D1154645

Climatism!

Science, Common Sense, and the 21st Century's Hottest Topic

STEVE GOREHAM

New Lenox Books

Climatism!
Science, Common Sense, and the
21st Century's Hottest Topic

ISBN: 978-0-9824996-3-4
Library of Congress Control Number: 2009912354

New Lenox Books
New Lenox, IL USA 60451
newlenoxbooks@comcast.net

PRINTED IN THE UNITED STATES OF AMERICA

Credits:

Sun photograph courtesy of SOHO (ESA & NASA)

Earth photograph courtesy of NASA

Iceberg photograph courtesy of National Science Foundation,
U.S. Antarctic Program Photo Library

Tornado and lightning photograph by Fred Smith

Book design and jacket design by Janice Phelps Williams

CONTENTS

PART I
GLOBAL WARMING ALARMISM DEBUNKED

The Medieval Warm Period and Little Ice Age ~ Medium Term: The 1500-Year Cycle ~ Short Term Cycles: El Niño, PDO, and AMO ~ Natural Cycles Explain Global Warming

Evidence Points to the Sun ~ The Link between Cosmic Rays and the Sun ~ New Evidence: Solar Activity Affects Clouds ~ Nature Is Not Cooperating ~ Headed for Global Cooling?

Icecap Melting Causes Sea Level Rise ~ An Increase in Reported Natural Disasters ~ Devastation from Hurricanes and Storms ~ Famine from Droughts and Floods ~ Increase in Temperature-Related Deaths ~ The Polar Bear and Species Extinction ~ Melting Glaciers and Water Shortages ~ Coral and Acidification of the Oceans ~ Gulf Stream Shut Down and New Ice Age

PART II
CLIMATISM AND "SNAKE OIL" REMEDIES

United Nations and Environmental Activism ~ Ozone Hole and Montreal Protocol ~ Theatrics in the Summer of 1988 ~ The IPCC Gathers a Host of Support ~ Score Decided before the Game Was Played ~ 1992 Earth Summit and the FCCC ~ Increasing "Certainty" of IPCC Reports ~ Re-writing History with the Mann Curve ~ Consensus Science from the IPCC ~ The Kyoto Treaty: Lesson in Futility ~ Alarming IPCC Temperature Forecasts

Malthus and the Roots of Climatism ~ The New Malthusians Adopt Climatism ~ The Improving State of the World ~ The Environment and Kuznets Curves

ACKNOWLEDGMENTS AND DEDICATION

I wish to acknowledge the integrity and courage of realist climate scientists of the world. These men and women have stood their ground in the face of pressures and attacks from Climatism as humanity adopted a misguided belief in man-made global warming. I also acknowledge the contribution of realist energy experts and their efforts to promote sensible energy policy in the face of renewable energy mania. *Climatism!* stands only on the shoulders of these scientists and experts. Special thanks go to David Archibald, Dennis Avery, Timothy Ball, John Coleman, Joseph D'Aleo, Michael Economides, Viv Forbes, William Happer, Howard Hayden, and Jay Lehr for their criticism, inputs, and encouragement. Thanks also to my editors, Janice Phelps Williams and Janet Weber for their patience and professionalism.

This book is dedicated to my wife and life partner Susan, and also to my family and friends, whose valuable inputs and inspiration helped me to push forward to completion.

Finally, this book is dedicated to the end of Climatism, to the restoration of sound climate science, and to the rebirth of common-sense energy policy.

FOREWORD

On December 7, 1972, the Apollo 17 crew took that famous first picture of the full sunlit Earth, the "blue marble." To me and millions of others it was an awesome sight. Our Earth is so beautiful. But that picture also showed how we are so alone here in space and our little planet seems so vulnerable. This is all we have. It is our only home. We must protect it for ourselves and our children. This picture is credited by many as the catalyst for the environmental movement.

I love this planet; it is a wonderful home for all six and one-half billion of us Earthlings. Clean air and clean water are my highest priority and leaving the beautiful blue marble as nice as it was when I was born is my goal. But, I have come to a parting of company with the extreme environmentalists who equate environmentalism with an anti-fossil fuel agenda that would sacrifice our modern society to protect the environment. Steve Goreham brands their agenda-driven extreme environmentalism as Climatism.

This book, however, is not a rant against environmentalists; not by a long shot. Steve loves this planet and wants to protect our ecosystem as much as you and I do. This is a serious book that carefully examines the issues that have been used to create the current climate change/global warming crisis that Al Gore, the United Nations Intergovernmental Panel on Climate Change and the media say threatens our planet.

The science is not settled; Al Gore take note. There is not a consensus among scientists; UN IPCC take note. The icecaps are not vanishing and the coasts are not in flood. Goreham carefully documents all of the claims and counterclaims and the science that should settle the debate. In this book common sense trumps nonsense.

I endorse *Climatism!* for its easy-to-read, well-illustrated presentation of complex science. That is an accomplishment you will appreciate. You will not get lost or confused.

And, in the end I think you will agree with me that our blue marble is not in crisis or in danger from our civilization. Our luck to be living in an interglacial period when the Earth is warm and life thrives is the ultimate in good fortune. If we must worry about the future, it seems concern about the onset of nature's next ice age would be a better concern that any about uncontrolled heat. There is no significant man-made global warming, there has not been any in the past and there is no reason to fear any in the future.

John Coleman

AUTHOR'S NOTE

It was June of 1995, a beautiful day in Colorado, and an excellent day for whitewater kayaking. My paddling companions and I were taking a brief rest in a river eddy on the Arkansas River at the bottom of Royal Gorge in Colorado. The blue-green river had narrowed to about 50 feet wide as it roared between the reddish-brown granite cliffs of the gorge. We leaned back in our kayaks to look up more than a thousand feet to the suspension bridge, which appeared tiny at the top of the canyon. What a magnificent world this is.

For more than three decades, when family and work allowed, I've enjoyed kayaking many of the whitewater rivers of the United States and Canada. I've navigated the rapids of several rivers on the Cumberland Plateau in eastern Tennessee, the home of Al Gore. I've camped by tent and trailer throughout North America with my family and friends. I've enjoyed many national parks and forests. These experiences are among my favorite memories.

No one wants to see Earth's air polluted, rivers dirtied, or oceans fouled. We *all* want to pass a better world on to our children. But our policies must be based on logic and sound science, not propaganda and fear.

For several years, I've watched the global-warming debate from afar. An occasional news report one decade ago has grown into a blizzard of news broadcasts, books, YouTube video clips, and movies. Most of this communication has proclaimed that Earth is warming;

that our industrial civilization is the cause; and increasingly, that we are on the road to climate catastrophe.

Despite the avalanche of publicity decrying man-made global warming, I wondered whether it all really made sense. Most news stories pointed to a weather event, such as Hurricane Katrina or catastrophic fires in Australia, and then jumped to the conclusion that the event was caused by human industry. Discussion of the science behind such conclusions was rarely included in the story.

Predictions announced that, by the year 2100, icecaps will melt, seas will inundate, and catastrophic hurricanes will destroy, among many other disasters destined to plague us all. But then my local meteorologist would be wrong on his five-day weather forecast. How is it possible that predictions for the year 2100 can be accurate, but not forecasts for the weekend?

As I read alarmist books on human-caused global warming, a pattern began to emerge. Most books spend 10–20 pages on the science of climate change, and then hundreds of pages on the impacts and possible remedies. These books rely almost entirely on findings of the Intergovernmental Panel on Climate Change (IPCC) of the United Nations for their science, and *assume* that the IPCC is correct.

At the same time, I discovered a growing group of scientists who, I believe, are climate realists, but are called "skeptics" (or worse) by the news media. These technologists are increasingly convinced that our globe is warming, but that the warming is due to natural, rather than man-made causes. Their arguments proved persuasive and moved me decisively into the skeptic camp. Most astonishing, the science clearly shows that global warming is due to natural causes, despite the tidal wave of world belief in man-made climate change.

Once convinced that mankind was heading down the wrong track, I felt compelled to add my voice to the debate. Exaggerations by Al Gore, James Hansen, and others added fuel to my fire to write. This book is the result.

Whether global warming is being caused by humanity should be decided by science, not by politics. The climate and energy policies of the nations of the world should be based on valid science, sound economics, and common sense. Unfortunately, as we shall see, this is currently not the case. Let me state up front that I am not in the pay of any energy company, or any other organization with a vested interest in the debate.

INTRODUCTION

Humanity is in the grip of a madness. It's a madness driven by fear and based on mistaken science. It's a madness now adopted by most of the governments of the world, trumpeted by the press, and taught in our schools and universities. It's a madness that, if left to run its course, will steal our freedoms, destroy our standard of living, and result in the deaths of millions of people in the developing world.

A new ideology—Climatism—drives this madness. **Climatism is the belief that man-made greenhouse gases are destroying Earth's climate.** Climatism is an extreme form of environmentalism that is using the natural climatic changes of Earth to re-define our societies. Climatism has adopted the mantle of science and uses fear of global catastrophe to alarm the public and achieve its objectives. Backed by academia and with millions in funding from foundations, Climatism is using modern public-relations techniques to mold public opinion. The attack is powerful and aimed at every part of our way of life.

The alarmists warn of climate catastrophe if we fail to act. Al Gore, former vice president of the United States, and now chief advocate for the position of catastrophic climate change, stated in testimony before the United States Senate:

> We have arrived at a moment of decision. Our home—Earth—is in grave danger. What is at risk of being destroyed is not the planet itself, of course, but the conditions that have made it hospitable for human beings.[1]

1

James Hansen, director of NASA's Goddard Institute for Space Studies, warns:

> ...coal is the single greatest threat to civilization and all life on our planet. The climate is nearing tipping points. Changes are beginning to appear and there is a potential for explosive changes, effects that would be irreversible, if we do not rapidly slow fossil-fuel emissions over the next few decades.[2]

Tony Blair, former Prime Minister of the United Kingdom, also waves the red flag:

> Climate change is the world's greatest environmental challenge. It is now plain that the emission of greenhouse gases, associated with industrialization and economic growth...is causing global warming at a rate that is unsustainable.[3]

The demands of the Climatists are extreme: eliminate carbon dioxide (CO_2) emissions from all human industrial processes. They define carbon dioxide as a pollutant and label it as dirty. They deplore our addiction to fossil fuels and attack carbon polluters with a religious fervor. They pressure us to jump on the Climatist bandwagon as it moves ever faster toward the utopia of a green world.

Climatists seek to control every aspect of our lives. Their remedies include forced replacement of coal-fired power plants with solar and wind power. They propose "Cap and Trade" legislation, which will impose carbon taxes on all industrial companies. They seek to reduce air travel by imposing heavy taxes. Their recipe for automobile travel is to "drive less, but pay more": downsize your SUV, pay more for gasoline, and drive fewer miles. They even point out that it's more green to be a vegetarian, since the production of meat causes a higher level of CO_2 emissions.

Many prominent persons have accepted the belief system of Climatism. These include almost all heads of state and 95% of the world's politicians. These include most of the mainstream media and the entertainers of Hollywood. These include most of the Fortune 500 companies. These include most of the major scientific organizations of the world. With such overwhelming support, how can Climatism be wrong?

All scientists agree that Earth's temperatures have increased by a small amount during the 20th century, on average. Most scientists agree that man's activities have had some impact upon Earth's climate. Most scientists also agree that CO_2 emissions from industry do contribute to the greenhouse effect and to increasing global temperatures. *The key question is: What is the primary and significant cause of global warming?*

This book takes a realistic look at the global warming crisis. Our conclusion will be that the global warming of the 20th century is primarily due to natural causes and not due to man-made emissions of carbon dioxide. The discussion is divided into three sections.

Part I will examine the science for global warming. This will start with a discussion of the leading global warming Climatists and their forecasts of catastrophe. We will then analyze the case for man-made global warming. Next, we'll show carbon dioxide is a trace element in our climate system, and that climate change is both natural and continuous, and much larger natural forces are at work with regard to Earth's temperature. We will then discuss the true driver behind the natural cycles of our climate system. Finally, we'll spend a chapter on what science tells us about the disasters predicted by Climatism. Part I contains a number of graphs and quite a bit of scientific data. We urge you to stick with us through the science, which is critical to understanding the global warming debate.

Part II will explain Climatism and proposed remedies for stopping global warming. First, we'll discuss the Intergovernmental Panel on Climate Change (IPCC), a United Nations organization regarded as the scientific authority on global warming. Climatism, the ideology behind man-made global warming alarmism, will be discussed next. Then we'll spend two chapters on the beliefs, objectives, and tactics of Climatism. We'll discuss Al Gore's claim that "the science is settled" and the impact of the news media on this misdirection of mankind. Finally, we'll cover the draconian proposals of Climatism to save the planet.

Part III will explain the economics of Climatist clean-energy remedies in depth and the results of Climatist policies where they have already been implemented. We'll talk about how convenient, low-cost energy is the lifeblood of economic prosperity. Since renewable energy from solar, wind, and bio-fuel sources are fundamental to Climatist solutions for a green world, we'll evaluate these energy solutions. Discussions of impacts on developing nations follow next. Examples of Climatist solutions in action in Europe and the United States will follow next. Finally, we'll look at specific possible impacts of Climatism in the U.S.

Despite what you've heard from the news media, from your political leaders, or from your college professors, we invite you to read on with an open mind. *If global warming is from natural causes, then all efforts to stop the Earth from warming are not only futile, but destructive to our way of life and economic prosperity of the developing nations.*

This is the real story.

Part I

Global Warming Alarmism
Debunked

"Nature, not human activity, rules the climate."
Dr. S. Fred Singer

CHAPTER 1

A TSUNAMI FOR GLOBAL-WARMING ALARMISM

"Nature never deceives us; it is always we who deceive ourselves."
Jean Jacques Rousseau (1762)

The scene is the annual meeting of the American Association for the Advancement of Science (AAAS) at the Fairmont Hotel in Chicago on February 14, 2009. Al Gore stepped to the lectern to a standing ovation from an estimated 3,000 attendees to address the crowd on "A Call to Action on Climate." During the next 50 minutes, Mr. Gore expertly entertained the audience with jokes and exhorted each member to get involved in "this historic debate."[1]

In his presentation, Mr. Gore spent little time on the science of how humanity is causing Earth to warm up. Using visual aids, he explained the greenhouse effect that produces warming due to carbon dioxide. Gore then pointed out that temperatures on Venus are 475°C (887°F), and attributed this to high levels of carbon dioxide in the atmosphere. His final evidence consisted of a graph illustrating temperature rise since 1900 along with the assertion that the ten hottest years on record (since 1850) occurred during the most recent eleven-year period.[2] For the benefit of the reader, leading U.K. and U.S. government agencies report a global temperature increase of 0.7°C (approx. 1.3°F) since the year 1900.

Having concluded his case that humanity is causing global warming, Mr. Gore went on to discuss shrinking arctic ice caps,

7

melting ice in Greenland, sea-level rise, a "negative ice balance" in Antarctica, shrinking glaciers, floods in the U.S., droughts in Georgia and Australia, and worldwide fires. All of these, he implied, are due to climate change caused by man. After this essentially political message, Gore left the stage to a standing ovation from the AAAS, an organization with a self-proclaimed commitment to the advancement of science.[3]

Gore's presentation is typical of global-warming alarmism. There is feeble discussion of the science of climate change, but an avalanche of discussion attributing every weather event on Earth to human carbon dioxide emissions.

THREE CLIMATIST PROPHETS OF DOOM

Former Vice President Al Gore is certainly the leading Climatist for man-made global warming. He has written three books on the subject, including *An Inconvenient Truth: The Planetary Emergency of Global Warming and What We Can Do About It*, which reached number one on *The New York Times* paperback best-seller list in July 2006. The book was made into the movie documentary *An Inconvenient Truth*, which opened in May 2006, earning $49 million at the box office worldwide, and winning two Academy Awards, including Best Documentary Feature.[4] Mr. Gore also shared the 2007 Nobel Prize with the Intergovernmental Panel on Climate Change (IPCC) of the United Nations. Sources estimate that he has earned $100 million in his global-warming crusade.[5]

Mr. Gore is no stranger to exaggeration. One week after his presentation to the AAAS, he removed a dramatic visual on disaster trends that stated "there have been four times more weather-related disasters in the last 30 years than in the previous 75 [years]," because of sharp criticism from the scientific community.[6] Both the book and the documentary, *An Inconvenient Truth*, have been heavily criticized for exaggerations and scientific inaccuracies. In fact, in 2007 Justice Burton of the London High Court of Justice found that the documentary contained at least nine scientific errors. He ruled that *An Inconvenient Truth* could not be shown in England's secondary

---

Scientific Errors in *An Inconvenient Truth*

In October of 2007, Justice Burton of the Royal Court of London ruled on a case brought against England's Secretary of State for Education and Skills. The Education Secretary proposed to distribute the documentary film *An Inconvenient Truth* as part of the curriculum for secondary students. Justice Burton ruled that the film could be used, but that guidance notes for teachers must point out scientific errors in the film.

Specifically, the court ruled that the film "promoted partisan political views" and contained nine "errors," which are:

1. Mr. Gore's forecast of sea level rise of up to 20 feet (7 metres) is not in line with scientific consensus.
2. Mr. Gore's claim that low lying pacific islands are being inundated due to global warming is not supported by evidence.
3. Mr. Gore's forecast that the North Atlantic Ocean current, the Gulf Stream, will shut down, is very unlikely.
4. Mr. Gore's claim that ice core data from over 650,000 years proves that rising CO_2 causes rising temperatures is incorrect.
5. Mr. Gore's claim that loss of snow on Mt. Kilimanjaro is due to man-made global warming is not scientific consensus.
6. Mr. Gore's claim that the drying up of Lake Chad is a result of global warming is not supported by the evidence.
7. Mr. Gore's assertion that the devastation of New Orleans from Hurricane Katrina was due to global warming is not supported by the evidence.
8. Mr. Gore's claim that polar bears are drowning due to loss of Arctic ice is not supported by any scientific studies.
9. Mr. Gore's claim that coral reefs are bleaching due to global warming is not supported by the evidence.

However, the court also ruled that "climate change is mainly attributable to man-made emissions of carbon dioxide" and that "the IPCC represented the current scientific consensus."[7]

---

schools without a discussion of these inaccuracies (see text box).[8]

Al Gore is not a scientist, but he is a political publicity machine. In addition to his more than 1,000 presentations on global warming, Gore's website, algore.com, furthers the global-warming crusade. Among Gore's listed projects on the site is "The Climate Project."[9]

The Climate Project (TCP) is an effort to spread the man-made global-warming message throughout the world. TCP trains presenters, provides them with materials (based on Al Gore's slideshow from *An Inconvenient Truth*), and sends them out to educate and convince the public. At last review, The Climate Project boasted more than 3,000 volunteers trained to sound the global warming alarm.[10]

A second leading Climatist, and the most outspoken, is Dr. James Hansen. Dr. Hansen is director of the NASA Goddard Institute for Space Studies in New York City and Adjunct Professor of Earth Sciences at Columbia University. He has also been a senior advisor to Al Gore and consulted for Gore on *An Inconvenient Truth*.[11] Dr. Hansen rose to national prominence from his testimony to the U.S. Senate on global warming in 1988.[12] His first testimony to Congress was to the House of Representatives in 1982, which was organized by then Representative Al Gore.[13]

Dr. Hansen is considered the leading scientist for man-made global-warming science. He was recently awarded the 2009 Carl-Gustaf Rossby Research Medal, the highest honor bestowed by the American Meteorological Society for "outstanding contributions to climate modeling, understanding climate-change forcings and sensitivity, and for clear communication of climate science in the public arena."[14] Dr. Hansen has called for closing all coal-fired power plants, which generate about one-half of the world's electricity, stating:

> The trains carrying coal to power plants are death trains. Coal-fired power plants are factories of death.[15]

He believes that special interests have blocked our transition to renewable energy:

> CEOs of fossil energy companies know what they are doing and are aware of long-term consequences of continued business as usual. In my opinion, these CEOs should be tried for high crimes against humanity and nature.[16]

Dr. Hansen is the most ardent spokesman for the concept of "tipping points" in climatology. A tipping point is a theoretical point at which the climate of the earth becomes unstable. According to Hansen, once a tipping point is passed, the earth's climate would become increasingly extreme regardless of any measures mankind could take:

The climate is nearing tipping points. Changes are beginning to appear and there is a potential for explosive changes, effects that would be irreversible, if we do not rapidly slow fossil-fuel emissions over the next few decades. As arctic sea ice melts, the darker ocean absorbs more sunlight and speeds melting. As the tundra melts, methane, a strong greenhouse gas, is released, causing more warming. As species are exterminated by shifting climate zones, ecosystems can collapse, destroying more species...The greatest danger hanging over our children and grandchildren is initiation of changes that will be irreversible on any time scale that humans can imagine.[17]

What if Dr. Hansen is right? Many things are possible, but Dr. Hansen's theories, despite his awards, are far outside the bounds of accepted climate science. Even the IPCC is much more conservative regarding sea-level rise and tipping points than Dr. Hansen. Apollo 7 astronaut and physicist Walter Cunningham says: "Hansen is a political activist who spreads fear even when NASA's own data contradict him."[18] As we shall see in Chapter 3, Dr. Hansen's alarmist assertions are based on projections from flawed computer models inconsistent with observed scientific data.

A third leading alarmist is Sir Nicholas Stern, Professor of Economics and Government at the London School of Economics and Political Science. Tony Blair commissioned Sir Nicholas to lead a team of economists at His Majesty's Treasury to publish the *Stern Review: The Economics of Climate Change* in 2006. The *Stern Review* stated: "The scientific evidence is now overwhelming: climate change presents very serious global risks, and it demands an urgent global response." The report detailed a wide spectrum of severe impacts, including water-supply shortages, declining crop yields, more widespread disease, extreme weather, flooding from rising seas, and extinction of species. The report warned that "business as usual" will result in 20% consumption losses per person by the end of the century. The report recommends CO_2 "stabilisation measures" that will cost the "relatively small" amount of 1% of world GDP.[19] In June 2008 Sir Nicholas revised this cost estimate upward:

...evidence that climate change was happening faster than had been previously thought meant that emissions needed to be reduced even more sharply...[this] would cost around 2% of GDP.[20]

Stern's predictions of doom have recently become even more alarmist, including world war and migrations of billions of people. In February 2009 he stated:

> … what we are talking about is extended world war… People would move on a massive scale. Hundreds of millions, probably billions of people would have to move…[21]

The hysterics of our three prophets, Mr. Gore, Dr. Hansen, and Sir Stern, continue to increase in volume and magnitude. However, as we discuss in Chapters 5 and 6, nature is not cooperating. Global temperature, sea levels, and other alarmist projections are running well below their dire predictions.

EIGHT "DISASTERS" OF GLOBAL WARMING

For every one word in the news media written or spoken about fact-based climate science, we find hundreds of words proclaiming the coming "disasters" of global warming. This is because the scientific evidence for man-made global warming is thin and also because global warming has become a political agenda driven by fear. Let's look at these eight forecasted disasters and the short answer to why these are incorrect.

1. Icecap Melting Causes Flooding from Sea Level Rise

The biggest projected disaster from manmade global warming is sea-level rise from the melting of the icecaps of the Arctic, Greenland, and Antarctica. Each of our prophets is forecasting about a 20-foot (6–7 meter) rise in the oceans, which would flood Bangladesh, the Maldives Islands, and many of the world's seaside cities. *An Inconvenient Truth* shows simulated pictures of the resulting flooding of South Florida, the Netherlands, Bangladesh, Calcutta, New York City, Beijing, and San Francisco Bay. The alarmists claim that millions of people will be displaced from their homes.[22]

However, sea-level rise has been a steady 6–7 inches per century for the last 5,000 years. With recent warming temperatures, the rise has been 7 inches per century since 1850.[23] Even the mid-level ocean-rise forecast of the IPCC is only 15 inches (0.39 meters) by 2100.[24]

2. Devastation from Hurricanes and Tropical Storms

Climatists have attributed the New Orleans devastation by Hurricane Katrina to global warming.[25] Alarmist forecasts predict the strength and frequency of tropical storms will increase with global temperature. Yet, studies from weather experts indicate storm frequency and strength have *not* increased over the last 100 years.

3. Famine and Death from Droughts and Floods

Alarmists predict an increase in the frequency of droughts and floods, leading to crop failures and global famine. However, droughts and floods have always been part of our climate, such as the dust-bowl droughts of the Plains States in the U.S. in the 1930s. Historical evidence shows dry and wet periods of the 1900s are well within the variability of natural conditions. While the frequency of drought and flood events has not increased, a decline in deaths from natural disasters during the 20th century shows improved ability by mankind to adapt to extreme events.

4. An Increase in Temperature-Related Human Deaths

The alarmists and the IPCC are forecasting a significant increase in heat-related deaths and increased vector-borne disease, such as malaria. However, each year the number of deaths from cold is many times larger than deaths from heat, so any temperature increase will reduce, rather than increase, human mortality. In addition, medical pathologists point out that window screens and insecticides are far more important in controlling malaria and other diseases, and that temperature is an insignificant factor in the spread of disease.

5. "Plight" of the Polar Bear and Species Extinction

The polar bear has become the poster child for the global-warming alarmist crowd. The cover of the book *The Hot Topic* shows swimming polar bears.[26] Al Gore uses pictures of polar bears in all of his presentations. The alarmists claim that man-made global warming will melt the arctic icecap, drive the polar bear to extinction, and also kill millions of other species. However, polar bear populations have more than doubled since 1950 and maybe *one* animal species has become extinct due to temperature increase. In addition, science tells us that biodiversity increases with modest temperature increase.

6. Water Supply Shortages from Melting of Glaciers

In *An Inconvenient Truth*, Gore shows many pictures of retreating glaciers. He forecasts that "40% of the world's people may well face a very serious drinking water shortage, unless the world acts boldly and quickly to mitigate global warming."[27] It is true that water is getting more expensive and shortages are occurring, but these are not due to man-made global warming. Studies show that our glaciers have been retreating since about 1800, long before any significant CO_2 was emitted from human industry.[28]

7. Acidification of the Oceans and the Death of Coral Reefs

Alarmists claim that increased CO_2 in the atmosphere will be absorbed by the seas, resulting in increased acidification of the ocean, causing bleaching of coral and eventually death. Bleaching is a natural process for coral reefs to adapt to either warming or cooling conditions. Contrary to fears, experimental data indicates that the recent increase in ocean surface temperature and atmospheric CO_2 concentration is having a positive effect on coral growth.[29]

8. Shut Down of the Gulf Stream Triggers a New Ice Age

In the disaster movie *The Day After Tomorrow*, global warming causes the Greenland ice sheet to melt, causing the Gulf Stream current to shut down, resulting in abrupt global cooling and the next ice age.[30] Scientists believe an event like this, the "Younger Dryas," may have occurred about 12,000 years ago at the end of Earth's last ice age. Fresh water from the melting of vast ice sheets in Canada, Greenland, and Siberia is believed to have flooded the North Atlantic, slowing the Gulf Stream. However, no such vast ice sheets exist today and the IPCC and most scientists think this scenario very unlikely.[31]

MEDIA CHEERLEADERS

There is a symbiotic relationship between the news media and global-warming alarmists. Climatists believe that the planet is warming, man has caused it, and so they are on a crusade to save the planet. Members of the media often think in the same way. They tend to be more activist than the general population, wanting to use their influence and play a part in moving humanity in the right direction. The media is also driven to boost ratings, sell copy, and increase circulation. Which title do you think would attract more viewers: "Global Warming Causes Hurricanes," or "No Long-Term Shift in Our Storm Patterns?"

THE CONSENSUS APPEARS UNANIMOUS

The alarmist theory of man-made global warming has a tsunami of support. The vast majority of national governments, local governments, scientific organizations, universities, mainstream press, and major corporations support this disaster theory. Communications in the news media and on the Internet run ten-to-one in favor of man-made global warming versus skeptical views.

In 1992, 40 nations and the European Economic Community adopted the United Nations Framework Convention on Climate Change, a global treaty with the stated purpose:

> ...to achieve...stabilization of greenhouse gas concentrations in the atmosphere at a level that would prevent dangerous anthropogenic interference with the climate system. [32]

"Anthropogenic" is used in climate discussions and means "caused or produced by humans." By greenhouse gases, the alarmists primarily mean carbon dioxide. In 1997, the Kyoto Protocol was adopted as an amendment to the 1992 treaty, calling for mandatory reductions in greenhouse gases to reach negotiated targets. Today, 184 nations have ratified the Protocol. [33] The main exception is the United States, which is also considering legislation to reduce CO_2 emissions.

The Kyoto Protocol is based on questionable science provided by the IPCC, a group founded by the UN in 1988. The IPCC is a powerfully effective political organization that has successfully passed itself off as the scientific authority on climate change. Actually, it is highly biased toward the dogma of man-made global warming. The vast majority of political leadership of the world has accepted the misguided warnings of the IPCC.

Dr. Václav Klaus, President of the Czech Republic, is the only head of state who disagrees with man-made global warming. He is also a past president of the European Union. President Klaus tells of his attendance at a closed session of the World Economic Forum in Davos in February, 2009. The topic of the session was "How to Prepare for the New Kyoto, the December, 2009, United Nations Copenhagen Summit." Attendees included heads of state of the major nations of the world, and "climate experts" such as Al Gore, Tony Blair, and Kofi Anon. He recalls:

> I looked around in vain to find at least one person who would share my views—there was no one. All the participants of the meeting took man-made global warming for granted. They're convinced of its dangerous consequences…[the following discussion was only about] one topic…whether to suggest a 20%, 30%, 50%, or 80% emissions cut as a worldwide project.[34]

The major scientific organizations of the world have also adopted the man-made global-warming theory (see text box). It's interesting to note that, while the official statements of the scientific organizations support man-made global warming, the views of the members do not always agree.

THE GLOBAL DELUSION

The shocking part is that **the theory of catastrophic man-made global warming from CO_2 emissions is increasingly shown to be incorrect.** As we will see in the next five chapters, mounting evidence from climate history, satellite data, and new experimental data shows natural causes, rather than man-made causes, are responsible for

---

**Some Major Scientific Organizations Supporting
Man-Made Global Warming**

United States
American Association for the Advancement of Science (AAAS)
American Geophysical Union (AGU)
American Meteorological Society (AMS)
American Physical Society (APS)
National Academy of Sciences (NAS)
National Oceanic and Atmospheric Administration (NOAA)
National Aeronautics and Space Administration (NASA)

International
Australian Meteorological and Oceanographic Society (AMOS)
Canadian Meteorological and Oceanographic Society (CMOS)
European Academy of Sciences and Arts (EASA)
European Science Foundation (ESF)
Intergovernmental Panel on Climate Change (IPCC)
Network of African Science Academies (NASAC)
Royal Meteorological Society (RMetS)
The Royal Society (UK)
Royal Society of New Zealand (NZ)

---

climate change. The world, driven by Climatism, is moving to implement extreme solutions for a greenhouse gas emissions problem that *doesn't* exist.

CHAPTER 2

THE THIN SCIENCE FOR
MAN-MADE GLOBAL WARMING

"Extraordinary claims require extraordinary evidence." Carl Sagan

Weather is defined as the general condition of the atmosphere at a particular time and place. Climate is an average of weather conditions over a period of years. Mark Twain once said: "Climate is what you expect, weather is what you get."

Earth's climate is powerful. A large hurricane is said to release heat energy at a rate of one exploding 10-megaton nuclear bomb every ten minutes.[1] Jet-stream speeds can reach 220 mph (350 Km/hr).[2] Lighting blasts an electrical path through the air at temperatures of 50,000°F—heat approaching five times the surface of the sun.[3] On a longer time scale, the great deep-water current known as the Global Ocean Conveyor Belt moves a huge volume of water through all five oceans on a 1,000-year cycle. Chicago was covered with an ice sheet an estimated *one mile thick* 20,000 years ago, during the last Ice Age.[4]

Earth's climate is complex. Warming radiation from the sun, cooling radiation from the Earth back into space, atmospheric currents (winds), and ocean currents interact in a chaotic interdependent system. These patterns change with night and day, with each season, and with every cycle of the planets in our solar system. Water in rivers, oceans, and polar regions (as ice), as well as in the atmosphere as vapor, clouds, and precipitation, is a major force in

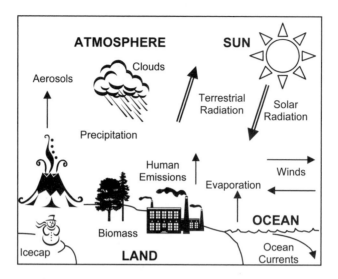

Figure 1. Elements of the Climate System.

shaping Earth's climate. Volcanoes play their part with significant emissions of dust and compounds thrown high into the atmosphere (aerosols). Even plants, animals, and man play an important role. A very simplified diagram is shown in Figure 1.

Earth's climate is only partly understood. The impacts of radiation from the sun, the formation of clouds and precipitation, ocean currents, and aerosols, biomass, and human activities remain largely unknown, despite the claims of global-warming alarmists. Prior to discussing the case for man-made warming, let's briefly review some basic elements of our climate system. These will help us in our discussion of global warming in this and succeeding chapters.

ENERGY BALANCE AND WEATHER

More than 99% all of energy received by Earth comes from the sun. It provides a continuous source of high-energy radiation in the form of sunlight. About 30% of sunlight is reflected back into space, by clouds and by Earth's surface. The remaining 70% of sunlight is absorbed (Figure 2).

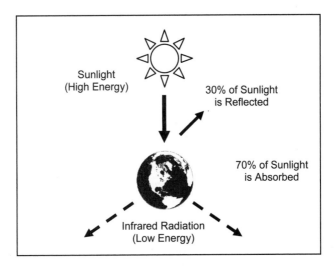

Figure 2. Energy from the Sun. Incident sunlight is reflected and absorbed. Earth radiates absorbed energy as infrared energy.

Earth emits low-energy radiation, called infrared radiation. In fact, all substances with heat, such as plants, animals, land, sea, clouds, and even ice, continuously emit infrared radiation. The infrared radiation emitted by any object is a function of its temperature. Therefore, the amount of infrared radiation emitted by Earth tells us our planet's average temperature, which is about 15°C or 59°F.[5]

Our eyes see reflected visible light when we look at the world around us. We are not able to see emitted infrared radiation, but it's there. It's continuously emitted from warm objects, even during night hours. Think of a common scene in action movies, where the bad guys have night-vision goggles. These devices sense infrared radiation, amplify it, and convert it to visible light for the goggles wearer.

Over the long term, Earth is said to be in "energy balance." That is, the incoming energy from the sun is equal to the outgoing energy from infrared radiation. Therefore, if the energy from the sun increases, Earth will absorb more energy, increasing its temperature. If Earth's temperature increases, the laws of physics tell us our planet will radiate more energy because of its higher temperature. This will increase the amount of outgoing infrared radiation until the balance is once again restored.

As you may recall from school, the tilt of Earth's axis provides our

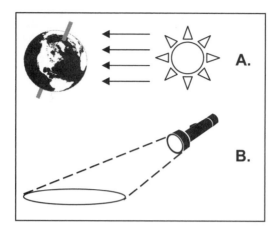

Figure 3. How the Sun's Energy Hits Earth. A. The tilt of Earth's axis and incident sunlight. B. Like a widened flashlight beam, reduced intensity of sunlight in polar regions causes the poles to be colder than regions near the equator.

different seasons as Earth orbits around the sun. But the tilt of Earth's axis and the curvature of its surface are also the driving force behind weather (Figure 3). When you shine a flashlight on a flat surface, and then move the light to reduce the angle of the beam to the surface, the intensity of the beam falling on the surface will decrease (Figure 3B). This is similar to the way sunlight falls with reduced intensity on polar regions relative to areas near Earth's equator where sunlight is more direct. Therefore, regions near the equator become hotter from absorbed sunlight than do polar regions. *Weather is the result of this uneven absorption of sunlight.* Earth uses weather and ocean currents to redistribute heat from the tropics to the poles. Hurricanes, monsoons, trade winds, warm and cold fronts, the Gulf Stream, and the North Pacific Current are all a result of heat imbalance between the poles and the equator. With this background, let's move on to the greenhouse effect and the science for man-made global warming.

EARLY RESEARCH ON CARBON DIOXIDE

A number of scientists in the 19th and early 20th centuries paved the way for today's global-warming alarmism. These men made solid contributions to climate physics without any other agenda. They contributed theory and experimental data to define what is now known as the greenhouse effect.

In 1824, Jean Fourier, the French mathematical physicist who developed Fourier Analysis, hypothesized that Earth's atmosphere trapped heat like the glass of a greenhouse.[6] John Tyndall, an English physicist, demonstrated in 1861 that "invisible gases and vapors," such as water vapor, carbon dioxide, ozone, and hydrocarbons, were able to absorb and emit radiant energy and affect Earth's climate.[7]

The first scientist to propose that increasing the amount of carbon dioxide in the atmosphere would increase Earth's temperature was Svante Arrhenius, a Swedish scientist. In 1896, he hypothesized that doubling the "carbonic acid" (CO_2) in the atmosphere would result in a temperature increase of 3–4°C. Arrhenius thought volcanoes were the primary source of CO_2 to the atmosphere, but also stated "the ocean plays an important role as a regulator of the quantity of carbonic acid in the air."[8]

In 1938, Guy Stewart Callendar, a British engineer and scientist, empirically connected temperature rise with man-made carbon dioxide emissions. Because of his work, the theory that global warming is due to hydrocarbon-fuel combustion by mankind has been called the "Callendar Effect." Callendar used data from weather stations around the world to estimate an average temperature increase of 0.25°C during the "past half-century" and estimated that 60% of the increase was due to "fuel combustion" by man. It's interesting to note that Callendar also stated in his landmark paper:

> ...the combustion of fossil fuel...is likely to prove beneficial to mankind in several ways...For instance,...small increases of mean temperature would be important at the northern margin of cultivation and the growth of...plants is directly proportional to the carbon dioxide pressure.[9]

THE SIMPLE GREENHOUSE EFFECT

Figure 4 shows a simple diagram of the greenhouse effect. As explained earlier, the sun provides a continuous source of energy for Earth in the form of sunlight. Sunlight is high-energy radiation.

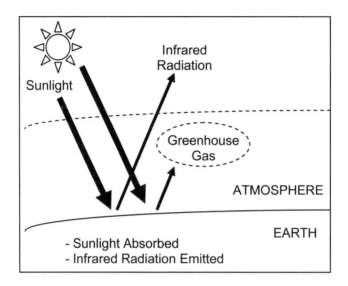

Figure 4. The Simple Greenhouse Effect. Emitted infrared radiation is absorbed by greenhouse gases in the atmosphere.

Sunlight's high energy allows most of it to pass through the atmosphere to be absorbed by Earth's surface. As we also discussed, Earth emits a lower-energy infrared radiation. Most of the infrared radiation passes through the atmosphere, but because it has a lower level of energy than sunlight, some of it can be absorbed by greenhouse gases. *This absorption of infrared radiation is called the greenhouse effect.*

Greenhouse gases strongly absorb infrared radiation. The most important of these gases are water vapor and carbon dioxide, but methane, sulfur dioxide, and chlorofluorocarbons are also greenhouse gases. Most of the greenhouse gases in the atmosphere are created by natural processes, so the greenhouse effect is a natural climatic process. However, our burning of hydrocarbon fuels adds to the carbon dioxide in the atmosphere and increases the greenhouse effect.

Recall that an energy balance is maintained between incoming solar radiation and outgoing infrared radiation. The energy balance assumption also means that Earth's surface temperature will warm as more infrared radiation is absorbed by greenhouse gases. If all other climatic factors remain the same, our burning of hydrocarbon fuels will cause Earth to warm. This is the theoretical basis Al Gore and others use for warming caused by CO_2 emissions.

However, all other factors are not the same. This is an over-simplified model, since evaporation, clouds, and precipitation play an even greater role in the process that converts heat energy at Earth's surface to outgoing infrared radiation.

INCREASING CO_2 AND TEMPERATURE

In 1956, Dr. Roger Revelle, a scientist at the Scripps Institution of Oceanography in San Diego, believed carbon dioxide was accumulating in the atmosphere, so he set out to measure the carbon dioxide concentration (Figure 5). He set up a research station on top of Mauna Loa, on the island of Hawaii, and hired a young researcher named Charles Keeling. Keeling started measurements in 1957, which eventually led to the "Keeling Curve."

The curve shows that the concentration of carbon dioxide in our atmosphere has risen from about 315 parts per million (ppm) in 1957 to about 385 ppm today. Others scientists hypothesize that the pre-industrial levels of CO_2 were about 280 ppm. If so, then the

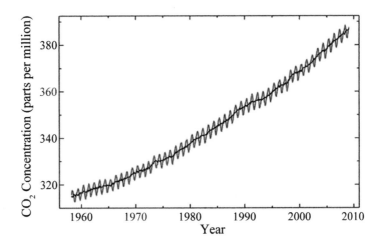

Figure 5. Atmospheric Carbon Dioxide Concentration. Atmospheric CO_2 concentration has increased from about 315 ppm in 1957 to about 385 ppm today. Data from the Earth System Research Laboratory in Mauna Loa, Hawaii. (Adapted from NOAA, 2009)[10]

atmospheric concentration of carbon dioxide has increased about 38% during man's Industrial Age.

Methods of global temperature measurement have improved since Callendar's work in 1938. Figure 6 shows global temperature data from 1850–2008 from the Hadley Centre Meteorological Office in the United Kingdom. Temperature "anomaly" in Figure 6 means difference from the average. The graph shows that global temperatures have warmed about 0.7°C since 1850, or about 1.3°F. Note that Earth's warming has not been steady, but cools from about 1880 to about 1910, warms from 1910 to about 1940, cools again from 1940 to about 1975, and then rises again from 1975 to 1998. We will discuss these variations later when we address Earth's natural temperature cycles.

A CLOSER LOOK AT TEMPERATURES

Determining an "average global temperature" for any given year is not easy. It requires recording and averaging of daily high- and low-temperature measurements from land and ocean stations at thousands of points distributed around the globe. Constructing the temperature record requires consistent long-term participation from governments

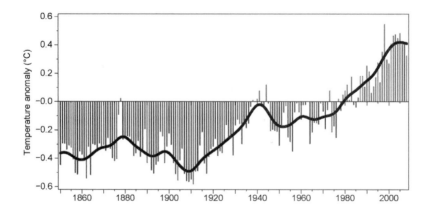

Figure 6. Global Surface Temperatures 1850–2008. Average global surface air temperatures from land and surface measuring stations (HadCRUT3). Data from Climatic Research Unit and the U.K. Meteorological Office Hadley Centre.[11]

of the world year after year, despite political disruptions that may occur. This is an important factor, considering the fact that two-thirds of the world's land stations (most of them from the former Soviet Union) stopped reporting around 1990.[12]

A growing temperature difference has become apparent between data sets that include land stations, such as Hadley Centre data in Figure 6, and temperature data measured by satellites. Satellite data, which has only been available since 1979, shows a 30-year increase of just less than 0.3°C, compared to the Hadley data that shows a rise of 0.5°C.[13] If Earth is warming due to atmospheric carbon dioxide, and if CO_2 is uniformly distributed throughout the atmosphere as the IPCC says, how can the surface station and satellite measurements differ by so much? It appears that this difference is due to "Urban Heat Island Bias" in the surface measurements.

When viewed from space, cities show up as bright areas not only in visible light photographs, but also in infrared pictures that capture heat energy. Human structures such as roads and buildings tend to absorb heat from sunlight each day and warm up to a higher temperature than areas with natural ground cover. Human structures also slowly release heat at night, making urban areas warmer than rural areas on average. Temperature bias occurs as cities grow and nations urbanize. Measurement stations that were once in remote areas end up next to parking lots, beside buildings, or other structures. Such stations end up measuring heat from air conditioners or automobile radiators, rather than Earth's temperature. These stations then record artificially higher temperatures because of Urban Heat Island Bias.

Numerous studies dating back to the 1930s have documented the existence of Urban Heat Island Bias. An example is a study by J. D. Goodridge in 1996. Goodridge plotted temperature data from 104 stations in California from 1909 to 1994 in three county groups: 1) counties with a population of more than one million persons, 2) counties with a population of 100,000 to one million, and 3) counties with a population of less than 100,000.

The results are striking. Over the 85 years of the study, temperatures in counties with large populations (urban) showed a temperature rise of over 3°F, while the counties with small populations (rural) showed a negligible temperature rise.[14] This is

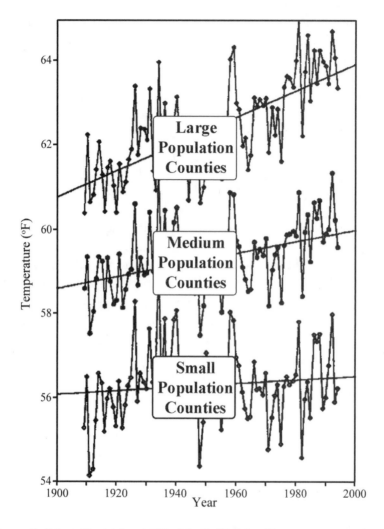

Figure 7. Urban Heat Island Effect in California. Temperature trends at 104 temperature stations in California from 1909 to 1994, grouped by population size. Counties of over one million in population showed greater than a 3°F rise (top curve), counties with populations between 100,000 and one million showed a 1°F rise, and counties with populations of less than 100,000 showed a negligible temperature rise. (Adapted from Goodrich, 1996)[15]

strong evidence that Urban Heat Island Bias may be causing error in our temperature stations (see Figure 7).

The 1,221 temperature stations of the United States Historical Climatology Network (USHCN) are meant to be the "high quality network" used to determine temperature trends in the U.S.[16] The USHCN is administered by the National Climatic Data Center (NCDC), which is part of the National Oceanic and Atmospheric Administration (NOAA). The NCDC published guidelines for sites, including recommendations on location, layout, and maintenance. They also have a five-level rating system for site temperature accuracy.

Meteorologist Anthony Watts set out to measure the effect of different paint types on temperature stations of U.S. Climate Reference Network Sites in early 2007. In the process, he found that many temperature sites did not meet proper site guidelines. Figure 8 shows two such poorly-sited temperature stations.

Since the NCDC did not have a site-audit program, Watts then started a volunteer project to photographically survey all 1,221 stations in the CRN network, which is discussed on the web site surfacestations.org.[17] Temperature stations were rated using NCDC guidelines, from CRN 1 (excellent), which was a site with a "grass/low vegetation ground cover" that was located "at least 100 meters from an artificial heating or reflecting surface" to CRN 5 (very poor), which was a site with a "temperature sensor located next to/above an artificial heating surface such as a building, roof top, parking lot, or concrete surface."[18]

By June, 2009, the project had surveyed 78% of the CRN sites. The project found that the CRN sites contained a significant warm temperature bias. Many sites were found to be located in parking lots. Some were next to air conditioner exhaust vents. Fully 69% of the

Figure 8. Poor Siting of U.S. Temperature Stations. The left picture shows a station located in an asphalt parking lot in Tucson, Arizona. The right picture shows a station in Waterville, Washington that is located over a cinder ground cover and next to a car radiator. (Watts, 2009)[19]

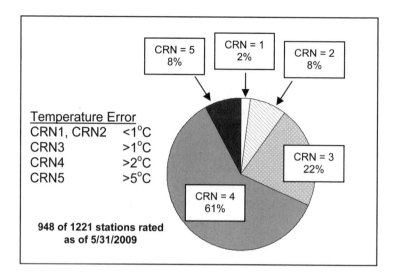

CRN = 5
8%

CRN = 1
2%

CRN = 2
8%

Temperature Error
CRN1, CRN2   <1°C
CRN3   >1°C
CRN4   >2°C
CRN5   >5°C

948 of 1221 stations rated
as of 5/31/2009

CRN = 3
22%

CRN = 4
61%

Figure 9. Quality of Climate Reference Network Sites. Quality of temperature data from 948 surveyed USCHN sites according to the CRN rating guidelines of the National Climatic Data Center. Sixty-nine percent of sites show a warm temperature error of over 2°C. (Adapted from Watts, 2009)[20]

sites were rated poor (CRN 4) or very poor (CRN 5), meaning they were located less than 10 meters from an artificial heating surface (see Figure 9). Therefore, most of the U.S. temperature-measuring sites *show a temperature error of over 2 °C too hot.* The National Climatic Data Center has used adjustments to attempt to eliminate the effects of heat bias, but their adjustments are too small to account for the large bias observed.

It's clear that U.S. land measurement stations are corrupted by Urban Heat Island Bias, which is completely unconnected from any greenhouse gas emissions. International land temperature stations are even less controlled and likely have the same bias. *Therefore, Hadley Centre temperature data and other temperature series that include land station data probably overstate global warming.* Dr. Joseph D'Aleo reports "half a dozen peer reviewed papers" estimate 50% of the measured global temperature increase since 1900 may be due to Urban Heat Island Bias.[21] For this reason, satellite data should be regarded as the more reliable indicator of global temperature change.

GLOBAL WARMING DOGMA

By the end of the 1950s, a simple basis was in place for establishment of the "Global Warming Dogma" that man-made emissions of carbon dioxide were causing catastrophic global warming. This basis included three elements: 1) greenhouse gas theory, 2) rising concentrations of atmospheric carbon dioxide, and 3) increasing global temperatures. A number of scientists, including Roger Revelle and Bert Bolin (the future first chairman of the IPCC) hypothesized that CO_2 was accumulating in the atmosphere and would upset the equilibrium of Earth's climate. Revelle voiced his concern over the burning of hydrocarbon fuels:

> Thus, human beings are now carrying out a large scale geophysical experiment of the kind that could not have happened in the past... Within a few centuries we are returning to the atmosphere and oceans the concentrated organic carbon stored in sedimentary rocks over hundreds of millions of years.[22]

Al Gore, a college student at the time, was heavily influenced by Dr. Revelle in the 1960s. The greenhouse gas warming hypothesis was pushed on by Gore, James Hansen and the other climate modelers, and eventually proclaimed as science fact by the geopolitical IPCC of the United Nations. To maintain this Global Warming Dogma, the Climatists sacrificed the laws of chemistry and physics, and ignored scientific and historical observations.

RISE OF THE CLIMATE MODELS

The Earth's atmosphere can be thought of as a dynamic fluid system in constant motion. During the 1940s and 1950s, climate scientists developed theories of the general circulation of Earth's atmosphere. At about the same time, the age of digital computers brought a new tool to science. A new class of climate scientists arose, dedicated to modeling the climate system on increasingly powerful computers. These models came to be known as General Circulation Models

(GCMs).

Norman Phillips, of Princeton's Institute for Advanced Study, is credited with developing the first true GCM. In 1955, he simulated a hemisphere of Earth as a cylinder with the crude computers of the time.[23] By the late 1950s, meteorologists in the U.S. and Europe had incorporated computer simulations into their regional forecasts on a regular basis.

By the 1960s, the growing power of computer systems allowed climate modelers to move from regional models to global models. Improved atmospheric data was increasingly available as inputs to the models. A number of modelers noted that carbon dioxide was accumulating in the atmosphere and wanted to estimate the effects. One of these was Dr. Syukuro Manabe, at the Geophysical Fluid Dynamics Laboratory in Washington, D. C.

Manabe developed a model that included cooling effects from both infrared radiation and convection of water vapor from Earth's surface. He then looked at what would happen if the concentration of CO_2 doubled in the atmosphere. Based on some previous work by Telegadas and London in 1954, Manabe assumed that the relative humidity of the atmosphere remained constant as it heated up.[24] This meant that the atmosphere would retain more water vapor, which is also a greenhouse gas, as Earth warmed. This is what scientists call a "positive feedback."

The concepts of positive and negative feedback are important, so let's pause to explain. Dr. John H. Sununu, former Governor of New Hampshire and former Chief of Staff for President George H. W. Bush, provided the following excellent example. When you step into the shower and the water is too hot, you reduce the temperature by turning the faucet toward "cold" until the stream of water cools. You have provided negative feedback to the system to cool the shower. However, suppose that shower knob was incorrectly reversed. Then, if the shower is too hot, your turn of the knob will inadvertently increase the temperature of the stream, an unpleasant situation. This is positive feedback. Such a system can be viewed as unstable, since an increase in temperature will lead to feedback causing a further increase in temperature.[25]

Manabe's assumption that water vapor would increase with increased atmospheric temperature served to amplify the greenhouse effect. The claim is that increased man-made emissions of CO_2 increase Earth's temperature, causing increased humidity and water vapor, causing a further increase in Earth's temperature. This positive feedback from water vapor became an important part of the Global Warming Dogma. Without positive feedback from water vapor, temperature increases from the greenhouse effect are small and calamity forecasts are impossible. This water vapor feedback theory was adopted by the IPCC and is present in all GCMs today. Positive feedback from water vapor results in more than two-thirds of the temperature rise forecasted by the alarmists. As we shall see in Chapter 3, this assumption is looking increasingly doubtful. See the text box for more discussion on models.

Most renowned of the modelers is Dr. James Hansen. Hansen's team developed an advanced GCM at NASA's Goddard Institute for Space Studies in New York City in the 1970s. In 1981, they published a landmark paper concluding that Earth was warming due to carbon dioxide emissions, and that "The global warming projected for the next century is of almost unprecedented magnitude." Hansen estimated a temperature rise of 2.8°C for a doubling of atmospheric CO_2 concentration, including the positive feedback from water vapor increase.[26]

Many of Hansen's ideas were eventually adopted by the IPCC, after it was established in 1988. These include 1) his idea of a range

---

How General Circulation Models (GCMs) Work

General Circulation Models are representations of the climatic behavior of the Earth's atmosphere and ocean. The building blocks of the GCMs are nonlinear mathematical equations based on the classical laws of physics for motion and thermodynamics. For example, the energy balance assumption we discussed regarding the greenhouse effect is part of these laws. Added to this core model is a wide array of sub-models describing cloud formation, interaction between the atmosphere and Earth's surface, and the effects of aerosols (dust and chemicals from human and volcanic activity), etc.

Modelers provide this body of equations with starting data (called initial and boundary conditions) and then run the model on supercomputers. The model outputs are temperature, air speed and direction, air pressure, and other parameters at all points across the globe. Each run of the model is for a single time interval, such as one simulated day. The model is run over and over in succession, to simulate long time periods and to forecast climatic results far into the future.

of temperature forecasts based on different world energy growth rates, 2) his chart on potential "radiative perturbations," and 3) his discussions on possible melting of icecaps and sea level rise.[27]

It's important to note that Dr. Hansen made his predictions even though his model did not incorporate other possible significant drivers of Earth's climate. These include the effects of solar variability, clouds and precipitation, and ocean currents. Today's models incorporate these effects, but have the same weaknesses because these important factors are not well understood.

THE VOSTOK ICE CORES

Model simulations are not measured data. Without some measured data showing that CO_2 was causing the Earth's warming, Dr. Hansen and the Climatists would have nothing but computer simulations. Enter the Vostok ice core data.

In 1987, J.M. Barnola and others published results from chemical analysis of ice cores retrieved from Vostok Antarctica (text box). Barnola's analysis showed a strong correlation between Earth's temperature and carbon dioxide.[28] Figure 10 shows Vostok data from some years after Barnola's early work. The top curve on the chart is temperature and the bottom curve is atmospheric carbon dioxide concentration. Scientists tell us that the temperature changes shown track the freeze and warm cycles of Ice Ages of Earth.

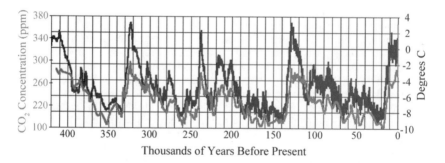

Figure 10. CO₂ and Temperature for the Last 400,000 Years. Temperature (top graph) and carbon dioxide levels (lower graph) from ice core analysis at the antarctic Vostok station. Present day at right. (Adapted from Petit et al.)[29]

Ice Cores and Atmospheric Data

Geologists have recently been able to learn about Earth's history from the antarctic and Greenland ice caps. Among the most important of this data is from the Vostok Ice Station in Antarctica. Since the 1970s, scientists from Russia, France, and the United States have drilled holes in the ice to a depth of more than two miles (3.2 kilometers). The ice cores retrieved from these holes provide a historic record of snowfalls deposited as ice over the last 500,000 years.

By analyzing oxygen and hydrogen molecules in the ice, scientists are able to construct a history of the temperature of Antarctica, and therefore of Earth in general. This allows a view of the ice ages and warming periods over thousands of years of Earth's history. By analyzing trapped air bubbles in the ice, researchers are able to measure the concentration of carbon dioxide in the atmosphere at points in the past. Some have questioned the accuracy of such measurements, but the results obtained are generally accepted by the scientific community.[30]

It wasn't long before James Hansen used this data as support for his theory that increased carbon dioxide causes climate change. He co-authored a paper in 1990 with C. Lorius and others, concluding that "the contribution of greenhouse gases to the Vostok temperature changes can be...between a lower estimate of 40% and a higher estimate of 65%." He supported this conclusion with computer simulations from his model.[31]

Al Gore has repeatedly used the Vostok graph showing temperature and carbon dioxide correlation as "powerful evidence" that CO_2 is causing climate change. It's used in both the book and movie *An Inconvenient Truth*, and Gore uses it in his lectures. Both Dr. Hansen and Mr. Gore *assume* that CO_2 is causing or amplifying the temperature increase. This all sounds plausible until one asks: If there was no human industry to emit carbon back in history, then how did the atmospheric CO_2 increase?

The answer is that the CO_2 increases are from natural causes, primarily out-gassing from the oceans. New research completed by the year 2000 showed CO_2 increases *lagging* Earth's temperature increase by centuries. Dr. Hubertus Fischer of the University of California demonstrated in 1999 that:

> High resolution records from Antarctic ice cores show that carbon dioxide concentrations increased...600±400 years *after* the warming...[our emphasis][32]

This means *temperature increase likely causes atmospheric carbon dioxide to increase, rather than the reverse.*

THIN SCIENCE FOR MAN-MADE GLOBAL WARMING

We can now summarize the argument for man-made global warming, which is simple and straightforward, but increasingly disproven. Here is the Climatist case:

1. Atmospheric CO_2 concentration is increasing. It has increased from about 315 parts per million since 1957 to 385 parts per million today (Figure 2). Most scientists agree.

2. Global temperatures are increasing. They have increased about 0.7°C since 1850 (Figure 3). These temperatures are swelled by Urban Heat Island Bias, but most scientists agree Earth has warmed.

3. The greenhouse effect tells us that increased CO_2 will cause some global temperature increase. Most scientists agree.

4. CO_2 concentrations and global temperatures are both at historic highs. *Many scientists and much evidence disagree.*

5. No known natural causes can account for Earth's temperature increases. *Many scientists and an increasing body of scientific evidence disagree.*

6. The models show that CO_2 is causing global warming. *Many scientists disagree and believe the models are in error.*

7. The Vostok ice core data shows a strong correlation between temperature and CO_2 concentration. *Scientific data now shows that CO_2 increases lagged temperature increases by an average of 600 years.*

The alarmist conclusion is that, since all of these are true, CO_2 must be responsible for global warming.

What we really have is a case of correlation, rather than causation. Just because global temperature and carbon dioxide are increasing, does not mean that CO_2 increase is the primary cause of global temperature increase. Argument number five is particularly important. The modelers have taken a great leap of faith that carbon dioxide is causing the warming *because they don't have another explanation*. **There is no direct experimental evidence that man-made CO_2 emissions are the primary cause of global warming.** As we'll show in Chapters 3, 4 and 5, other natural climatic forces, although less understood, offer more powerful explanations for the current warming.

Let's discuss a simple example that's closer to our everyday experience. As we all know, on average, more people are overweight today, so we agree that obesity is increasing in our society. Suppose as scientists, we develop a theory that eating peanut butter causes obesity. We have scientific evidence that peanut butter is high in calories and, if eaten in sufficient amount, will produce fat in the body (this is a theoretical basis). We have graphs that show that peanut butter consumption is increasing in the same way as obesity in humans (this is correlation). Suppose we also develop a supercomputer model that shows eating peanut butter also "causes" a person to eat more foods of all kinds, therefore leading to greater weight gain (this is positive feedback). We therefore proclaim that a ban on peanut butter will solve the problem of increasing obesity in society.

We expect you will say that other factors are more important in causing people to be overweight. These may be richer diet, greater consumption of meat, less exercise than in the past, the "supersizing" of restaurant portions, etc. The point is that peanut butter consumption is only a small factor in the cause of human obesity. So it is with man-made carbon dioxide emissions and global warming. As we will show, other climatic factors are much more important.

SUMMARY

By the late 1980s, the Climatist modelers had rushed to the judgment that emission of man-made carbon dioxide was the primary cause of global warming. They cobbled together rising temperatures, increasing atmospheric CO_2 concentration, Vostok ice core data, and simulation outputs from their computer models. At this point they left the path of science and chose the path of politics. As we'll see in Chapter 7, a United Nations–sponsored organization called the Intergovernmental Panel on Climate Change (IPCC) drove worldwide adoption of the Global Warming Dogma, requiring the bending of science to fit the dogma. We'll discuss the scientific evidence and show that man-made greenhouse gases are small compared to major natural climatic forces in determining Earth's temperature.

CHAPTER 3

CARBON DIOXIDE: NOT GUILTY

"All the rivers run into the sea; yet the sea is not full."
Ecclesiastes 1:7, The Bible

C arbon dioxide is not a pollutant. It is an odorless, colorless, harmless gas that is naturally produced by nature. It is an essential part of plant photosynthesis. In fact, most life on Earth would not exist without carbon dioxide.

Carbon dioxide is not poisonous to humans, plants, or animals. It does not cause cancer. We drink it in our beer and soft drinks. In fact, it's created by humans and animals, and 96% percent of atmospheric CO_2 is produced by nature's processes. The air we inhale has only a trace of carbon dioxide, but humans exhale air containing 4% carbon dioxide, produced from normal body processes. Yet, the Climatists and news media now call carbon dioxide a pollutant.

But, can't too much of anything become a pollutant? Even excessive drinking of water or an atmosphere full of carbon dioxide can be harmful. But the concentration of carbon dioxide in our atmosphere is very, very small.

CARBON DIOXIDE—A TRACE GAS

Carbon dioxide is a trace gas. Methane (CH_4), Sulfur Dioxide (SO_2),

39

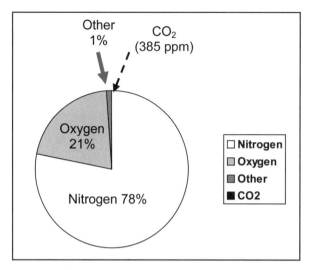

Figure 11. Gases in Our Atmosphere. Percentages shown are for dry air. Water vapor can be up to about 2% of the atmosphere for some local conditions. The CO_2 portion is within the thin vertical line on the graph.

Carbon Monoxide (CO), Ozone (O_3), and other hydrocarbon gases are also greenhouse gases discussed by the IPCC, but the atmosphere contains even less of these gases. A breakdown of gases in our atmosphere is shown in Figure 11. Can't see the carbon dioxide? That's because 385 parts per million isn't very much.

To picture how small 385 ppm is, imagine yourself at a packed indoor sports arena with 10,000 spectators. A capacity crowd at a Duke University or Notre Dame basketball game is about the right size. Or, for international arenas, consider an event at Yoyogi National Gymnasium in Tokyo, the Hovet in Stockholm, or Madrid Arena at Casa de Campo. If the attendance at the event represented a 10,000-molecule sample of our atmosphere, then the amount of carbon dioxide would be only four people of the ten thousand attending. *The total man-made CO_2 emissions in history would amount to only one of those 10,000 spectators.*

THE CARBON CYCLE

Carbon is found in the atmosphere, oceans, and Earth's crust. There is a continuous exchange of carbon between the atmosphere and

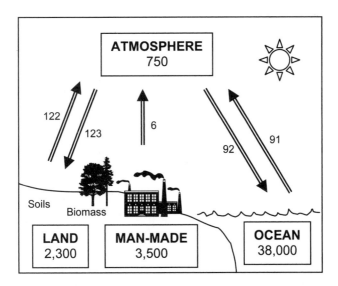

Figure 12. The Carbon Cycle. Numbers in billions of tons of carbon. The numbers shown within the boxes are estimated carbon totals residing in each climate subsystem. The numbers next to the arrows are estimates of annual transfers of carbon in the form of CO_2. (Adapted from IPCC 2007)[1]

the surface systems. Plants capture CO_2 from the atmosphere and convert it to carbon-based biomass as part of photosynthesis. Decomposing vegetation forms CO_2, which is returned to the atmosphere. Carbon dioxide is absorbed by the oceans and emitted from oceans to the atmosphere, depending upon surface temperature, and other factors. These CO_2 processes are called the "carbon cycle," as shown in Figure 12. The data in the figure are from the IPCC.

Once alarmists adopted the dogma that man-made CO_2 was causing global warming, they then developed pseudo-scientific assumptions to support this dogma. Two of these assumptions involved the carbon cycle. First, the IPCC assumed man-made carbon dioxide was impacting the carbon cycle:

> The additional burden of CO_2 added to the atmosphere by human activities…leads to the current 'perturbed' global carbon cycle… These perturbations to the natural carbon cycle are the dominant driver of climate change because of their persistent effect on the atmosphere.[2]

Second, the IPCC assumed that atmospheric carbon dioxide must stay in the atmosphere a long time:

> About 50% of a CO_2 increase will be removed from the atmosphere within 30 years, and a further 30% will be removed within a few centuries. The remaining 20% may stay in the atmosphere for many thousands of years.[3]

But both of these assumptions are refuted by science. Let's discuss the "long-lifetime" assumption first.

The idea that carbon dioxide stays in the atmosphere for hundreds of years runs afoul of Henry's Law, a law of chemistry. Henry's Law is one of the laws governing the properties of gases, developed in 1803 by William Henry. It states:

> For a given temperature, the amount of gas dissolved in a solution is directly proportional to the pressure of the gas in the air above the solution.[4]

We can observe Henry's Law in action when we open a bottled soft drink. Prior to opening, most of the gas in the neck of the bottle is carbon dioxide, with CO_2 also dissolved in the drink. Upon opening the bottle, a loud pop is heard when the CO_2 escapes into the atmosphere. If someone drinks the soda immediately, dissolved carbon dioxide bubbles in the mouth of the drinker. However, if the drink is left open for an hour before drinking, its taste becomes "flat." Carbon dioxide dissolved in the soda has escaped into the atmosphere, bringing the level of dissolved CO_2 in the bottle into equilibrium with the lower CO_2 vapor pressure of the air.

Henry's Law means that dissolved CO_2 in the ocean is closely connected to CO_2 in the atmosphere. Therefore, if atmospheric CO_2 increases from man-made emissions, the vapor pressure of CO_2 will increase, causing oceans to absorb more CO_2 from the atmosphere. Figure 12 shows the oceans have about 38,000 billion tons of carbon dissolved as CO_2, or over 50 times as much as the 750 billion tons of carbon in CO_2 in the atmosphere. So, if man-made CO_2 adds six billion tons of carbon per year to the atmosphere and Henry's Law holds, this increase should soon be absorbed by the ocean.

So what to do if faced with Henry's Law? The alarmist scientists assumed a "buffer factor," which is also sometimes called the "Revelle factor."[5] Bert Bolin postulated in 1958 that the surface layer of the ocean acts a buffer to defeat Henry's Law and allow accumulation of CO_2 in the atmosphere:

> ...less than 10% of the excess fossil CO_2 in the atmosphere should have been taken up by the mixed layer. It is therefore obvious that the mixed layer acts as a bottleneck in the transport of fossil CO_2 into the deep sea.[6]

However, experimental evidence shows that water adjusts quickly to equalize CO_2 vapor pressure and dissolved CO_2, which contradicts the buffer factor. This is well summarized by Dr. Tom Segalstad, of the University of Oslo:

> Experimentally it has been found that CO_2 and pure water at 25°C reaches 99%...equilibrium after 30 hours and 52 minutes; after shaking (like wave agitation) 99% equilibrium is reached after 4 hours and 37 minutes.[7]

If Henry's Law holds, then man-made emissions of carbon dioxide will not be able to accumulate in the atmosphere and must be absorbed by the oceans.

Let's look back at Figure 12. The arrows are the IPCC estimates for flows of carbon between oceans, land, and atmosphere. These flows are shown to be approximately in balance, except for the six billion-ton arrow from man-made emissions, which, according to the IPCC, causes carbon dioxide to accumulate in the atmosphere. The figure shows that each year the oceans absorb about 92 billion tons of carbon as CO_2 with plants absorbing an additional 123 billion tons of carbon. This means that about 28% of all atmospheric carbon is absorbed every year. This is much larger than the six billion tons emitted by industry each year, which forms only about 1% of the CO_2 in the atmosphere. Have we missed something? If over 20% of the carbon dioxide is being absorbed every year, how can human emissions remain for 30 years or more, per the IPCC?

In fact, the prevailing science of the last 50 years, prior to the re-write by the IPCC, places CO_2 atmospheric lifetime at between two and twelve years, with a mean of about five to six years. More than 30 papers have experimentally measured short CO_2 lifetimes, including one from Bert Bolin.[8] Yet the IPCC mentions none of this research and sticks to assumed values for carbon dioxide lifetimes of 30 years to hundreds of years or more.

The other carbon cycle assumption claimed by the IPCC is that the Keeling Curve (Chapter 2, Figure 5) is driven by man-made CO_2 emissions to the atmosphere (rather than natural CO_2 emissions). However, the analysis of isotopes of carbon that are part of CO_2 in the atmosphere challenges this assumption. See the text box on page 46 for more detail on carbon isotopes.

The carbon dioxide emitted from combustion of hydrocarbon fuels is an isotope that is lighter than the CO_2 that is naturally found in nature. That is, CO_2 from hydrocarbon fuels contains less heavy carbon atoms. This is because such fuels are believed to come from plants, which prefer CO_2 with lighter carbon atoms for photosynthesis. Therefore, an analysis of the carbon isotopes in atmospheric carbon dioxide can show whether the atmosphere is being filled with CO_2 from man-made or natural sources.

As we have discussed, the IPCC and global warming alarmists claim the increase in atmospheric carbon dioxide over the last century is *entirely* due to man-made CO_2 emissions. The isotope ratios of carbon in naturally occurring CO_2 and in man-made CO_2 are well known. Using an estimated rise of 105 parts per million from the "natural background level" of 280 parts per million of atmospheric CO_2, the atmospheric mix of naturally occurring CO_2 and man-made CO_2 should yield a $^{13}C/^{12}C$ isotope value of -12 as shown in Figure 13 if the IPCC is correct. However, the actual measured atmospheric value is -8. This means that the current atmospheric portion of man-made carbon dioxide is only 4%, not 27% per the IPCC. *Therefore, the measured carbon isotope composition of the atmosphere is different than predicted by IPCC assumptions.*

Carbon isotope ratios can be measured today with a high degree of accuracy. If the atmosphere is actually filled with man-made CO_2, this should be apparent from the isotope measurements. Dr. Segalstad, formerly an expert reviewer for the IPCC, brought this evidence to the attention of the IPCC. But the IPCC ignored his analysis.[9] When a scientific theory (increasing atmospheric CO_2 due to man-made emissions) does not agree with experimental results (the carbon isotope composition of the atmosphere), the theory should be modified or abandoned. The IPCC has done neither.

In addition, the actual measured atmospheric isotopic carbon ratio

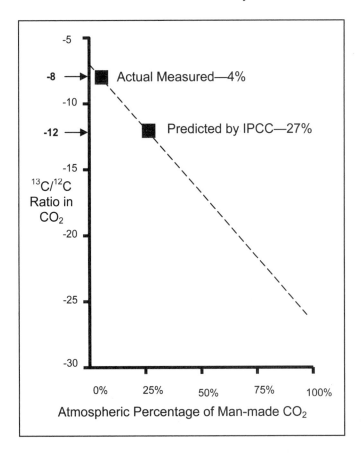

Figure 13. Atmospheric Carbon Isotopes of CO₂. The diagram shows the percentage of atmospheric CO_2 along the dotted line, depending upon the mix of naturally produced CO_2 and man-made CO_2 as of 2008. Experimental measurements show a carbon ratio of -8 vs. the IPCC predicted -12. Units are "per mil" relative to the Pee Dee Belemnite carbon isotope standard. The graph uses methodology of Segalstad (2009).[10]

has increased only very slightly since 1980, while the IPCC predicted carbon ratio for an atmosphere filled with man-made CO₂ increases much faster. *Therefore, both the isotopic composition and rate of change are wrong for IPCC assumptions.* This means that Henry's law is working and emitted man-made CO₂ is rapidly being absorbed by surface systems. This further means that most of the increase in atmospheric CO₂ concentration on the Keeling curve is due to natural processes rather than man-made emissions. Ian Plimer, Australian geologist and author of *Heaven and Earth: Global*

Carbon Isotope Analysis

Nature's elements come in varying atomic weights called isotopes. Isotopes are atoms of the same chemical element that have different atomic weight. In the case of carbon, the most common carbon atoms found in nature have an atomic weight of 12, which is called Carbon 12 or ^{12}C. Carbon is also found in lesser concentration as Carbon 13 (called ^{13}C). The extra atomic weight of Carbon 13 is due to an additional neutron in the nucleus of the atom. Carbon 12 is about 99% and Carbon 13 is about 1% of carbon found in nature, with other isotopes found only in very small amounts.[11]

Isotope analysis can be very useful for tracing climatic and other processes. Isotope analysis was used to analyze the Vostok ice cores we discussed earlier to correlate atmospheric CO_2 to atmospheric temperature. It's also used in radiocarbon dating to determine the age of organic materials.

Today's scientists use sophisticated equipment to measure the ratio of carbon isotopes in liquids, gases, and solids. These include elemental analyzers and isotope ratio mass spectrometers as part of computerized systems. Carbon isotope ratios can be measured with a high degree of accuracy.

Since carbon isotopes are part of carbon dioxide molecules, carbon ratio analysis can be used to determine the composition of atmospheric CO_2. Carbon ratios are measured relative to the Pee Dee Belemnite standard, which is a belemnite fossil formation of South Carolina. Units are expressed in molecules per thousand, or "per mil" relative to the Pee Dee standard.

Carbon ratio measurements of naturally occurring CO_2 show a $^{13}C/^{12}C$ ratio of -7 per mil. Similar measurements of man-made CO_2 from combustion of hydrocarbon fuels show a ratio of -26 per mil. Yet, measurements on mixed air samples show -8 per mil, a much lower ratio than the atmospheric mix predicted by the IPCC, and their claims that the atmosphere is filled with man-made CO_2.

Warming, the Missing Science, argues that CO_2 ejected from undersea volcanoes is a major source of emissions to the atmosphere, not considered by the IPCC.[12] It's likely that most of the rising CO_2 in the atmosphere is due to emissions from the oceans over the last several hundred years.

CARBON DIOXIDE IS GREAT FOR PLANTS

Carbon dioxide is to plants as oxygen is to humans. Plants "breathe" CO_2 and use it to make sugars in photosynthesis. It's ironic that environmentalists, sometimes called "tree-huggers," have now become Climatists with a mission to eliminate carbon dioxide, *a gas*

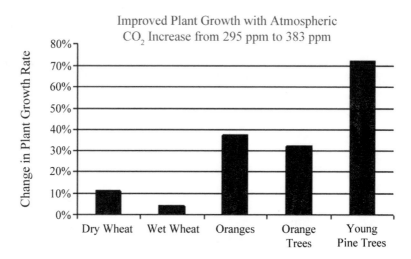

Figure 14. Plant Growth with Increased CO₂. Calculation of improved plant growth rates from increased atmospheric CO_2 concentration in 2007 compared to the pre-industrial CO_2 level. (Adapted from Robinson et. al., 2007)[13]

beneficial to trees!

Trees and food crops grow faster with more CO_2 in the air. They lose less water vapor from their leaves and are better able to grow and survive in drier conditions. This makes plants more drought resistant and able to prosper in more varied climatic conditions.

Figure 14 shows calculated growth rates of wheat and trees from Earth's recent increase in atmospheric CO_2. Multiple studies have shown similar positive effects from increased carbon dioxide. These studies show increased forest growth in the United States and Amazon rain forests over the last 50 years. Should atmospheric CO_2 increase to 600 ppm, it's estimated that the rate of tree growth will increase over 100% from the IPCC's "natural level" of 280 ppm.[14] Plants live and grow by using carbon dioxide in the air. Animals and humans get their carbon by eating plants. **Carbon dioxide is essential for life and certainly not a pollutant.**

THE CORE ISSUE: THE GREENHOUSE EFFECT

Suppose we accept the dogma of the Climatists that man-made CO_2 is accumulating in the atmosphere. This means we'll need to ignore the carbon cycle evidence that we've just discussed. But this will allow us to get to the main question, which is "Will the greenhouse effect from carbon dioxide cause catastrophic global warming?"

It's important to keep in mind that no one really knows the certain answer to this question. Despite the assertions of Al Gore and James Hansen, large areas of climate science, including the effects of clouds, precipitation, ocean currents, aerosols, and even the radiation of the sun, remain only partly understood.

Forecasts by the General Circulation Models and IPCC are usually based on doubling atmospheric CO_2 concentration, from the background level of 280 ppm to 560 ppm. They claim this doubling will cause a certain amount of "radiative forcing," resulting in increased global temperatures. Radiative forcing is defined as the net change in radiation (both sunlight and from Earth's surface) at the top of the troposphere, which is Earth's lowest layer of atmosphere.

It's estimated that, on average, each square meter of Earth's surface receives about 240 watts of solar energy, resulting in an average surface temperature of about 15°C (or 59°F). For a doubling of CO_2, the modelers put forward the following three claims:

Claim #1: A radiative forcing of 3.7 watts per square meter will occur, resulting in a surface temperature increase of 1.2°C (or 2.2°F) from the CO_2 increase alone.[15]

Claim #2: An additional greenhouse effect will occur from increased water vapor (a positive feedback), so that the total surface temperature rise is about 3°C (or 5.4°F).[16]

Claim #3: Additional temperature increases may occur from other positive feedbacks, such as ice cap melting or Amazon rain forest deforestation. For these additional events to happen, Claims #1 and #2 must occur.

This 3°C increase is the basis for all of the calamity scenarios. Additional events, such as ice cap melting in Claim #3, are needed for some of the greatest disasters. Let's look first at radiative forcing due to carbon dioxide itself and then alleged additional temperature increase from positive feedback.

CARBON DIOXIDE GREENHOUSE IMPACTS

Most scientists agree that the greenhouse effect from CO_2 is non-linear. This means that each additional amount of emitted carbon dioxide has less effect on temperatures than the previous amount. This effect is shown in Figure 15, reproduced from geologist David Archibald. As can be seen, most of the temperature increase comes with the first 20 ppm of carbon dioxide in the atmosphere.

Doubling CO_2 from 280 ppm to 560 ppm causes less than a 2% increase in radiative forcing. The concentration would need to be doubled again to over 1100 ppm to get another 2%. Dr. Lindzen, Alfred E. Sloan Professor of Meteorology at MIT, points out:

> ...greenhouse gases added to the atmosphere through man's activities since the late 19th Century have already produced three-quarters of the gross radiative forcing that we expect from a doubling of CO_2.[17]

In other words, because of the non-linear effect, the rise in carbon

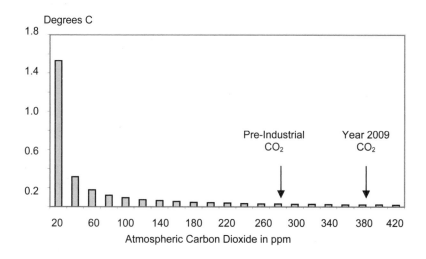

Figure 15. Diminishing Warming Effect of Atmospheric CO₂. Change in global temperature for each 20 ppm increase in atmospheric carbon dioxide. Most temperature change is caused by the first CO₂ added to the atmosphere. (Adapted from Archibald, 2008)[18]

dioxide concentration from 280 ppm to 385 ppm has already provided most of the forcing that the IPCC expects from a doubling of CO_2 in Claim #1. As you recall, we have seen about a 0.7°C increase in global temperature, which is clearly not a catastrophic situation. The Climatists need positive feedback added to CO_2 forcing to get their forecasted calamities. So the key issue is: Will the CO_2 increase result in additional temperature increase from positive water vapor feedback, per IPCC Claim #2?

IT'S ALL ABOUT THE FEEDBACK

You may ask, "Isn't carbon dioxide the most important greenhouse gas in our atmosphere?" Well, not by a long shot. Most scientists estimate that more than 90% of greenhouse gas warming is due to water vapor, with CO_2 accounting for less than 10% of the effect.[19] So feedback from water vapor, as well as weather in general, is vital to estimating the effects of global warming.

We've modified our greenhouse gas model to discuss some of these effects in Figure 16. As before, high energy sunlight enters our atmosphere and is absorbed by the Earth. Earth continuously emits lower energy infrared radiation. Most of the infrared radiates out into space, but some is absorbed by greenhouse gases. We have agreed that Earth is in radiation-energy balance, with the incident solar radiation equal to the outgoing infrared radiation. The Keeling data shows that CO_2 concentration is growing in the atmosphere since 1957, so Earth's temperature rises by a small amount to maintain the energy balance (Claim #1). The models of the Climatists then *assume* the amount of water vapor in the atmosphere rises as a positive feedback, further increasing the greenhouse effect (Claim #2).

There are two major problems with the water-vapor positive-feedback assumption. First, this assumption was made without any supporting experimental data. It was first assumed by modelers in the 1970s, providing results showing large global temperature increases from CO_2 forcing. Dr. William Gray, Emeritus Professor of Atmospheric Science at Colorado State University and leading hurricane forecaster, has remarked:

...so of course Hansen got warming. Because he assumed as the

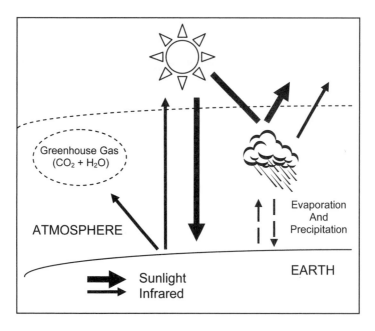

Figure 16. Greenhouse Effect with Weather Feedbacks. Cloud formation and precipitation are major forces involved in the greenhouse effect.

CO_2 built up, he put all this extra water vapor in the model. Well, they all do this [the models].[20]

Second and equally important, the assumption ignores other larger processes of our weather systems. We've added a cloud to our greenhouse model in Figure 16 to show the involvement of weather. Weather and clouds (along with ocean currents) have the largest effect on the surface temperature of Earth. Much of the 30% of sunlight reflected from Earth is reflected by clouds. A small increase in cloud cover, resulting in a 1% increase in Earth's reflectivity, would more than compensate for any CO_2 greenhouse warming. A 1% increase in rainfall efficiency would also compensate for CO_2 warming. A small decrease in low-level cloudiness or an increase in high-level cloudiness could increase warming.[21] Climate scientists continue to learn more about these processes, but they are still poorly understood. But the modelers assumed carbon dioxide would drive Earth's temperature, relegated weather to a minor position in their simulations, and made their alarming predictions anyway.

Let's consider the size of forces involved. To look at the positive feedback assumption another way, *the Climatists have essentially claimed that build up of a trace gas, carbon dioxide, is controlling the weather, a system of huge forces with thousands of times more energy.* This is more like the flea wagging the dog than the tail wagging the dog. Dr. Roy Spencer, Principle Research Scientist at the University of Alabama, puts it in perspective:

> Al Gore likes to say that mankind puts 70 million tons of carbon dioxide into the atmosphere every day. What he probably doesn't know is that mother nature puts 24,000 times that amount of our main greenhouse gas—water vapor—into the atmosphere every day, and removes about the same amount every day. While this does not "prove" that global warming is not manmade, it shows that weather systems have by far the greatest control over the Earth's greenhouse effect, which is dominated by water vapor and clouds.[22]

The point is that weather processes of evaporation, cloud formation, and precipitation are the primary cooling agents of Earth's surface. The greenhouse gas effect from carbon dioxide is a secondary process which adjusts to Earth's weather, not the primary driver of the weather, as the modelers have assumed.

Papers have been written on both sides regarding whether positive feedback is occurring or not, but we'll not pursue these here. Let's close the topic of positive feedback with a statement by Dr. William Happer, Professor of Physics at Princeton University:

> With each passing year, experimental observations further undermine the claim of a large positive feedback from water. In fact, observations suggest that the feedback is close to zero and may even be negative. That is, water vapor and clouds may actually diminish the already small global warming expected from CO_2, not amplify it. The evidence here comes from satellite measurements of infrared radiation escaping from the earth into outer space, from measurements of sunlight reflected from clouds...[23]

GARBAGE IN, GARBAGE OUT

Model simulation results are not experimental data. No matter how many different models get similar results, no matter how many times models are run, if the underlying physical basis of the assumptions is wrong, the models will give the wrong result. And the modelers have been finely tuning the wrong result for thirty years.

The IPCC puts great weight on simulation results from the General Circulation Models (GCMs) to project catastrophic global warming. They devote an entire chapter to the climate models in the Fourth Assessment Report, released in 2007. The report talks about how "confidence is higher" and "resolution is better" in model results since the Third Assessment Report (TAR) published in 2001. The IPCC also discusses how the complexity of the models has "increased over the last few decades." [24]

While the complexity of the GCMs has increased, they also contain two fundamental flaws. First, every model includes CO_2 forcing of global temperatures, and second, they assume increasing water vapor with increasing temperatures. These two assumptions are the quicksand foundations of the early Manabe and Hansen models, and continue to be fundamental errors in today's GCMs.

The GCMs are basically large regression models that rely on "curve fitting" to build model parameters. That is, values are assumed for physical factors such as cloud formation and sunlight reflection from ice, computations are done by supercomputers, and then model results are compared to actual weather history. The models are run over and over, each time changing parameter values, until the "right" values are determined to make model results match climate history.

Once models are built according to the dogma that man-made carbon dioxide is causing global warming, it's relatively easy to get simulations to match past climate. The great mathematician John von Neumann reportedly said:

Give me four adjustable parameters and I can simulate an elephant. Give me one more and I can make his trunk wiggle. [25]

Figure 17 shows an IPCC graph of different groups of model simulations that is a favorite of the alarmists. The top gray band of model results includes greenhouse-gas forcings that closely track historical temperatures (the black line), compared to the bottom gray band with the man-made effects removed, which no longer tracks

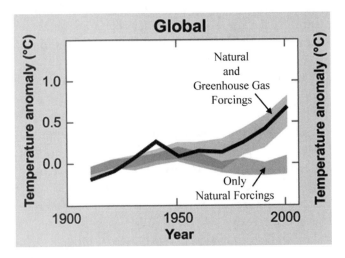

Figure 17. IPCC Model Claims of Man-made Warming. The graph shows two gray bands of model simulations, the top with both natural and man-made forcings and the bottom with only natural forcings. The top simulation tracks the black historical temperature line. (Adapted from IPCC, 2007)[26]

temperature. The IPCC claims this as evidence that climate change is man-made. This is complete rubbish. Of course the curves based on the complete models track better! *The models were curve-fitted to historical temperatures with the greenhouse effects included.* If you curve-fit a model to temperature and then remove the key factors, it's obvious that the curve will no longer fit the record. But the IPCC puts this right on page 40 of their 2007 Synthesis Report as "evidence" of man-made global warming.[27] This same graph is used over and over again in many published books by climate alarmists.[28]

An example of this curve fitting is the way GCMs have modeled aerosols. Aerosols are the dust and miscellaneous materials that nature expels into the atmosphere. Values have been *selected* for aerosols to make the models work and to match the past historical record, without sufficient experimental data to justify the selection. Dr. Theodore Anderson of the University of Washington has commented about model handling of aerosols:

> ...the possibility of circular reasoning arises—that is, using the temperature record to derive a key input to climate models that are then tested against the temperature record.[29]

The net effect of aerosols is to cool the planet. Recent studies show that the effect of aerosols is larger than expected, possibly great enough to offset any greenhouse effect from carbon dioxide. Studies show that plankton in the oceans and plants on land emit large quantities of chemicals and organic particles (such as pollen in the case of plants). These particles make their way into the atmosphere, forming nuclei that stimulate cloud formation. Increased levels of cloud formation can result in greater reflection of sunlight, which reduces global temperatures. Studies further show that increased atmospheric carbon dioxide increases the production of organic aerosols from both the ocean and land biospheres, providing a negative feedback to the climate system.[30]

But without any experimental data, the modelers have just tuned their aerosol model values to match model results to historical temperature records. It's fairly easy to track history. Since modelers know the answer first (the historical record), by adjusting parameters such as aerosol effects they can build a model that matches the historical record and supports the man-made global warming dogma.

Yet, the greatest weakness of the models continues to be their inability to effectively model clouds and precipitation. Weather is complicated. Updrafts and downdrafts of cumulus cloud convection, the organization of clouds into fronts, the turbulence of chaotic cloud systems, and complicated cloud formation mechanisms are difficult to model. The GCMs attempt to do this by "parameterization," or selecting numerical parameters to represent real-world effects. Dr. David Randall of Colorado State University is one of many climatologists who has written about model limitations. He states:

> There is little question why the cloud parameterization problem is taking a long time to solve: It is very, very hard…a sober assessment suggests that with current approaches the cloud parameterization problem will not be "solved" in any of our lifetimes.[31]

How can we tell if the models are right? This will require a little explanation, so please stick with us. The IPCC has used the term "fingerprint" to describe an observed pattern of warming with a pattern calculated by the GCMs. Chapter 9 of the IPCC's 2007 Fourth Assessment Report shows such a fingerprint of temperature increases predicted by the models (See Figure 18).

The graph shows model-predicted temperature increases in the atmosphere due to greenhouse gas warming over the last 100 years. The right axis shows height above Earth's surface in kilometers. The horizontal axis shows Earth's latitude. "Eq" in the middle is for the equator, with the North Pole on the left (90N) and the South Pole on the right (90S).

Recall that the IPCC defines forcing as occurring at the top of the troposphere, which is at a height of about 12 km. The models *unanimously predict* increasing temperature from Earth's surface to the top of our troposphere and centered over the equator, due to man-made warming. This forms a "hot spot," which is the large dark spot in the center of the chart in Figure 18. Scientists believe that only greenhouse warming will produce this warming fingerprint.[32]

But satellite and weather balloon (radiosonde) data do not show this signature warming. Figure 19 shows actual observed temperature change data from Hadley Center weather balloon data from 1979–1999. The data shows little or no warming of the tropical troposphere, in stark contrast to the warming which must occur if greenhouse gases are forcing the warming. United States MSU satellite data confirms the lack of a hot spot in the balloon data.

In 2000, the National Academy of Sciences attempted to resolve this problem, but found:

> …The various kinds of evidence examined by the panel suggest that the troposphere actually may have warmed much less rapidly than the surface…due both to natural causes and human activities…[33]

The US Climate Change Science Program published a report in 2006 with the expressed purpose of resolving this problem. In line with the Climatist dogma, the executive summary states:

> Specifically, surface data showed substantial global-average warming, while early versions of satellite and radiosonde data showed little or no warming above the surface. This significant discrepancy no longer exists because errors in the satellite and radiosonde data have been identified and corrected.[34]

This conclusion in the front of the study is surprising, because it conflicts with conclusions in the document. Chapter 5, the key chapter of the report that analyzes the difference between model projections and observed temperatures states:

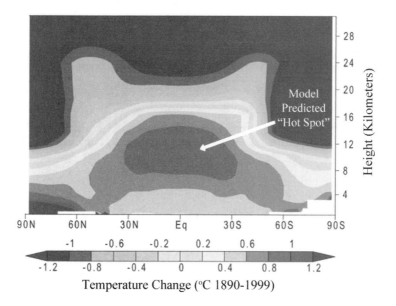

Figure 18. Model-Predicted Temperature Change. The models predict a "hot spot" in the troposphere centered over the equator. (Adapted from IPCC, 2007)[35]

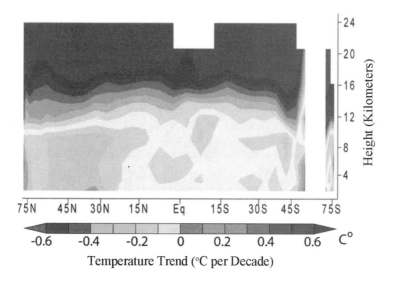

Figure 19. Actual Observed Temperature Change. Weather balloon data from Hadley Centre from 1979–1999 shows no significant warming in tropical latitudes, in conflict with model predictions. (Adapted from CCSP, 2006)[36]

Tropical...trends in both radiosonde data sets and UAH satellite data are always positive (larger warming at the surface than aloft), and lie outside the range of model results.[37]

The report attributes the differences to "net spurious cooling" that corrupts the radiosonde data. In any case, there is no experimental evidence to support the signature temperature rise in the tropics due to greenhouse warming predicted by the models. **This is powerful evidence that the models are wrong and do not accurately portray our climate.**

SUMMARY—CARBON DIOXIDE: NOT GUILTY

An understanding of the science is critical to determine the cause of global warming. We have more science in Chapters 4 and 5, but our story is less technical thereafter. If you have stayed with us thus far, we can conclude:

A. Carbon dioxide is not a pollutant, but essential for life on Earth.

B. The IPCC carbon cycle model has serious discrepancies. Isotopic evidence shows that man-made CO_2 is rapidly being absorbed by Earth's surface systems.

C. The greenhouse effect of CO_2 is accepted science and may raise the Earth's temperature by $1.2\,°C$ for a CO_2 doubling, but this is far from a catastrophic outcome.

D. The Climatist assumption of water vapor greenhouse forcing, driven by the trace gas carbon dioxide, is neither probable nor supported by experimental data.

E. The General Circulation Models are fundamentally flawed. Tropospheric warming predicted by the models is not found in measured atmospheric data.

If CO_2 is not the primary cause of global warming, what is the cause?

CHAPTER 4

CLIMATE CHANGE IS CONTINUOUS

"We need a proper understanding of the past to correctly judge the present if we ever are to foretell the future." Dr. Craig Idso, 2009

"Climate change is real." This is probably the most frequently used phrase in the global warming debate. A recent Google search on this phrase turned up 200,000 hits. But it's also a meaningless phrase. Of course climate change is real. Global warming is real. Global cooling is real. *Climate change is not only real, but continuous.*

A trip to most any national or state park in the U.S. will provide a lesson in climate change. The history of the glaciers that dug the "kettles" in Kettle Moraine State Park in Wisconsin is an example. The walls of the Grand Canyon show thousands of years of changing climatic conditions. Geologic studies of the Southern California coast show more than a dozen climatic changes, *each* with a sea level change of 400 feet.[1] All of these changes involved natural adjustments to the Earth's temperature, without man-made influences.

OUR CLIMATE IS DOMINATED BY CYCLES

Our lives run in cycles. Our heart beats in a continuous cycle about 70 times per minute. We inhale and exhale about 20 times per minute. Anyone who has flown over the ocean has experienced "jet lag," in which the body must adjust to the change in our daily cycle of sleep and wake.

As we know, our weather is also dominated by cycles. During a long winter, it may seem that spring will never come, but it does, every year. We have four seasons as part of Earth's revolution around the sun. These seasons cause major changes in not only temperature, but also in many elements of the weather, such as wind current direction and magnitude, the location and size of high-pressure areas, the amount and frequency of precipitation, and the formation of major storms. Similar weather patterns tend to repeat year after year.

Ocean forces are also driven by cycles. Tides rise and fall in excess of 30 feet in a single day on some shores, acted on by the gravitational pull of both the moon and the sun. The orbit of the moon around Earth provides a monthly recurring cycle of tidal activity. The Meridional Overturning Circulation, also called the great Global Ocean Conveyor Belt, moves huge quantities of water on a path that stretches from the North Atlantic Ocean to the Antarctic Ocean to the North Pacific Ocean and back again, over a cycle of roughly 1,000 years (Figure 20).

Earth's climate is also dominated by cycles. Climatic cycles are driven by energy arriving from the sun, Earth's revolution around the sun, and regular cycles of ocean currents and other natural forces.

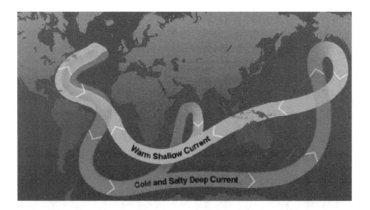

Figure 20. Meridional Overturning Circulation. The great ocean conveyor belt makes a journey from the North Atlantic to the Antarctic to the North Pacific and back on a cycle of about 1,000 years. The light-grey paths show the warm surface water portion of the circulation and the dark-grey paths show the cold deep water portion of the circulation. (USGCRP, 2009)[2]

Some of these repeat on a period of tens of thousands of years, such as the Milankovich Cycles (discussed below). Others repeat on a 1,500-year cycle or shorter time periods. A recognized short cycle is the El Niño Southern Oscillation that changes Pacific Ocean weather every three to seven years. El Niño operates as part of the Pacific Decadal Oscillation, which cycles on a time period of decades.

Climatism chooses to downplay many of the natural cycles of Earth in the belief that mankind is responsible for global warming. Climatists believe the climate of Earth was optimum and unchanging prior to the growth and industrialization of humanity. On this mistaken belief, they conclude that the temperature of Earth was mostly constant prior to man-made greenhouse gas emissions. The IPCC adopted the pseudo-science of the infamous Mann "hockey-stick curve" to declare temperature unchanged for 1,000 years prior to human industrialization. We will show this assertion is incorrect. *Natural cycles have dominated the Earth's climate for all of history.* These same cycles explain most of the recent warming of our planet.

LONG-TERM: THE MILANKOVITCH CYCLES

The Milankovitch Cycles are three long-term cycles in the Earth's axis and orbit motion. They are named after work of Milutin Milankovitch, a Serbian astronomer, who developed theories in the 1920s to explain Earth's Ice Ages. It's generally accepted that Milankovitch's theories explain much of the long-term climatic cycles of Earth, but uncertainties remain.

The Earth's rotational axis forms an angle of about 23° relative to the plane of our orbit around the sun. This angle remains almost fixed relative to the plane of Earth's orbit over the short-term. As the Earth orbits around the sun, the lean of the axis changes position relative to the sun. In the northern hemisphere, the axis leans toward the sun during summer, but the axis leans away in the winter, giving us our seasons. Southern hemisphere seasons are reversed. Today the Earth's axis points toward Polaris, the North Star, on our yearly journey around the sun.

However over the long-term, the axis of Earth is actually rotating, which is called "precession." Precession is the first of the Milankovich Cycles. This motion is similar to that of the axis of a spinning

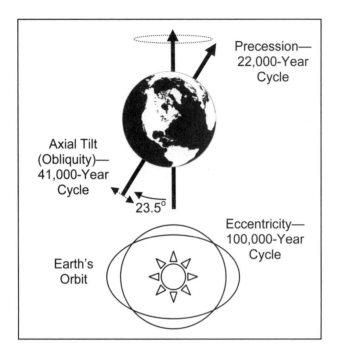

Figure 21. The Milankovich Cycles. Cyclical changes in the pointing of the Earth's axis relative to the fixed stars (Precession), the axial tilt of the axis (Obliquity), and the elliptical shape of the Earth's orbit (Eccentricity). (Milankovitch, 1941)[3]

gyroscope that leans slightly from perpendicular (see Figure 21). A full cycle of Earth's axial precession takes about 22,000 years. About 10,000 years from now, our axis will no longer be pointed at Polaris, but instead at the star Alpha Draconis, as it was 10,000 years ago.[4]

The second cycle involves changes in the angle of Earth's axis, called "obliquity." The angle is about 23.4° today and varies between about 21.6° and 24.5°.[5] A complete cycle of the Earth's axial tilt is about 41,000 years.

The third Milankovitch Cycle involves the shape of the Earth's orbit, called "eccentricity." Earth's orbit around the sun is almost circular, but periodically becomes a more flattened ellipse and then back to a circle again. The eccentricity of the Earth's orbit moves through a complete cycle in about 100,000 years.[6]

Astronomers have added another cycle to the three that Milankovich studied. This is the inclination of the Earth's orbital plane which drifts up and down (tilts) relative to its present orbit. The inclination cycle of Earth's orbit is about 70,000 years.[7]

Precession, obliquity, eccentricity, and inclination have a powerful long-term impact on Earth's climate. These forces combine to provide climatic cycles in subsets and supersets of 22,000, 41,000, 70,000, and 100,000 years. These cycles are believed to be responsible for temperature changes resulting in the freezing and thawing of the Earth's Ice Ages, including the glaciers that covered much of North America and Eurasia 20,000 years ago. The Ice Ages are believed to have lasted as long as 90,000 years, with temperatures dipping 7 to 12°C lower than the intervening warmer periods.[8]

The Milankovitch Cycles are primarily caused by the interaction of the gravitational forces of the Sun, Moon, Jupiter, and Saturn with Earth's rotation and orbit. Therefore, our climate is continually being pushed and pulled by major forces of our solar system. The Climatists believe in the actions of the Milankovitch Cycles, but dismiss the medium-term and short-term cycles we'll discuss next.

MEASUREMENT OF PAST TEMPERATURES

As evidence to support man-made global warming, Al Gore states:

> The ten hottest years in the atmospheric record, going back only 160 years, have been in the last eleven years.[9]

Following Mr. Gore's lead, the alarmist crowd sounds two notes of concern, often without the 160-year qualification: 1) it's warmer today than ever in history, 2) the 0.5°C temperature warming since 1970 is unprecedented. Therefore global warming must be due to hydrocarbon emissions! This seems to make sense on the surface. I recall years from my youth in Chicago when skating rinks were frozen for the whole month of January. Now it's sometimes tough to find snow for skiing.

Consider a man who lives all of his life in New York City, who has no knowledge of geography, and who takes a driving trip to San Francisco. Upon reaching the Continental Divide in Colorado, he was heard to say: "I've driven 2,000 miles and this is the highest elevation I've reached. I must be on the highest mountain on the planet!" The point is that 40 years, or even 160 years, is not sufficient time for "hottest ever" or "fastest warming" conclusions to be drawn. Climate and temperature trends operate over cycles that are hundreds and thousands of years long.

So how can we compare today's temperatures to those that prevailed hundreds of years ago? While the Greeks made simple thermometers about 2,000 years ago, the first mercury thermometer was made in 1714 by Gabriel Fahrenheit, the German physicist.[10] But it was not until the 1800s when global temperatures began to be recorded on a regular basis.

Past temperatures can be estimated by: 1) examining written human history and 2) analyzing physical indicators from nature's geologic record, which are called temperature "proxies" (see text box).

Temperature Proxy Analysis

Temperature has an effect on most of nature's chemical processes. For example, trees grow in annual cycles, adding a ring to the trunk for each year of growth. Warmer atmospheric temperatures result in wider and denser tree rings (over some range of temperatures). By measuring temperatures today and comparing to today's tree ring growth, scientists can calibrate a process for measuring past temperatures.

Ratios of oxygen isotopes are often used as a temperature proxy. Recall that isotopes are atoms of the same chemical element with a different atomic weight. The ratio of Oxygen 16 (called ^{16}O) and Oxygen 18 (called ^{18}O) is a key indicator of atmospheric or ocean temperature.

The Greenland Icecap was built season-by-season by snowfall over thousands of years. Ice cores recovered from holes drilled in the icecap show annual variation in the $^{18}O/^{16}O$ ratio. More ^{18}O is contained in the snow during the summer, so scientists can count each historical year by the summer ^{18}O peaks with good accuracy. Long term trends in the size of the peaks show changes in atmospheric temperatures.[11]

Oxygen isotope analysis is also used to measure historical sea surface temperatures. Warmer oceans mean less ^{18}O (heavy oxygen). Plankton build calcium carbonate shells from oxygen atoms in sea water. The $^{18}O/^{16}O$ ratio in the ocean is then reflected in the shells of plankton. Plankton die and are captured in sediment on the ocean floor. By recovering sediment cores and using oxygen isotope analysis, scientists can estimate ocean temperatures for past years.[12]

Proxy analysis comes with significant uncertainty. Tree-ring growth is also affected by rainfall, tree age, atmospheric CO_2 concentration, and other factors. Changes in ocean saltiness cause errors in sea temperature estimates from oxygen isotope analysis. Ice-core data can be corrupted by melting and re-freezing or other contaminations. With these uncertainties, it's important to look at a number of studies before drawing conclusions.

Historical information for the last 1,500 years is sometimes available in written records, literature, and art. Scientists are able to develop proxies for temperature, such as tree-ring growth, oxygen isotope ratios from ice cores, and the composition of oxygen isotopes in calcite shells of plankton.

THE MEDIEVAL WARM PERIOD AND LITTLE ICE AGE

The Medieval Warm Period (MWP) is a well-known period of warmer temperatures from about 900 to 1300 AD. Even the IPCC acknowledges the existence of the Medieval Warm Period, but they claim today's temperatures are warmer. They state in their Fourth Assessment Report: "the warmth of the last half century is unusual in at least the previous 1,300 years."[13] However, both historical writings and temperature proxy studies support the importance of the MWP.

Eric the Red and the Vikings established a colony in southern Greenland in 982 (see text box). Taking advantage of the mild climate, the settlement grew to 5,000 inhabitants by 1300, but failed in the next 100 years as Earth's climate moved into the considerably cooler Little Ice Age period. Leif Ericsson, the son of Eric, is credited with being the first European ever to set foot in North America. He

---

Greenland and the Vikings

Eric the Red was an infamous Viking adventurer with a fiery temper. History records that he was exiled from first Norway and then Iceland for killing some of his fellow men. He sailed west from Iceland in 982 and discovered a land of fjords and green pastures, which he named Greenland. He then returned to Iceland with news of his discovery.

In 985 he set out again for Greenland, leading a fleet of 25 ships and 500 settlers. Only 14 of the ships survived the voyage, but nevertheless a community was established. By the year 1000, the settlement had grown to 3,000 inhabitants and 300 to 400 farms.[14]

The settlers were able to enjoy the mild climate of the Medieval Warm Period and warm temperatures until about the year 1300. The population grew to 5,000 inhabitants with an economy based on livestock such as cattle, pigs, goats, and sheep. Trade of polar bear skins and walrus tusks, as well as fishing, boosted the economy.[15]

However, after about 1300 the climate cooled as Earth entered the Little Ice Age. The settlement declined rapidly due to a shorter grazing season for livestock and a sea ice increase that blocked trade. The last written evidence of the settlement was recorded in 1408.[16]

arrived in northwest Newfoundland in 1000, which he named Vinland, reportedly because of "the profusion of grapes that grew there."[17] The history of Viking efforts in Greenland and Newfoundland points to a Medieval Warm Period with temperatures at least as warm as those of today.

The climate change that doomed the Viking settlement on Greenland was the transition to the Little Ice Age. The Little Ice Age is named for the period from 1300 to 1850, when Earth's temperatures were about 1–2°C cooler than both warm temperatures of the Medieval Warm Period and temperatures of today. Written history recounts agricultural consequences of the Little Ice Age in Europe, including shorter growing seasons, reduced crop yields, and higher prices for grains. As a result, European populations suffered increases in famine, malnutrition, disease, and social unrest.[18] It's interesting that the only known instance of a missing oak tree ring occurred in 1816, which is also known as the "Year Without a Summer." In 1816, some oak trees in Europe were dormant due to cold summer temperatures from a combination of the Little Ice Age and a volcanic eruption in 1815.[19]

Further historical evidence for the Medieval Warm Period and the Little Ice Age is provided by Dr. Quansheng Ge of the Chinese Academy of Sciences. Dr. Ge's team compiled records for 2,000 years of history in eastern China to construct a historical temperature record. Data compiled included plant and crop information, dates for frosts and snowfall, lake and sea freezes and thaws, and distribution of crops and farming systems. A graph of Dr. Ge's data is shown in Figure 22. Dr. Ge and his team found strong evidence for the Medieval Warming Period:

> Starting from AD 510s, the temperature rose rapidly and entered the warm period of the AD 570–1310s. In this epoch, climate was dominantly warmer…The 30-year mean temperatures of two warm peaks were generally 0.3-0.6 degrees higher than present day.[20]

In addition to written history, an increasing number of scientific studies have used temperature proxies to verify the warm climate of the Medieval Warm Period. Dr. Lloyd Keigwin of the Woods Hole Oceanographic Institution estimated surface temperatures in the Sargasso Sea, south of Bermuda in the Atlantic Ocean. He examined

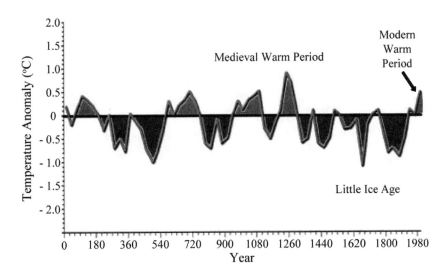

Figure 22. Temperature History of Eastern China. Graph constructed from written history records of agriculture and climate. Note the peak of the Medieval Warm Period and cooler cycle of the Little Ice Age. (Ge et al., 2003, Adapted from CO_2Science)[21]

oxygen isotope ratios in shells of plankton remains that were retrieved in sediments from the ocean floor. Dr. Keigwin found:

> Sea surface temperature was ~1° C cooler than today ~400 years ago (the Little Ice Age) and 1700 years ago, and ~1°C warmer than today 1,000 years ago (the Medieval Warm Period). Thus, at least some of the warming since the Little Ice Age appears to be part of a natural oscillation.[22]

Dr. Keigwin further concludes "the warming of the 20th century (0.5°C) *is not unprecedented* [our emphasis]."[23] Dr. Arthur Robinson of the Oregon Institute of Science and Medicine has graphed and extended Dr. Keigwin's data. The graph shows that sea surface temperatures were warmer than today's temperatures during several periods in the last 3,000 years, including the Medieval Warm Period (see Figure 23).

Dr. Haken Grudd of the University of Stockholm found similar results in the Tornetrask area of northern Sweden. He used tree-ring width and wood density of the Scots pine as a proxy to estimate atmospheric temperatures over the last 1,500 years. Dr. Grudd

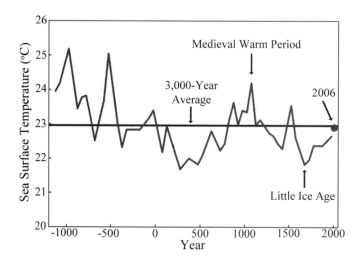

Figure 23. Sea Surface Temperatures in the Sargasso Sea. Temperature data was derived from oxygen isotope ratios in shells of plankton remains that were retrieved in sediments from the ocean floor. The Sargasso Sea is located in the Atlantic Ocean, south of Bermuda. Adapted by Robinson et al., including addition of a 2006 temperature data point. (Keigwin, 1996, Adapted from Robinson et al., 2007)[24]

summarizes:

> The late-twentieth century is not exceptionally warm in the new Tornetrask record…The warmest summers in this new reconstruction occur in a 200-year period centered on AD 1000. A "Medieval Warm Period" is supported by other paleoclimate evidence from northern Fennoscandia, although the new tree-ring evidence from Tornetrask suggests that this period was much warmer than previously recognized.[25]

"Fennoscandia" includes the area of Scandinavia, plus Norway, Sweden, Finland, and the Russian peninsula of Kola. A graph of Dr. Grudd's temperature reconstruction is shown in Figure 24. The graph shows that temperatures of the Medieval Warming Period were more than 1 °C warmer than current temperatures.

Since Tornetrask is north of the Arctic Circle, this data directly contradicts the statement of the IPCC in 2007 that "the last time the polar regions were significantly warmer than present" was about

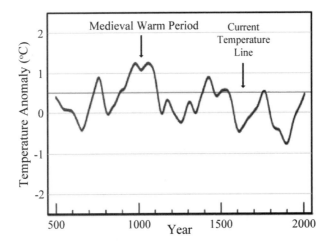

Figure 24. Temperature History from Sweden Tree Rings. Temperatures constructed from tree-ring width and density of Scots pines found in Tornetrask, Sweden. (Grudd et al., 2008, adapted by CO_2Science)[26]

125,000 years ago.[27] Climatists admit there was a Medieval Warm Period, but that it "only occurred in the Northern Hemisphere." But this is not the case. There is a growing body of peer-reviewed scientific evidence that the MWP was a *global* climatic period that was as warm as, or warmer than today's climate.

The Center for the Study of Carbon Dioxide and Global Change, also known as CO_2Science, directed by Dr. Craig Idso, provides a website (www.co2science.org) that has captured much of the evidence for the Medieval Warm Period. Their Medieval Warm Period Project is an ongoing effort to assemble scientific studies that verify the MWP. Over 200 peer-reviewed studies are listed on the site, including more than 30 studies from the Southern Hemisphere.[28]

MEDIUM TERM: THE 1500-YEAR CYCLE

The Medieval Warm Period and the Little Ice Age are part of a 1500-year cycle of the Earth's climate. This is fully discussed by Dr. Fred Singer and Dennis Avery in their book *Unstoppable Global Warming Every 1,500 Years.* Dr. Singer is an atmospheric physicist at George Mason University and founder of the Science and

Environmental Policy Project, a think tank on environmental issues. Dennis Avery is a senior fellow at the Hudson Institute. They conclude:

> The Earth is warming but physical evidence from around the world tells us that human-emitted CO_2 (carbon dioxide) has played only a minor role in it. Instead, the mild warming seems to be part of a natural 1,500-year climate cycle (plus or minus 500 years) that goes back at least one million years.[29]

Singer and Avery point to a 1984 report by Dansgaard and Oescher as strong experimental evidence for the 1500-year cycle (Figure 25). Dansgaard and Oescher were research scientists at the Geophysical Isotope Laboratory in Copenhagen who retrieved ice cores from the Crete site in central Greenland. By using oxygen isotopes as a proxy for temperature, they were able to reconstruct the temperatures of Greenland over several thousand years.[30] The data clearly shows the Medieval Warming Period and the Little Ice Age, as well as other warm and cool eras in the 1500-year cycle. Singer and Avery hypothesize that solar activity is the primary driver for the 1500-year cycle of climate change on Earth. But, regardless of the cause, they are certain that today's global warming is primarily due to natural causes, and not man-made carbon dioxide emissions.

It's clear from the evidence that the climate of Earth is continually driven by long-term and medium-term natural cycles. It's also clear that today's temperatures are not abnormally high, as the Climatists claim. As an additional bit of evidence, see Figure 26. It shows the stump of a white spruce tree which has been radiocarbon dated at 4,900 years. *The stump is located 100 kilometers north of the current arctic tree line.*

SHORT-TERM CYCLES: EL NIÑO, PDO, AND AMO

News Hour with Jim Lehrer, February 4, 1998:

> Strong storms socked California this week, causing heavy damage yesterday. Some meteorologists tied the first winds and rain to the weather pattern known as El Niño. Fourteen counties in California declared local states of emergency. Northern California has been hit

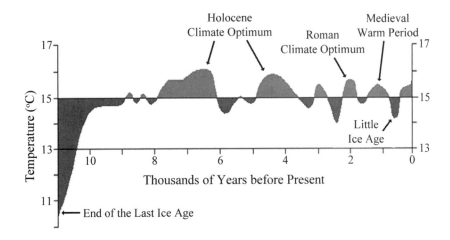

Figure 25. Temperature Cycles of the Last 12,000 Years. Reconstructions show temperatures in Earth's Northern Hemisphere varied about 4°C over the last 10,000 years. Data from oxygen isotope analysis of ice cores at Crete, Greenland. (Dansgaard et al., 1984, Adapted by Avery, 2009)[31]

Figure 26. White Spruce 100 km North of Tree Line. This stump is from a white spruce tree carbon-dated to have died about 4,900 years ago. It's located in northern Canada above the Arctic Circle, 100 kilometers north of the current tree line. (Ritchie, 1962)[32]

particularly hard since Monday night. More than 11,000 people were evacuated because of 80-mile-an-hour winds and floods. At least one death has been attributed to the storm…More than 400,000 people around the state lost power for at least part of yesterday.[33]

In addition to the California impacts, the very strong El Niño of 1997–1998 was also blamed for:

• droughts and forest fires in Mexico and Central America,
• droughts in Australia and the Midwest United States,
• ice storms in New England and Canada,
• economic disaster to Peruvian fishing companies, and
• mild winters in the Eastern United States.[34]

El Niño is more than a weather pattern. It's the best known of the Earth's short-term climatic cycles. In addition to the long-term Milankovitch Cycles and the 1500-year cycle identified by Singer and Avery, a number of short-term cycles drive the climate of Earth, including El Niño, the Pacific Decadal Oscillation (PDO), and the Atlantic Multidecadal Oscillation (AMO). No one knows exactly what causes these cycles, but they have recently been well-documented by scientists. Evidence shows that they have probably been oscillating for a least a thousand years.[35]

El Niño means "Little Boy" or "Christ Child" in Spanish, so named by the people of Peru because it usually arrives in December. El Niño is only part of the cycle. La Niña (Little Girl) is the other part of the cycle, which is more properly called the El Niño Southern Oscillation (ENSO) by climatologists. The ENSO cycle takes place in the southern Pacific Ocean (see Figure 27 for a diagram and discussion of El Niño and La Niña).

During La Niña, the normal condition, trade winds blow west, pushing warm water to the western Pacific. According to NASA, the ocean at Indonesia is usually about one-half meter higher then at Ecuador, due to these winds.[36] When El Niño occurs, every three to seven years, the winds reverse and blow to the east, pushing warm water into the eastern Pacific. Cold water from the Thermocline layer is pushed deeper near Peru, ruining local fishing conditions.

Dr. Fred Goldberg, professor at the Royal Institute of Technology in Stockholm, Sweden, reminds us that the oceans store more than

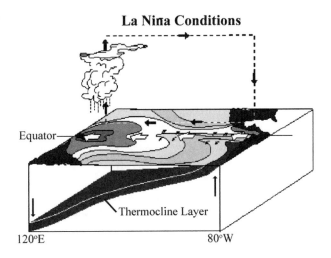

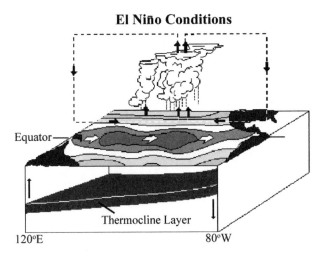

Figure 27. The El Niño Southern Oscillation (ENSO). The figure shows the La Niña (top) and El Niño (bottom) cycles of ENSO, operating in the central Pacific Ocean. La Niña, the normal condition, is characterized by trade winds that blow to the west, pushing warm water to the western Pacific. Storms form above the warm pool to the west. During La Niña, cool water from the Thermocline layer wells up near Peru, bringing deep-water nutrients to feed Peruvian fishing waters. During El Niño, which occurs every 3–7 years, trade winds reverse and blow to the east, pushing warm water and storms to the eastern Pacific. Less cold water wells up near Peru, reducing fishing harvests. (NOAA, 2009)[37]

30 times as much solar energy as the atmosphere.[38] This makes the Pacific Ocean, the home of El Niño, Earth's largest heat reservoir. When El Niño occurs, a huge amount of heat is pushed thousands of miles from the western Pacific to the eastern Pacific. This has a major impact on weather all over the world. El Niño has been shown to reduce rainfall and increase drought conditions in South Africa, the Indian sub-continent, Indonesia, and Australia. Increased rainfall often occurs in South America, North America, and even eastern Africa. El Niño also results in warmer conditions in North America and Japan and reduced hurricane activity in the Atlantic Ocean.

The El Niño Southern Oscillation operates in concert with the Pacific Decadal Oscillation. The PDO was named in 1996 by scientist Steven Hare who studied connections between Alaska salmon production and Pacific climate.[39] The PDO is a 30 to 40 year-long cycle of warming and cooling that operates throughout the Pacific Ocean. The weather impact of the PDO is primarily in the North Pacific, in contrast to the impact of the El Niño Southern Oscillation, which is primarily in the tropics.

The Pacific Decadal Oscillation is shown in Figure 28. The graph is a measure of sea-surface temperature change for the Pacific Ocean. The PDO includes the effects of ENSO. As the figure shows, the PDO was in warm cycles from about 1920 to 1940 and from about 1975 to 1998, with a cooler cycle from about 1940 to 1975. Recently, it appears the PDO has entered a cooling phase. Along with

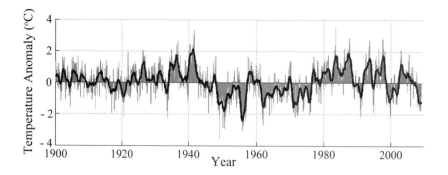

Figure 28. The Pacific Decadal Oscillation 1900–2008. The graph plots monthly average sea surface temperature variation for the Pacific Ocean for the last 108 years. The PDO Index incorporates effects from El Niño events in 1993, 1998 and other years. (JISAO Washington, 2009)[40]

cooler temperatures, fishermen are noticing increased catches of salmon in the Pacific Northwest as spawning runs move south.

The last short-term cycle we'll discuss is the Atlantic Multidecadal Oscillation (Figure 29). This cycle has less impact on Earth's temperatures than El Niño or PDO cycles, but still shows a variation in sea-surface temperature of 0.25°C on 60-year cycle. Remember that alarmists are warning of disaster from model simulations and a temperature rise of only 0.5°C from 1970 to 2000. If man-made greenhouse gases caused the rise in Atlantic Ocean temperatures from 1970 to 2000, what caused the AMO peaks in 1880 and 1940 when there was *negligible man-made* CO_2 in the atmosphere?

Our apologies for burying you with data on the cycles of Earth's climate, but this is vital. It should be clear by now that temperatures on our planet are continuously changing as the result of long-term, medium-term, and short-term cycles. The alarmists are unable to show that CO_2 causes the 1,500-year cycle, El Niño, the PDO or the AMO. They claim El Niño is amplified by man-made emissions, as they claim for every other weather event. But the General Circulation Models do not account for, and cannot explain, these cycles. In short, *Climatism minimizes the strong scientific evidence for Earth's natural cycles and their impact on global temperatures.*

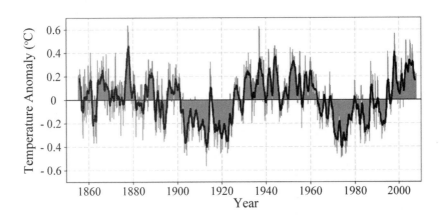

Figure 29. The Atlantic Multidecadal Oscillation. The graph shows the variation in average monthly sea-surface temperatures in the Atlantic Ocean for the last 152 years. The alarmists claim the temperature rise from 1975 to 2000 is due to carbon dioxide emissions, but have no explanation for past temperature cycles. (Adapted from NOAA, 2008)[41]

NATURAL CYCLES EXPLAIN GLOBAL WARMING

After discussion of Earth's short term cycles, we're now in a position
to show how these cycles can account for global temperatures over the
last century. We'll follow a line of reasoning by Dr. Syun Akasofu,
Professor of Physics at the University of Alaska, to show that current
global warming is well explained by natural causes.

The global temperature curve is shown in Figure 30, along with
the annual increase in man-made carbon dioxide emissions. Recall
that Earth's 20th century temperature rise has been irregular, with
periods of warming and also of cooling. We've had a decline in global
temperatures from 1880 to 1910, a rise from 1910 to 1940, another
cooling from 1940 to about 1975, and then the recent warming from
1975 until about 1998. Compare this with the steady increase in
man-made carbon dioxide emissions, which were at a low level in
1940 and then increased steadily until 2000. We can draw three
conclusions from Figure 30. First, global temperatures are irregular,

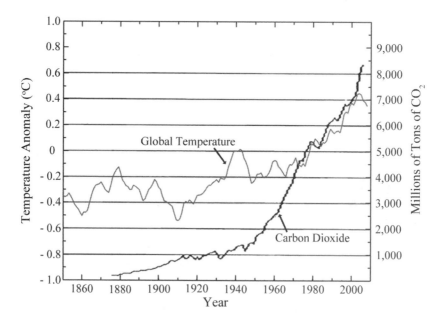

Figure 30. Global Temperature and CO₂ Growth.   Globally averaged
temperature plotted with annual man-made CO₂ emissions from 1850 to 2008.
(Adapted from Akasofu, 2009)[42]

while CO_2 rise is mostly continuous. Therefore, some global temperature variation must be due to natural causes. Second, the temperature rise from 1910 to 1940 must be due to natural causes, since CO_2 emissions were very small until after 1940. Third, global temperatures were flat to declining from 1940 to 1975, despite an exponential growth in emissions. It appears man-made emissions really don't match temperatures very well. Is there a better way to explain temperature variations other than greenhouse gas emissions?

Dr. Akasofu hypothesizes that global temperatures of the 20th century can be explained by the sum of two natural factors: 1) a linear temperature rise as Earth recovers from the Little Ice Age, and 2) a decadal oscillation in the Earth's temperature (Figure 31).[43] As we

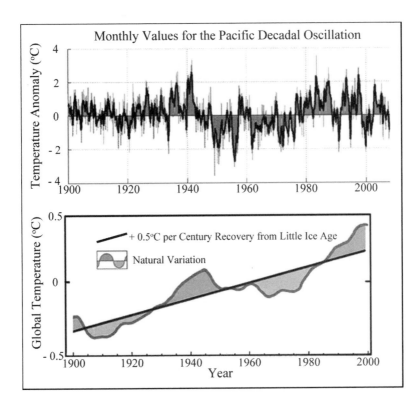

Figure 31. Causes of 20th-Century Global Warming. Lower graph shows warming as the sum of: 1) a 0.5°C per century temperature rise as recovery from the Little Ice Age and 2) a multi-decadal oscillation. Top graph is the Pacific Decadal Oscillation, which fits well with natural fluctuations seen in global temperatures. (Adapted from Akasofu, 2009 and NOAA, 2009)[44, 45]

learned earlier, global temperatures were cool in the period of the Little Ice Age until about 1850. Earth is now warming at a rate of about 0.5°C per century, which is close to the 0.7°C identified by the IPCC. This can be explained as a recovery from the Little Ice Age and part of the 1500-year temperature cycle identified by Singer and Avery. According to Dr. Akasofu, on top of this 0.5°C rise is the short-term temperature cycle of the Pacific Decadal Oscillation, which includes El Niño variation.

The graph of the PDO Index is shown on the top graph of Figure 31, matching up well with the natural fluctuations observed in global temperatures. Akasofu concludes:

> ...it is quite obvious that the temperature change during the last 100 years or so includes significant natural changes, both the linear change and fluctuations. It is very puzzling that the IPCC reports state that it is mostly due to the greenhouse effect.[46]

Dr. Roy Spencer finds similar results, pointing out that "the PDO can explain three quarters of 20th-century warming."[47] It's surprising that the IPCC discounts the short-term cycles in Earth's climate. But then, the cycles don't fit the dogma.

SUMMARY

Climate change is continuous. Earth's climate is dominated by long-term, medium-term, and short-term cycles. Evidence for the Medieval Warm Period shows that global temperatures were as warm, or warmer, 1,000 years ago as temperatures today. **The warming of Earth in the 20th century is well explained by natural causes, such as medium-term recovery from the Little Ice Age and short-term variations of the Pacific Decadal Oscillation, rather than by the rise in human greenhouse gas emissions.**

Interplanetary forces on Earth's orbit and axial motion (the Milankovitch Cycles) appear to account for long-term climatic cycles, but what causes the short- and medium-term cycles in our climate? We'll propose an answer to this next chapter.

CHAPTER 5

THE SUN IS OUR CLIMATE DRIVER

"Truth is like the sun. You can shut it out for a time but it ain't goin' away." Elvis Presley

We've shown that cycles of nature, particularly the 1,500-year cycle combined with short-term variation of the Pacific Decadal Oscillation, can account for most of the modern day temperature rise. But what drives the 1,500-year cycle and the PDO? The Danish National Space Center in Copenhagen gives us one likely answer:

> The scientific results…indicate that the varying activity of the Sun is indeed the largest and most systematic contributor to natural climate variations.[1]

The sun is the big dog in our solar system. It contains 98% of the mass of the solar system by itself. Each second more than 5 million tons of matter in the sun's core is converted to energy.[2] All of Earth's weather is caused by the difference in sunlight received between the tropics and the polar regions. It seems likely that the sun also drives our climate, rather than four carbon dioxide molecules in every 10,000 air molecules, as claimed by Climatism. But let's examine the evidence.

EVIDENCE POINTS TO THE SUN

The term solar radiation includes visible sunlight, but also high-energy ultraviolet radiation and low-energy infrared radiation, which our eyes are unable to see. Physics text books describe the amount of solar radiation Earth receives as the "solar constant." The measured radiation received by Earth at the top of the atmosphere is about 1,366 watts per square meter. Since half of the Earth is not receiving sunlight at any one time, and since Earth is a sphere, this number is reduced to an average of about 342 watts of energy at the top of the atmosphere for each square meter of Earth's surface.[3] Historical measurements of solar radiation have not been able to detect any changes in this number, hence the name "solar constant." The IPCC accepts the historical concept of the solar constant and states:

> ...we conclude that it is very likely that greenhouse gases caused more global warming over the last 50 years than changes in solar irradiance.[4]

However, scientific evidence now shows the IPCC is incorrect, and that energy we receive from the sun is the primary driver of Earth's climate. We will show it's likely that a combination of solar radiation and other solar effects account for both long-term and short-term global climate cycles.

For many years, scientists did not have methods to accurately measure the energy received from the sun. It's only with the launch of satellites in 1979 that we have been able to better measure solar radiation. These new measurements show a variation of about 0.1% in solar radiation between the peaks and valleys of each 11-year solar cycle.[5] This small variation is not large enough to account for Earth's short-term temperature cycles. However, as you know by now, 30 years is not long enough to assess a driver of climate change.

How can we estimate received solar energy over the last few centuries, without a history of satellite measurements? Sunspots are one method to track solar activity. Since 1600, astronomers have been observing and recording the presence and absence of sunspots. The German astronomer Heinrich Schwabe looked at the sun each day from 1826 to 1843 and observed the number of "solar spots" to

Figure 32. The Sun and Sunspots. Huge sunspot area during a period of high solar activity on March 30, 2001 (left). The big sunspot group is 13 times as large as Earth. A period of low activity on July 15, 2009 (right). (SOHO, 2009)[6,7]

peak about every 11 years. In 1844, he published a paper describing what became known as the 11-year Schwabe sunspot cycle.[8]

Sunspots are dark spots that appear on the surface of the Sun, which are easily seen using a telescope (Figure 32). They are areas of intense magnetic activity that are often many times larger than Earth. Sunspots, along with solar flares and "coronal holes," can be regarded as measures of solar activity.

Sunspots also play an important role in the "solar wind." The sun continuously emits a plasma of atomic particles called the solar wind. Solar wind is best known to cause the Aurora Borealis, also known as the Northern Lights (Figure 33). The Northern Lights are streamers

Figure 33. Aurora Borealis. Aurora above Bear Lake in Alaska, caused by interaction of the solar wind with Earth's magnetic field. (USAF, 2009)[9]

of colored lights that appear in the sky in the Arctic. When present in the Antarctic, they are called the Aurora Australis, or Southern Lights. Auroras are caused as particles of the solar wind interact with Earth's magnetic field in the polar regions. When sunspots, solar flares, and other events on the surface of the sun occur, the intensity of the solar wind increases.

John Eddy, an astronomer at the NOAA High Altitude Observatory in Colorado, used sunspot data to provide a thorough study of long-term variations in solar activity and climate in 1976. Eddy found evidence for a solar "Grand Maximum" at the time of the "medieval Climatic Optimum of the 11th through 13th centuries" and also evidence for a "prolonged solar minimum with the excursion of the Little Ice Age."[10] The "medieval Climatic Optimum" Eddy mentions is the Medieval Warm Period we have discussed. Eddy hypothesized:

> The mechanism of this solar effect on climate may be the simple one of ponderous long-term changes of small amount in the total radiative output of the sun, or solar constant.[11]

In other words, Eddy proposed the "solar constant" increased gradually over many hundreds of years to cause global warming.

Sunspot cycles have been proposed by a number of scientists as a measurement of solar activity. Figure 34 shows such an analysis, as

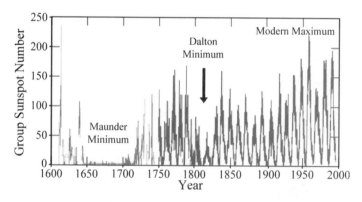

Figure 34. Sunspot Activity from 1600 to 2000. Sunspot cycles derived from the number of sunspot groups. The peaks shown are the maximum of each sunspot cycle every 11 years. (Hoyt et al., 1998)[12]

developed by Hoyt and Schatten in 1998. Note the Maunder Minimum, a period from 1650 to 1700 with an almost complete lack of sunspots, during the middle of the Little Ice Age. The Dalton Minimum of 1800 is another period of low solar activity. Note also the rise in sunspot activity from 1800 to 2000 corresponds with the warming of Earth's temperatures from the Little Ice Age.

Other scientists looked for a correlation between the Earth's temperatures and the solar activity cycle. One problem they encountered is that the thermal inertia of Earth's oceans causes a lag in temperature change from a change in solar activity. Also, the average period of the sunspot cycle is about 11 years, but the length of this cycle varies.

Dr. Eigel Friis-Christensen and Dr. Knud Lassen, geophysicists at the Danish Meteorological Institute in Copenhagen, showed a strong correlation between temperatures in the Northern Hemisphere and the length of the sunspot cycle in 1991. Figure 35 shows this relationship, along with a graph of atmospheric CO_2 concentration. The length of sunspot cycles has shortened during the 20th century. This means the cycles are coming faster, resulting in increased solar activity. As shown, the shorter cycle correlates well with increasing temperatures. Note that the increasing level of CO_2 does not track

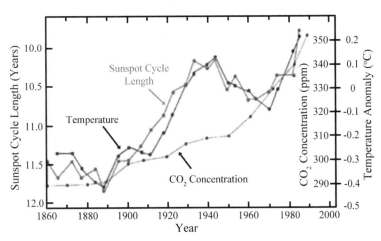

Figure 35. Sunspot Cycle Length and Temperature. Sunspot cycle length has decreased for 130 years, inversely matching rising temperatures. Northern Hemisphere land temperature data used. Rising CO_2 is unable to explain temperatures from 1900 to 1970. (Adapted from Friis-Christensen, 1991)[13,14]

temperature, except over the last 30 years. Friis-Christensen and Lassen published a follow-on paper in 1995, using proxy data to extend their analysis to 400 years, back to 1580. Good correlation was again found between solar cycle length and Northern Hemisphere temperatures.[15]

Researchers also find examples of strong correlation directly between solar radiation and temperature. Meteorologist Joseph D'Aleo has shown that a sum of the Pacific Decadal Oscillation and the Atlantic Multidecadal Oscillation closely tracks total solar radiation in the 20th century.[16] Dr. Willie Soon, geoscientist at the Harvard-Smithsonian Center for Astrophysics, compared total solar irradiance with air temperatures in the Arctic over the last 125 years. "Solar irradiance" is another term for solar radiation. Dr. Soon used temperature data from coastal land stations and drifting stations in the Arctic. His results are shown in Figure 36.

As can be seen, arctic air temperature closely tracks solar irradiance, but the rise in greenhouse gas emissions appears to be independent of arctic air temperature. These results are significant,

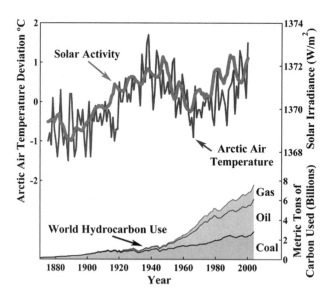

Figure 36. Solar Radiation and Arctic Temperatures. Received solar radiation matches arctic surface temperatures over 125 years. Hydrocarbon emissions have steadily risen since 1940 and do not track arctic temperatures from 1940-1970. (Adapted from Robinson et al., Soon, 2005)[17,18]

because the Climatists claim the Arctic Icecap is melting due to carbon dioxide emissions.   Dr. Soon points out that solar radiation:

> ...explains well over 75% of the variance for...Arctic...surface air temperatures...CO_2-dominated forcing of Arctic surface air temperatures is inconsistent with...Arctic-wide records.[19]

Soon believes that total solar radiation may have increased 0.3% to 0.4% over the last several hundred years, accounting for much of the warming since the Little Ice Age.[20]

In summary, **long-term solar activity and global temperatures, as well as 20th-century temperatures and solar activity are highly correlated.   In contrast, atmospheric carbon dioxide emissions do not match global temperatures, particularly before 1970.** While correlation does not assure causation, solar activity more closely accounts for recent global temperature rise than greenhouse gas emissions from hydrocarbon fuels.

THE LINK BETWEEN COSMIC RAYS AND THE SUN

Cosmic rays are subatomic particles that continuously bombard Earth from deep space. These particles are created from exploding stars far away in the Milky Way galaxy. Upon entering our atmosphere, they collide with atoms of atmospheric gases. These collisions create secondary cosmic ray particles that reach us on Earth's surface. Some of these secondary particles are passing through you just as you read this book. Earth's magnetic field and atmosphere are good barriers, shielding us from most cosmic rays.

Scientists have used instruments to measure cosmic rays since about 1935.[21] They soon discovered that cosmic ray intensity declines when solar activity increases, as measured by the sunspot cycle. A graph of sunspot count and cosmic ray flux, which is measured at the Climax neutron monitor station in Colorado, is shown in Figure 37. Note that the cosmic ray intensity, measured on the right scale, decreases as you move up the chart. Therefore, the data shows that high sunspot activity reduces cosmic rays

Physicists now agree that solar activity from sunspots helps to shield us from cosmic rays. This is because the solar wind and its

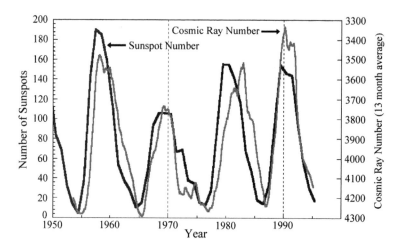

Figure 37. Sunspot Effect on Cosmic Rays. Inverted cosmic ray flux (right scale), plotted with sunspot number (left scale). When sunspots increase, cosmic ray count decreases. Cosmic ray count comes from the Climax monitor station in Colorado. (Svensmark et al., 1997)[22]

associated magnetic field act as a shield to block cosmic rays from reaching Earth. When sunspots, solar flares, and other events on the surface of the sun occur, the intensity of the solar wind increases, reducing the cosmic rays that strike Earth.

Another way to measure the intensity of past solar radiation is with a solar radiation proxy, similar to other proxies we have discussed before. One such proxy is the concentration of the beryllium 10 isotope (called ^{10}Be). When cosmic rays strike our atmosphere, they collide with atoms of nitrogen and oxygen and produce beryllium 10. Since the solar wind blocks cosmic rays, the strength of the solar wind affects the amount of ^{10}Be created. Therefore, the more sunspots, the stronger the solar wind, and the less ^{10}Be is created in the atmosphere.[23]

Dr. Jürg Beer, physicist at the Swiss Federal Institute of Aquatic Science and Technology, worked with others to analyze Greenland ice-core data for ^{10}Be and plotted it alongside a combined temperature proxy record. As shown in Figure 38, the proxies for solar activity and temperature moved closely together for the last 270 years, despite variability that comes with graphing two proxy measures. Beer stated:

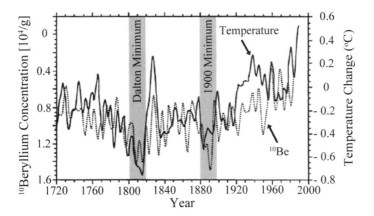

Figure 38. Temperature and Solar Activity 1720-1990. Beryllium 10 (^{10}Be) as a proxy for solar activity (dotted line) is graphed with a combined temperature proxy record (solid line). The temperature curve closely tracks solar activity, including the Dalton and 1900 Minimum periods of low solar activity. (Beer et al., 1994)[24]

Based on the analysis of historical data we conclude that solar forcing indeed plays an important role in past and present climate change.[25]

NEW EVIDENCE: SOLAR ACTIVITY AFFECTS CLOUDS

Scientists at the Danish Space Research Institute made remarkable discoveries in the mid-1990's that also point to the sun as our climate driver. They observed that, in addition to sunlight, the effects of sunspots themselves appear to have a second but major impact on Earth's temperatures through cloud formation. Satellite data that has only been available since 1979 was the basis for the new discovery. Unfortunately, this new evidence surfaced after the IPCC and Climatists had chiseled their man-made carbon dioxide dogma in stone.

In 1995, Dr. Nigel Marsh and Dr. Henrik Svensmark of the Danish National Space Center published work that examined satellite data from the International Satellite Climatology Project. They noted that for more than 20 years, satellite records of low-altitude clouds closely followed variations of cosmic rays, with a very high level of correlation. This connection is shown in Figure 39.

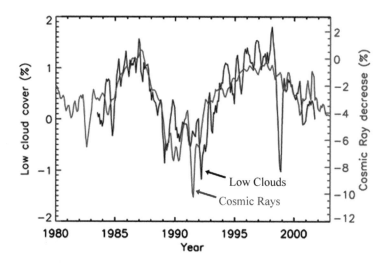

Figure 39. Variation in Low Clouds and Cosmic Rays. Cosmic ray variation closely tracks the global level and variation of low clouds from 1984 to 2005, according to satellite data. Cosmic ray variation data is from the Climax neutron monitor operations. (Adapted from Shaviv, 2009 and Svensmark)[26,27]

The data shown in Figure 39 is remarkable for several reasons. First, the amount of decrease in cloud cover due to cosmic ray reduction is large. Marsh and Svensmark found the variation to be more than 2% within one 11-year sunspot cycle. This is many times larger than any changes from solar radiation over the same period. Second, the researchers found the relationship with cosmic rays holds only for low-altitude clouds (below 3.2 kilometers), with no change for middle and high-altitude clouds. Recall from Chapter 3 that a small change in low-altitude cloudiness will cause a significant change in global temperatures. Marsh and Svensmark hypothesized that a high level of sunspot activity reduces the of level of cosmic rays and therefore low-altitude cloudiness, resulting in an increase in global temperatures.[28]

Recall from Chapter 3 that one of the weaknesses of the Global Circulation Models was their inability to model clouds. This discovery from the Danish National Space Center is completely unexplained by the models. The work by Marsh and Svensmark was greeted with much skepticism, largely because there was no mechanism known for cosmic rays to influence low-altitude clouds.

But shortly afterward, Svensmark and the team at the Danish National Space Center developed both a theoretical basis and experimental data to support their hypothesis. They set up the SKY experiment in Copenhagen, which is a simulation of Earth's atmosphere. SKY is a room filled with air, like our atmosphere, with controlled concentrations of gaseous elements such as water vapor, sulfur dioxide, and ozone. Cosmic ray particles continually enter the SKY chamber through its ceiling, as they bombard all areas of Earth's surface. The scientists found that when small amounts of sulfur dioxide are added to the chamber, sulfuric acid ions are created by the cosmic rays, resulting in large numbers of water droplets. As a result, Svensmark developed a new theory for cloud formation:

> The theory describes mathematically the early growth of sulphuric acid droplets in the atmosphere. These are the building blocks for the cloud condensation nuclei on which water vapor condenses to make clouds.[29]

Meteorologists have long known that clouds begin to form best around dust or other aerosol particles. Such particles provide a platform on which water vapor molecules can condense to start the cloud formation process. Cloud seeding using silver iodide crystals or dry ice is an effort to use this principle. The SKY experiments showed that sulfur dioxide ions created by cosmic rays are very effective condensation nuclei for forming clouds.

A summary of how sunspot activity causes global warming is shown in Figure 40 (next page). First, an increased level of sunspot activity causes an increase in the solar wind and its associated magnetic field, which allows fewer cosmic rays to penetrate to Earth's atmosphere. Fewer cosmic rays create fewer ions and therefore fewer condensation nuclei for the formation of clouds. Less low-altitude cloudiness causes less sunlight to be reflected, causing Earth to warm.

The European Organization for Nuclear Research (CERN) has initiated the CLOUD project (Cosmics Leaving Outdoor Droplets), as a follow-on effort to SKY. CLOUD is building a chamber to simulate atmospheric conditions in concert with CERN's particle accelerator in Geneva, Switzerland. Operation is planned to begin in 2010. More than 50 scientists from institutes in Europe and the U.S. plan to test and extend the results of Svensmark's experiments.[30]

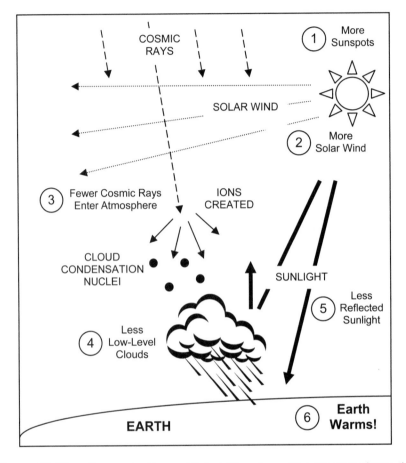

Figure 40. More Sunspots Cause Global Warming. The diagram shows the mechanism for global warming from increased sunspot activity. More sunspots increase the solar wind, reduce cosmic rays, and reduce low-level cloudiness, which reflects less sunlight, causing Earth to warm.

Dr. Svensmark and Nigel Calder describe their theories in their book *The Chilling Stars: A New Theory of Climate Change.*[31] It appears the sunspot-to-cloud-formation mechanism explains many of the short-term cycles of Earth's climate. Dr. Fred Singer states:

There is now little doubt that solar-wind variability is a primary cause of climate change on a decadal time scale. Once the IPCC comes to terms with this finding, it will have to concede that solar variability provides a better explanation for 20th Century warming than greenhouse effects.[32]

Svensmark and his team also believe that cosmic ray variability from sunspots can account for climatic changes over centuries. Svensmark uses another proxy (you're an expert in proxies by now) to discuss cosmic ray effects on climate. Like beryllium 10, cosmic rays also create carbon 14 (^{14}C) as they bombard our atmosphere, so ^{14}C is a proxy for cosmic ray intensity. After being created in the atmosphere, carbon 14 accumulates in ice cores, cave stalagmites, trees, and other deposit sites to give us a record of cosmic ray intensity throughout history. The more carbon 14 present in the atmosphere, the higher the level of cosmic rays, and according to Svensmark's findings, the higher the level of low-altitude cloudiness, resulting in lower global temperatures. Therefore high carbon 14 concentration means colder global temperatures.

A historical tracking of carbon 14 is shown in Figure 41. Note from the figure that the history of carbon 14, and therefore cosmic rays and solar activity, tracks the major climatic changes of Earth, including the Medieval Warm Period, the Little Ice Age, and the 20th century warming.

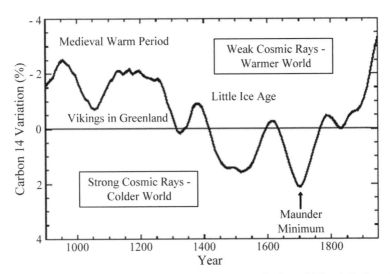

Figure 41. Carbon 14 over the Last 1000 Years. Carbon 14 is plotted upside down to match temperature changes over the last millennium. [14]C is a well known proxy for cosmic rays, but also shows the Medieval Warm Period, the Little Ice Age, and the current day warming. (Adapted from Svensmark, 1999)[33]

As we have shown, **the recent work of many scientists has provided solid scientific evidence that the sun is the primary driver of global temperatures.** Cloud-variation from cosmic rays, driven by changes in sunspot activity, appears to explain the short-term temperature variations of the 20th century. A combination of solar radiation and sunspot activity appears likely to be able to explain the 1500-year cycle, including the Medieval Warm Period and the Little Ice Age. The dogma of global warming from hydrocarbon emissions explains little of these cycles. Dennis Avery states:

> The sunspots and cosmic rays have a 79 percent correlation with our thermometer record since 1860. Meanwhile the CO_2 correlation is a mere 22 percent.[34]

NATURE IS NOT COOPERATING

In December of 2008, Climatist groups announced plans for a "mass non-violent civil disobedience" on March 2, 2009 to shut down the coal plant that powers the U.S. Capitol in Washington, D.C. But when March 2 rolled around, nature did not cooperate with the spirit of the protest. About 2,000 protesters were greeted by 24°F weather and a massive storm that set records for snowfall in dozens of northeastern cities in the United States.[35]

The protest was an example of what has come to be called the "Gore Effect" by some climate scientists. Almost every time a global warming protest is scheduled, nature intervenes with a cold-weather day. But, all joking aside, the warming of global temperatures in the last quarter of the 20th century has halted, despite continuing increases in atmospheric carbon dioxide concentration.

Figure 42 shows global temperature trends over the last eight years. University of Alabama Huntsville (UAH) satellite data shows a decline of almost 0.2°C for the period. Hadley Centre temperature station data shows a decline of over 0.1°C. It can be argued that these declines are not significant, but they contrast markedly with the steady rise in atmospheric carbon dioxide over the period. This is a shift in direction from the rising global temperatures measured from 1975 to 1998.

Another indicator is provided by the Argo ocean buoy system of

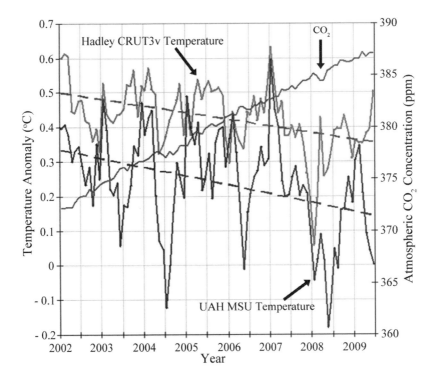

Figure 42. Declining Global Temperature 2002-2009. University of Alabama Huntsville (UAH) satellite data and Hadley Centre CRUT3v show declining global temperatures since 2002. Dotted lines are average temperature trends. Atmospheric CO_2 continues to increase over the same period. (D'Aleo, 2009)[36]

NASA's Jet Propulsion Laboratory. The Argo system consists of 3,000 robotic diving buoys designed to measure ocean temperatures to a depth of 700 meters.[37] Since its deployment in 2003, the Argo system shows that the trend of heat content of the world's oceans is flat, as shown in Figure 43 (next page). This appears to indicate a stabilization in ocean heat content, which was measured by other means as steadily rising since 1975.

These observations conflict with forecasts offered by the alarmists. Both IPCC predictions and climate models show accelerating global temperatures from 2000 throughout the 21st century. Since nature is not cooperating with Climatist predictions, do we have reason to expect a period of global cooling?

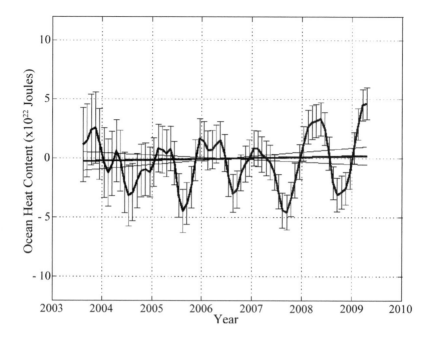

Figure 43. Stable Ocean Heat Content since 2003. Data from Argo robotic buoy system. The vertical bars show uncertainty in the plotted curve. (Willis, 2009)[38]

ARE WE HEADED FOR GLOBAL COOLING?

On April 21, 2009, the BBC reported:

> The sun is the dimmest it has been for nearly a century. There are no sunspots and very few solar flares—and our nearest star is the quietest it has been for a very long time. The observations are baffling astronomers, who are due to study new pictures of the Sun, taken from space, at the UK National Astronomy Meeting.[39]

Indeed by April of 2009, the sun had hit a 100-year low in sunspot activity and a 50-year low in solar wind pressure.

David Archibald, climate scientist from Perth, Australia, is concerned about this low level of solar activity. Another way to measure solar activity is to measure the strength of the Interplanetary Magnetic Field (IMF). More sunspots cause a stronger solar wind,

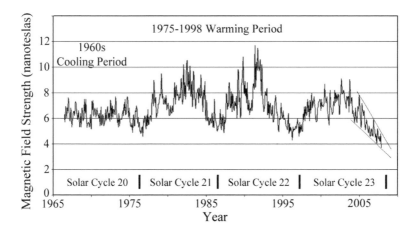

Figure 44. Interplanetary Magnetic Field 1966–2008. Interplanetary Magnetic Field strength, generated by solar wind from sunspots, is shown for the last 40 years. Sunspot cycles 20–23 are shown. Recent field strength has dropped below that of the 1960's cooling period. (Adapted from Archibald, 2008)[40]

which carries the sun's magnetic field out into the solar system, where it is called the IMF. The magnetic field strength of the IMF is graphed over the last four sunspot cycles in Figure 44.

Note the flatness of Solar Cycle 20, which occurred during the late 1960s global-cooling period. Archibald points out that the Interplanetary Magnetic Field was lower in 2008 than during the 1960s period. This means solar activity is now lower than it has been over the last several sunspot cycles.[41] Note also that Solar Cycles 21 and 22 show high solar activity, corresponding to the global warming period of 1975–1998.

Recall from our earlier discussion and Figure 35 that an upside-down graph of solar-cycle length strongly matches the global temperature curve. The longer the solar cycle, the lower global temperatures will be. Archibald estimates that Solar Cycle 24 began in the summer of 2009, but remains at a very low level of activity. This means Solar Cycle 23 was 13 years long, compared with 9.6 years for Solar Cycle 22, pointing to a coming period of cooler global temperatures.

Like realist climate scientists, I've begun a habit of looking at the daily pictures of the sun that are posted on weblogs. During most of 2009, we've seen the "cue ball" picture of the sun with few

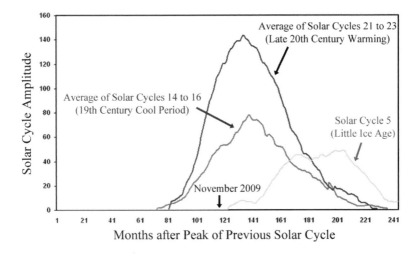

Figure 45. Solar Cycle Period in Months. Start and end times after the peak of the previous solar cycle and cycle amplitudes are shown. Solar Cycles 21–23 are from the warm period of 1975–2005. Solar Cycles 14–16 are from the cooler periods of the late 19th Century. Solar Cycle 5 is from the Maunder Minimum during the Little Ice Age. Solar Cycle 24 began in the summer of 2009, but its amplitude remains near zero. (Adapted from Archibald, 2008)[42]

sunspots. The quiet sun is a probable indicator of cooling temperatures ahead.

Figure 45 displays this concern. The figure plots solar cycle amplitude against "months after peak of previous solar cycle", which is a measure of how long it takes to start a new cycle. The solar cycles of the second half of the 20th century started an average of 81 months after the peak of the previous cycle. Solar Cycles 14–16 from the cooler period of the late 19th century began about 72 months after the previous peak, but only reached 60% of the amplitude of 20th century cycles. Solar Cycle 24 has started, but its amplitude remains close to zero. Solar Cycle 5 from the Little Ice Age is shown for comparison, when the Thames River froze hard at London.

The graph by Friis-Christensen in Figure 35 shows a global cooling of about 0.4°C for each year of longer solar cycle. Archibald estimates between 0.4°C and 0.7°C of cooling for each extra year, depending upon location.[43] The long Solar Cycle 23 and slow ramp-up of Solar Cycle 24 is strong evidence that we are entering a period of global cooling, contrary to IPCC predictions.

IT'S THE SUN!

Mounting scientific evidence shows the sun is our climate driver. Solar activity better explains both 20th century and multi-century temperature variations than man-made hydrocarbon emissions. The sun provides a "double whammy" effect on Earth's climate: 1) by long-term changes in solar radiation and 2) by short-term changes in sunspot activity that vary cosmic ray intensity, and the level of low-altitude cloudiness.

Global temperatures have been cooling since 2002, according to satellite and surface-station data. This cooling trend contrasts with steadily rising atmospheric carbon dioxide, and is contrary to both climate model and IPCC forecasts. The current low level of solar activity warns of a cooler period for Earth's global temperatures.

If you've stayed with us so far through all this science, congratulations; we've covered a lot of information. The science is important because, for climate change, it's not what you believe, but what the science tells you. Most of the remaining book employs fact-based common sense, rather than science and technology.

CHAPTER 6

GLOBAL WARMING DISASTERS DEBUNKED

"This is one of those cases when imagination is baffled by the facts."
Winston Churchill (1941)

The cries of alarmists grow more shrill each day. James Hansen considers President Obama to be our last chance to save the planet:

We cannot afford to put off change any longer…We have to get on a new path within this new administration. We have only four years left for Obama to set an example to the rest of the world.[1]

Another example is from *The Early Show* on CBS from February 16, 2009:

MAGGIE RODRIEGUEZ: A dire new warning from scientists this morning. Global warming is happening much faster than expected. CBS News science and technology correspondent Daniel Sieberg has the story. Good morning, Daniel.

DANIEL SIEBERG: Good morning, Maggie. Yes, a dire new warning from scientists says the amount of greenhouse gas emissions worldwide is higher than predicted, mostly from places like India and China. That's leading to warmer temperatures and more

99

extreme conditions all over the planet. For years, scientists have known Greenland's glaciers were melting and the seasonal sheet of ice covering the Arctic had been shrinking. But the thaw is literally just the tip of the iceberg.[2]

The Early Show segment continues with pictures of shrinking arctic sea ice, discussion about wildfires in Australia, and footage of hurricanes, floods, and blizzards.

Britain's Prince Charles has also boosted the level of alarm. He has been flying between different countries of the world in his carbon-emitting private jet, proclaiming an alarming message:

> ...the best projections tell us that we have less than 100 months to alter our behavior before we risk catastrophic climate change and the unimaginable horrors that this would bring.[3]

I'm not sure that "unimaginable" is the right word, because Climatist doomsayers can imagine a lot.

Weather is a great source of fodder for Climatism and the sensationalist media. Absolutely *any* weather event—including floods, droughts, heat waves, cold snaps, fires, snowfalls, you name it—is due to climate change. Recently Rachel Maddow of MSNBC attributed a second-time-ever snowfall in Dubai to climate change.[4]

In this chapter we'll refute eight categories of disasters forecasted by the alarmists. Many of these predicted catastrophes seem to be an exercise in "how to turn white into black." If one can imagine a disaster, a scientist can be found to generate a climate change cause for such a disaster. But as Sir Winston rightly put it: "This is one of those cases when imagination is baffled by the facts."

ICECAP MELTING CAUSES SEA LEVEL RISE

The *first* and biggest projected disaster from man-made global warming is sea-level rise from the melting of Earth's icecaps. Al Gore and James Hansen warn of an ocean level rise of 20 feet, or 6–7 meters, by the year 2100. Such a sea-level rise could flood Bangladesh, the Maldives Islands, and many of the world's coastal areas. This, the alarmists claim, will cause the dislocation and

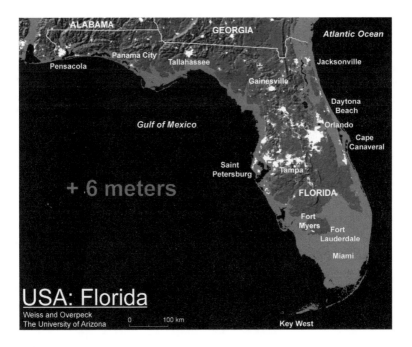

Figure 46. Florida Flooding from a 6-Meter Sea-Level Rise. Light gray areas show flooded areas from a simulated sea level rise of 6 meters. (Weiss and Overpeck, University of Arizona, 2009)[5]

migration of millions of people, as well as billions of dollars in damage. Figure 46 shows a simulation of flooding in Florida from a 6-meter sea-level rise. A similar picture was used by Mr. Gore in *An Inconvenient Truth*.[6]

Frozen water on Earth is mostly concentrated in three icecaps: the Arctic in the north, the Antarctic in the south, and Greenland. Antarctica contains about 90% of Earth's ice and about 60% to 70% of all fresh water. Greenland's icecap has about 8–9% of Earth's ice. The remaining 1–2% of the ice is held by the arctic pack ice and mountain glaciers.[7] Antarctica and Greenland are land masses with most of their ice deposited on top. The arctic ice pack floats on the Arctic Ocean. Arctic ice shrinks more than 50% every summer and then recovers during the winter, annually varying more than the icecaps of Greenland or Antarctica. We'll discuss the warming situation for each of the Arctic, Antarctic, and Greenland Icecaps.

The arctic ice pack has provided the most visible evidence of the

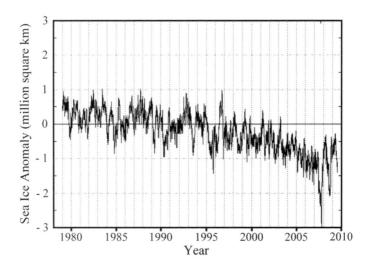

Figure 47. Arctic Sea Ice 1978–2009. The graph shows variation (anomaly) of arctic sea ice from a mean value for 1978–2000. Fall of 2007 was the lowest value in the last 30 years. Data from satellites. (University of Illinois, 2009)[8]

recent global warming period. Figure 47 shows a measure of arctic sea ice from 1979 to 2008 from satellite data. Anomaly means "variation from the average" for the period from 1978–2000. The average ice pack has declined by about 1 to 1.5 million square kilometers over the last 30 years, which is about 12% to 18% of the average annual amount.

According to the Climatists, the decline in arctic sea ice is the "canary in the coal mine" for global warming. After the sea ice hit a recent low in the fall of 2007, predictions were very aggressive. Professor Mark Serreze of the University of Colorado predicted that the Arctic Ocean will be completely ice free by the summer of 2030.[9] Wieslaw Maslowski of the Naval Postgraduate School was even more outspoken, projecting "2013 for the removal of ice in summer."[10] The implication of these statements is that the ice would be gone, never to return because of man-made global warming. However, just because ice is melting, doesn't mean carbon dioxide emissions are causing it.

It's important to note that melting of all the arctic sea ice *will not* cause the world's oceans to rise by a measureable amount. Try a simple experiment by filling a glass with water and floating ice. After all the ice melts, you'll see no change in the water level. There is

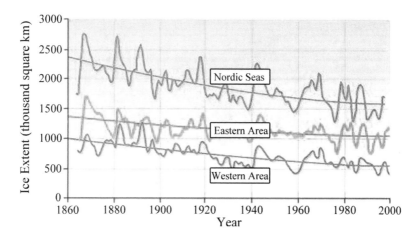

Figure 48. Ice Extent for the Nordic Seas 1860–2000. Ice extent decline over 135 years in the Nordic Seas, including the Greenland, Iceland, Norwegian, Barents, and Western Kara Seas. Data from ship logs and meteorological observations. (Adapted from Vinje, 2001)[11]

no volume gain when floating ice melts.

Once again, alarmist scientists are looking at data across 30 years and making conclusions about the climate, which operates over a timescale of many centuries. Dr. Torgny Vinje of the Norwegian Polar Institute in Oslo studied the sea ice in the ocean area north of Iceland, east of Greenland, and west of Norway from 1860 to 2000. Dr. Vinje found a long-term decline in the ice extent over 135 years, as shown in Figure 48. Vinje concludes:

> The extent of the ice in the Nordic Seas…has decreased by ~33% over the past 135 years…The time series indicates that we are in a state of continued recovery from the cooling effects of the Little Ice Age.[12]

If man-made global warming caused the arctic ice melt from 1979 to 2005, then what caused the melt from 1865 to 1940, before hydrocarbon emissions? The answer must be natural causes. Also, it appears that nature is not cooperating again. The graph in Figure 47 shows an uptick in arctic sea ice during 2008. This continues in 2009 and it appears that arctic ice will once again reach the 1978–2000 average.

Figure 49. Submarine at North Pole, March 17, 1959. The nuclear submarine *SSN Skate* at the North Pole in March, 1959. Note the open water at the time of year when polar ice should be thickest. (US Navy, 1959)[13]

As a last comment on arctic ice, see the picture in Figure 49, which was taken in March, 1959 when annual arctic ice is typically at its peak. It shows the United States nuclear submarine *Skate*, surfaced at the North Pole, with an apparent lack of ice. Arctic ice melt is neither new, nor caused by greenhouse gas emissions.

Antarctica has the coldest temperatures on Earth. Average monthly temperatures are -50°C in the interior and -5°C to -15°C on the coast.[14] *Monthly average antarctic temperatures are below the freezing point of water at all locations even during the summer.* The continent is often discussed in two sections, as shown in Figure 50. Eastern Antarctica is the major land mass of the continent. This section contains about 90% of the ice mass. Western Antarctica is a large area of land and frozen sea ice also called the West Antarctic Ice Sheet. International antarctic weather stations show on average no temperature increase or decrease for East Antarctica over the last 50 years. Data shows a 0.2–0.5°C per decade increase only for the small area of the Antarctic Peninsula.[15]

In addition to flat antarctic land temperatures, satellite data shows sea ice in the Antarctic Ocean slowly accumulating, as shown in Figure 51. This means the Antarctic Ocean has been cooling. Even the IPCC states that Antarctica has seen a "lack of rise in near-surface

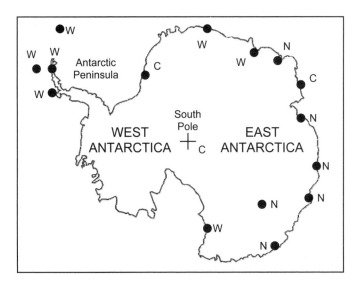

Figure 50. Antarctica and Temperature Trends 1956–2008. The chart shows Antarctica and temperature trends at 15 stations (W-warming, C-cooling, N-no change). East Antarctica temperatures are flat over the last 50 years. Antarctic Peninsula temperatures show a small increase of 0.2°C–0.5°C per decade. (Data from British Antarctic Survey, 2009)[16]

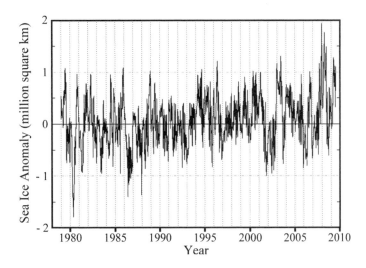

Figure 51. Antarctic Sea Ice 1978–2009. The graph shows variation (anomaly) of the antarctic sea ice from a mean value for 1978–2000. The sea ice has been slowly accumulating for the last 30 years and reached a high in 2008. Data from satellites. (University of Illinois, 2009)[17]

atmospheric temperatures averaged across the continent."[18]

So, little change in land temperature data, growing sea ice, and the IPCC agrees that temperatures are stable. Therefore, sea level rise from antarctic ice melt must be unlikely, right? Well, not quite. Climatists can still warn of catastrophe even if observed data does not show warming temperatures.

The National Science Foundation (NSF), which manages the U.S. Antarctic Program, provided $20 million in support for the Antarctic Geological Drilling (ANDRILL) program, along with an additional $10 million in support from other sources. Several of the ANDRILL projects were funded to investigate the effect of climate change on Antarctica.[19] This resulted in the sensational press release March 18, 2009 from the NSF:

> New Evidence from NSF-funded ANDRILL Demonstrates Climate Warming Affects Antarctic Ice Sheet Stability.[20]

Scientists from the ANDRILL program looked at ice cores that were 3–5 million years old and concluded that the Western Antarctic Ice Sheet (WAIS) collapsed millions of years ago. Computer models of the ice sheets were developed, which drew some alarming conclusions. Researchers found that warming 4 million years ago was primarily caused by changes in the Earth's axial tilt, but concluded that high levels of atmospheric carbon dioxide "amplified" the warming and caused ice sheet instability. Professor Ross Powell, a professor of geology at Northern Illinois University, uses "projected," "if," and "could" to paint a dire situation:

> The sedimentary record indicates that under global warming conditions that were similar to those *projected* to occur over the next century, protective ice shelves *could* shrink or even disappear and the WAIS would become vulnerable to melting. *If* the current warm period persists...This would result in a potentially significant rise in sea levels [our emphasis].[21]

This means that, *if* the IPCC computer models are correct, and *if* ice-core data from 4 million years ago is correct, and *if* the ice sheet computer modeling is correct, then we *could* have a rise in sea levels. They're asking us to trust computer model results built on top of other computer model results, rather than real temperature observations from land stations and satellites. The NSF press release also fails to mention that the West Antarctic Ice Sheet sits above an undersea ridge with at least one active volcano.[22] A volcano just might have more of an impact on the ice sheet than four molecules of CO_2 in every 10,000 in Earth's atmosphere.

As a last comment on the Antarctic Icecap, in April 2009 part of the Wilkins Ice Shelf split from the Antarctic Peninsula, formed a huge iceberg, and drifted out to sea. The climate alarmists jumped on this event in the press, claiming evidence of man-made global warming.[23] The stories failed to mention that the calving of ice shelves is a natural occurrence. Antarctica has lost more than seven ice shelves in the last 20 years without disastrous global effects.[24]

Greenland is probably the most photographed of the three ice caps by the Climatists. Global warming documentaries typically show a Greenland river from glacier melt roaring through a crevasse, and claim this to be an abnormal event. In fact, the Greenland ice cap is fairly stable. Most scientific studies show some melting at the edges of the ice cap, but also ice accumulation in the center of the Greenland ice mass from snowfall. Figure 48 showed us that ice extent in the Nordic Seas east of Greenland has been declining since about 1860 as Earth warms from the Little Ice Age.

Figure 52 shows temperatures at Godthab in southwest Greenland

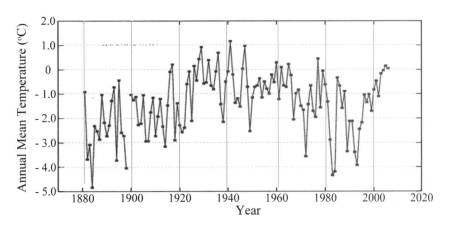

Figure 52. Temperatures at Godthab, Greenland. Graph of 128 years of mean annual temperatures at Godthab on southwest Greenland. Past temperatures were warmer than today. (Adapted from D'Aleo, NASA, 2009)[25]

from 1880 to 2008. The graph shows a warming trend of more than 1°C over the last 120 years. However, the graph also shows temperatures generally warmer during 1930–1940 than today, prior to significant man-made carbon dioxide emissions. Temperatures in Greenland during the Medieval Warm Period were also at least as warm as today. Dr. Joseph D'Aleo points out temperatures at Godthab track the Atlantic Multidecadal Oscillation, a natural temperature cycle which we discussed in Chapter 4.[26]

Historical sea level trends have been very consistent. Dr. Arthur Robinson of the Oregon Institute of Science and Medicine led a team that published a summary paper on climate change. They provided the graph in Figure 53 showing sea-level rise over the last 200 years. The figure shows that sea level has been increasing at 7 inches (18 centimeters) per century for the last 150 years. Intermediate trends over one or two decades have varied between 12 inches and zero inches per year. The 7-inch-per-year trend was the same prior to the ramp in greenhouse gas emissions in 1940 and during the period of rapidly increasing emissions after 1940.[27]

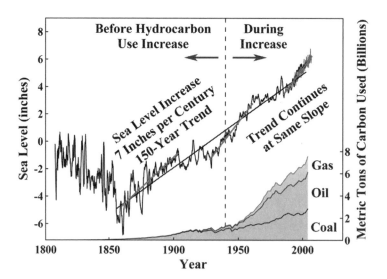

Figure 53. Sea-Level Rise for the Last 150 Years. The sea-level rise has been a steady 7 inches (18 cm) per century both before and after the rapid rise in hydrocarbon fuel emissions. Data is from surface gauges from 1807 to 2002 and satellites from 1993 to 2006. Carbon used numbers are annual amounts. (Robinson et al., 2009)[28]

Its interesting that sea-level rise projections from the IPCC have been falling with each assessment report. The First Assessment Report of 1990 projected an average sea-level rise of 188.5 centimeters (74 inches) by year 2100. By the Fourth Assessment Report in 2007, the average estimate was down to 38.5 centimeters (15 inches).[29] A rise of 15 inches by year 2100 is only 1.7 inches per decade, hardly cause for concern.

In summary, the specter of catastrophic sea-level flooding is very unlikely to happen. In contrast to computer-model projections, measured data for the Arctic, Antarctic, and Greenland Icecaps show only modest changes. Since the Arctic Icecap is floating in the ocean, a complete melt will not raise sea levels. Antarctic temperatures have been stable over the last 50 years and ice in the Antarctic Ocean is slowly growing. The Greenland icecap is also stable and not warmer today than in recent history. Measured sea-level rise has been a steady seven inches per century over the last 150 years and appears to be independent of the recent rise in atmospheric carbon dioxide.

AN INCREASE IN REPORTED NATURAL DISASTERS

The IPCC warns about climate change impacts due to natural disasters in its 2007 Fourth Assessment Report:

> Projected climate change-related exposures are likely to affect the health status of millions of people, particularly those with low adaptive capacity, through…increased deaths, disease, and injury due to heat waves, floods, storms, and droughts…[30]

The World Bank points to an increase in natural disasters in a 2007 report:

> In the aggregate, the reported number of natural disasters worldwide has been rapidly increasing, from fewer than 100 in 1975 to more than 400 in 2005.[31]

Indeed, the number of reported disasters has been increasing over the last century. Figure 54 shows a marked increase in the number of natural disasters reported and the damage caused.

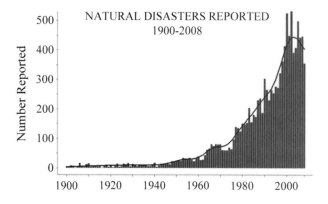

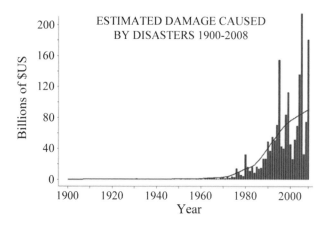

Figure 54. Number and Cost of Natural Disasters 1900-2008. The top graph shows the number of disasters reported and the bottom graph the estimated cost of damage in billions of U.S. dollars. (Adapted from EM-DAT, 2009).[32]

Al Gore, the United Nations, and the Climatists frequently use these graphs to blame global warming for human suffering from disaster events. Mr. Gore's book, *An Inconvenient Truth*, contains similar graphs.[33] We should ask two questions about this data. First, have the number of natural events *actually increased* over the last 100 years, or have other factors caused both the rise in the number and impact of reported disasters? Second, to what extent is the increase in natural disasters due to global warming?

The graphs in Figure 54 are provided by the International Emergency Disasters Database (EM-DAT). EM-DAT natural disasters include temperature-related floods, storms, droughts, wildfires and disease events that the IPCC warns will worsen due to global warming, but also earthquakes, volcanoes, and tsunamis. To qualify as an EM-DAT disaster, an event must cause one of these:

- ten or more people reported killed,
- 100 people reported affected,
- declaration of a state of emergency, or
- a call for international assistance.[34]

There are strong indications that both reporting and demographic factors have shaped the EM-DAT data over the last century. Reporting of disasters is more consistent worldwide today where it was lacking fifty years ago. Dr. Indur Goklany, an expert on economic development, put it this way:

> In the past, it is quite likely that unless there was significant loss of life (or property), the event would not be remarked upon, particularly if it happened in Asia, Africa or Latin America. As a result, fewer events would have been identified...[35]

The World Bank attributes the causes of disaster increase to population growth, urbanization, land use, and development of the coastal areas. World population has increased from approximately 2.7 billion in 1940 to over six billion today. As the population grew and economies developed, there was a shift from rural to urban for most nations of the world. By 2006, there were 25 cities of over 10 million in population, with 19 of these in the developing countries. Changes in land use increased vulnerability to fire, flood, and drought. Finally, the development of coastal areas, now with over 50% of the world's gross domestic product, has increased vulnerability to tropical storms and tsunamis. Each of these factors contributed to the number and cost of natural disasters.[36]

Despite the increase in reported events from 1900 to 2006, the number of deaths caused has fallen sharply, as shown in Figure 55 (next page). Mankind has been remarkably effective in reducing the deaths from disaster events. It appears that demographic trends explain much of the recent increase in reported disasters from natural events. Let's look first at storms and next at droughts and floods. We'll examine trends in the number and intensity of these events and

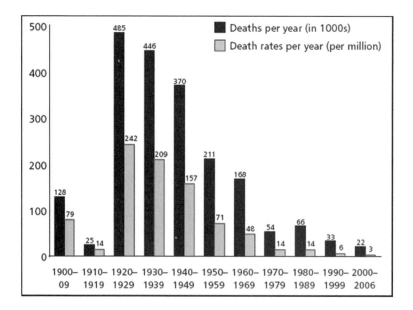

Figure 55. Global Decline in Deaths from Natural Disasters. Deaths from natural disasters have been declining over the last century, both in total and in per million of population. Data from EM-DAT. (Goklany, 2007)[37]

to what extent global warming plays a role.

DEVASTATION FROM HURRICANES AND STORMS

On August 29, 2005, Hurricane Katrina struck Louisiana with Category 3-force winds estimated at 125 mph. The subsequent failure of levees in New Orleans made Katrina the most destructive U.S. hurricane in history in terms of economic losses. The years 2004 and 2005 were the most active hurricane seasons of recent memory, with 17 and 31 identified storms, respectively.[38]

Increased devastation from stronger hurricanes and tropical storms is the *second* disaster projected by the alarmists. It's known that the strength of tropical storms increases as they travel over warmer ocean water. Building on this, some General Circulation Models project that warmer sea-surface temperatures from global warming will generate more and stronger tropical hurricanes and typhoons.

Kerry Emanuel of the Massachusetts Institute of Technology

produced a paper in 2005 that became the basis of storm alarmism. He examined tropical storms in both the Atlantic and Pacific Oceans from 1930 to 2005 and found that the frequency of storms was the same, but that the "potential destructiveness of hurricanes…has increased markedly since the mid-1970s." Emanuel defined a "power dissipation (PD) index" as a quantity that is proportional to the cube of the wind speed of a tropical storm. His paper claimed the PD had increased 75% over the last 30 years. Emanuel then found that increasing sea-surface temperature was well correlated to PD and attributed the cause to global warming.[39] Despite much criticism, Emanuel's results were adopted by the IPCC and cited frequently by Al Gore and the Climatists.

Emanuel's analysis suffered from a major problem: The historical record for both the number and strength of tropical storms is inaccurate. Tropical storm forecasting and measurement science has rapidly improved over the 20th century. Prior to 1944, information was gathered by ships, which generally tried to avoid contact with storms. Airborne reconnaissance was added in the 1940s, satellite measurements in the 1970s, and many improvements were made in techniques over the last 50 years. The Dvorak Technique to measure wind speeds from satellites was invented in 1972 and not used routinely until the 1980s. Prior to the Dvorak Technique, there was no systematic way to measure tropical storm wind speed.[40]

Dr. Christopher Landsea, research meteorologist for the NOAA National Hurricane Center in Orlando, Florida remarks:

> …the 1970 Bangladesh cyclone–the world's worst tropical-cyclone disaster, with 300,000 to 500,000 people killed—does not even have an official intensity estimate…[41]

Much of Landsea's work has involved reconstruction of databases for tropical storms, both in count number and storm intensity. He further states:

> Trend analyses for extreme tropical cyclones are unreliable because of operational changes that have artificially resulted in more intense tropical cyclones being recorded, casting severe doubts on any such trend linkages to global warming.[42]

Sea-surface temperature is only one of many parameters that impact the formation and strength of tropical storms. Wind shear is a more important factor. Wind shear is the difference in speed and direction between high-altitude winds and winds near the ocean surface. High wind shear tends to prevent formation of hurricanes. Natural variability due to the cycles of the Pacific Decadal Oscillation and Atlantic Multidecadal Oscillation is also a factor. Dr. William Gray identified reduced hurricane activity in the Atlantic Ocean during El Niño years. El Niños cause changes in wind patterns that increase wind shear in tropical areas where hurricanes are formed.[43]

Accumulated Cyclone Energy (ACE), which is similar to Emanuel's Power Dissipation index, is now the accepted measure of tropical storm intensity. Dr. Philip Klotzbach of Colorado State University analyzed ACE for tropical storms across the globe from 1986 to 2005. Klotzbach found:

> ...a large increasing trend in tropical cyclone intensity and longevity for the North Atlantic basin and a considerable decreasing trend for the Northeast Pacific. All other basins showed small trends, and there has been no significant change in global net tropical activity.[44]

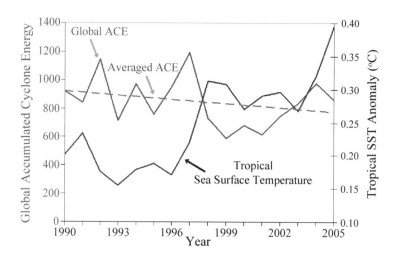

Figure 56. Tropical Storms and Ocean Temperature. Tropical storm activity measured by the Accumulated Cyclone Energy (ACE) index has been flat to slightly down while tropical Sea Surface Temperature has been rising from 1990 to 2005. (Adapted from Klotzbach, 2006)[45]

In other words, the hurricane activity in 2004 and 2005 in the Atlantic showed a big increase, but at the same time Pacific storm activity was considerably less. Dr. Klotzbach also found that global ACE was steady to slightly declining, while Sea Surface Temperatures increased by 0.3 °C over the period, as shown in Figure 56.

Other factors appear to be more important drivers to tropical storm activity than sea surface temperature. After the high hurricane activity years of 2004 and 2005, 2007-2009 have been low activity years. Dr. Ryan Maue of Florida State University maintains a global ACE index. In June, 2009 he reported that global tropical cyclone activity had plunged to a 30-year low.[46] Along with robbing Climatism of material for headlines, it appears that cyclone activity does not track the level of atmospheric carbon dioxide, which is still rising. Figure 57 shows Dr. Maue's ACE index from 1979 to 2009. Note the low level of tropical storm activity in 2009 compared to the higher activity in 2005.

Few scientists will attribute tornado activity to global warming, but that hasn't stopped politicians from adding to the alarm. Senator John Kerry appeared on MSNBC in February, 2008 to discuss a tornado event that killed more than 50 people in the

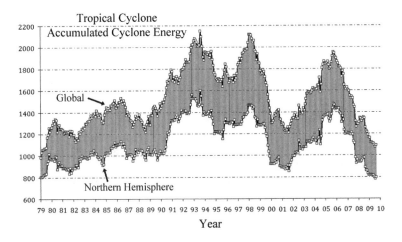

Figure 57. Global Accumulated Cyclone Energy 1979–2009. The top graph shows total global Accumulated Cyclone Energy for tropical cyclones and the bottom graph shows ACE for the Northern Hemisphere. The ACE index is the sum of the squares of the maximum surface wind speed measured every 6 hours for all storms with wind speeds exceeding 35 knots. (Adapted from Maue, 2009)[47]

Southeastern United States. According to Mr. Kerry:

> ...this is related to the intensity of storms that is related to the warming of the Earth. And so it goes to global warming and larger issues that we're not paying attention to. The fact is the hurricanes are more intensive, the storms are more intensive, and the rainfall is more intense...[48]

In fact, data from the National Climatic Data Center shown in Figure 58 shows no increase in the number of violent tornados in the U.S. over the last 50 years.

In summary, the data does not show increased ocean temperatures or global warming cause an increase in tropical storm frequency or intensity, or an increase in the number of violent tornados. Indeed, most tropical storm scientists do not believe that global warming is a significant contributing factor. Dr. Stanley Goldenberg, meteorologist at the Hurricane Research Division of NOAA, remarks

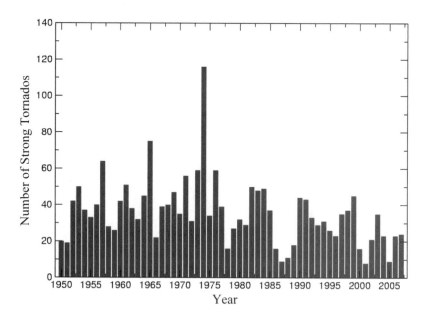

Figure 58. Number Strong to Violent Tornados in the U.S. The chart shows the number of F3 to F5 tornados that occurred in the U.S. from 1950 to 2008, as measured on the Enhanced Fujita Damage Intensity Scale. No increasing trend can be seen. (NCDC, 2009)[49]

about the views of his colleagues:

> To my knowledge, not a single scientist at the Hurricane Research Division, the National Hurricane Center, or the Joint Typhoon Warning Center believes…that there is any measureable impact on hurricane numbers or activity from global warming.[50]

FAMINE FROM DROUGHTS AND FLOODS

The *third* catastrophe projected by the climate alarmists is an increase in the frequency of droughts and floods, leading to worldwide deaths from famine. Of course, every drought and flood is blamed on global warming. In April 2007, Sir Nicholas Stern spoke to a group of environmentalists in South Africa, attributing past disasters to global warming:

> You'd see more floods like you've seen in Mozambique in 2000, you'd see more droughts like you saw in Kenya in the late 1990s, there would be a serious threat to the water flow down the Nile on which 10 countries depend.[51]

The summer of 2002 brought both drought and floods to India. Hot winds blocked the usual monsoon rains, bringing drought to Punjab, India's breadbasket. These same winds also caused a rapid melting of snow in the Himalayas, resulting in floods in the eastern states of Assam, Bengal, and Bihar. Rajendra Pachauri, chief of the IPCC, blamed climate change:

> There are strong reasons to connect the current drought to larger climate change, since what we are witnessing is a peculiar and sudden variation in climate as predicted by experts studying global warming.[52]

Another example was from Al Gore in his testimony to the United States House of Representatives in April, 2009:

> …2009 saw the eighth "ten-year flood" of Fargo, North Dakota, since 1989. In Iowa, Cedar Rapids was hit last year by a flood that

exceeded the 500-year flood plain. All-time flood records are being broken in areas throughout the world.[53]

Droughts and floods caused 87% of extreme weather deaths from 1900 to 2006, according to the EMDAT database.[54] But are these events abnormal, providing evidence for global warming? Or are they part of the natural cycles of Earth? Let's look at historical analysis.

Dr. Henry Lamb of the University of Wales, United Kingdom, led a team that analyzed sediments from the bottom of Lake Hayq in northern Ethiopia, in Eastern Africa. Lake Hayq is a closed basin lake, so the lake level varies primarily due to precipitation and evaporation. By analyzing oxygen and carbon isotopes from carbonates formed in the lake and also pollen evidence, they were able to construct a 2000-year precipitation record.[55] This data is shown in Figure 59. As can be seen, drought and flood conditions during the 20th century appear to be well within natural variability for Eastern Africa. In fact, the 20th century variation (the

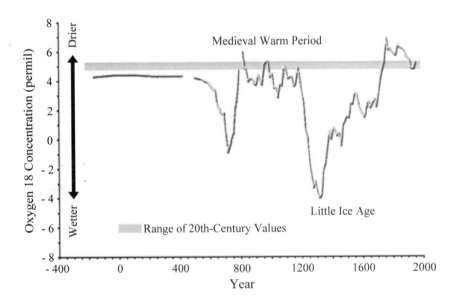

Figure 59. Historical Rainfall in Ethiopia. Estimate of rainfall in Ethiopia for the last 2000 years using an oxygen 18 proxy. The grey band shows the small variation over the 20th century. Note that conditions were dryer from about 1700 to 1900 and also much wetter during the Little Ice Age. (Lamb, 2007, Adapted by CO$_2$Science)[56]

thin gray bar on the chart) appears to have been abnormally stable in terms of variation in rainfall, during a period of maximum rise in atmospheric carbon dioxide.

Tree rings can be a good proxy for historical precipitation and are especially sensitive to periods of drought. Dr. Timothy Shanahan of the University of Arizona led efforts to analyze tree rings from submerged forests at the bottom of Lake Bosumtwi, Ghana in Western Africa. The team found that the lake completely disappeared for centuries at a time during drought periods over the last 3,000 years. The Sahel drought of the 1970s and 1980s killed an estimated 100,000 people, but Shanahan found that historical droughts appeared to be much worse:

> What's disconcerting about this record is that it suggests that the most recent drought was relatively minor in the context of the West African drought history…You have droughts that last 30 to 60 years, and then some that last four times as long.[57]

The researchers found that the most recent multi-century drought was from 1400 A.D. to 1750 A.D. They further found that the cycles of drought were linked to the natural cycle of the Atlantic Multidecadal Oscillation.[58]

Dr. Brendan Buckley of the Tree Ring Laboratory of Columbia University participated in a similar study. Tree rings from the mountains of Vietnam were analyzed to develop a 535-year historical record. Southeast Asia has a history of swings between flood and drought conditions. The researchers found major 30-year droughts in the 1300s, the 1400s, and the 1700s, believed to have contributed to the fall of the civilization of Angkor in what is modern day Cambodia. Dr. Buckley said that the tree-ring history showed a close match between dry spells and the El Niño pattern that brings drought to Southeast Asia and eastern Australia.[59]

U.S. Senator Debbie Stabenow recently proclaimed: "Global warming causes volatility. I feel it when I'm flying."[60] We all have feelings about the weather, but it's better to trust recorded data when trying to determine long-term trends. Despite the alarms of Al Gore and others, recent wet and dry conditions in the United States also appear to be within the range of normal variation. Figure 60 uses data from the National Climatic Data Center to chart the percentage of

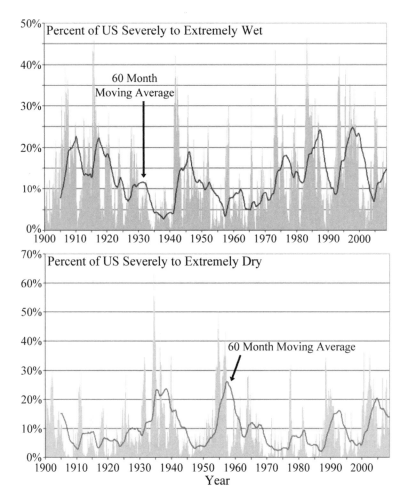

Figure 60. U.S. Wet or Dry Conditions from 1900-2008. The graphs show the percentage of the U.S. severely to extremely wet (top) and severely to extremely dry (bottom). Data is monthly and provided from the National Climatic Data Center. The black lines are 5-year moving averages. (Adapted from Meyer, 2009)[61]

the U.S. with wet conditions and dry conditions. The droughts of the 1930s and 1950s can clearly be seen. Wet periods occur several times during the century, centered at about 1907, 1915, 1942, 1974, 1984, and 1996. No trend toward more extreme droughts or floods is apparent.

In summary, data from Africa, Southeast Asia, and North America show that recent droughts and floods are within the range of natural variability over the last 500 to 1000 years. The proclamations of the Climatists show an ignorance of the natural weather cycles of Earth's climate. Remember this when news of the next flood or drought makes the headlines.

INCREASE IN TEMPERATURE-RELATED DEATHS

The *fourth* disaster projected by Climatism is an increase in temperature-related deaths from heat waves and diseases. Tony Blair, in a speech to the United States Congress in September, 2004, stated:

> There is good evidence that last year's European heat wave was influenced by global warming. It resulted in 26,000 premature deaths and cost $13.5 billion.[62]

The IPCC warns about heat waves in its 2007 Fourth Assessment Report:

> Hot days, hot nights, and heat waves have become more frequent...the observed higher frequency of heat waves is likely to have occurred due to human influence on the climate system...[63]

This last winter was a tough one for illness. My wife and I both suffered from a respiratory infection that we battled for almost a month. We're healthy people and rarely get sick, but when we do, it's usually during the winter.

Why is it that most illness seems to occur in the winter? According to the Center for Disease Control (CDC), the influenza (or flu) season in the United States is from November to April.[64] According to the World Health Organization of the United Nations, the flu season in the Southern Hemisphere is from May to October, during their winter months.[65] Could it be that cold climate has a greater negative impact on human health than warm climate?

In fact, many more persons die during cold weather than hot weather. The late Dr. William Keatinge, physiology professor at Queen Mary and Westfield College, led a team that studied temperature-related deaths for people aged 65 to 74 in Finland, Netherlands, England, Germany, Italy, and Greece. In 2000,

Keatinge's team found that deaths related to cold temperatures were more than *nine times* greater than those related to hot temperatures in Europe. Heart attacks, strokes, and respiratory illness were responsible for most of the cold-related deaths. Keatinge comments on the projected global warming:

> ...in the regions we have studied, the direct effect of the moderate warming predicted in the next 50 years would be to reduce, at least briefly, both winter mortality and total mortality.[66]

Dr. Bjorn Lomborg, adjunct professor at Copenhagen Business School goes even further:

> Winter regularly takes many more lives than any heat wave: 25,000 to 50,000 each year die in Britain from excess cold. Across Europe, there are six times more cold-related deaths than heat-related deaths...by 2050...Warmer temperatures will save 1.4 million lives each year.[67]

Other studies have found a similar pattern in the United States, even attributing increases in U.S. life expectancy to the migration of the elderly to warmer southern states.[68] Why do senior citizens in the U.S. retire to Florida or Arizona, instead of North Dakota or Alaska? Don't they know the IPCC claims warmer climates cause temperature-related death?

The IPCC briefly acknowledges these facts, but they state *without evidence* that rising temperatures will be worse:

> Studies in temperate areas have shown that climate change is projected to bring some benefits, such as fewer deaths from cold exposure. Overall it is expected that these benefits will be outweighed by the negative health effects of rising temperatures worldwide, especially in developing countries.[69]

It can be argued that heat-related deaths will be worse in the developing nations, which don't have the means to cope with rising temperatures. But the IPCC's own scenarios show per capita income of the developing nations reaching the 1990 per capita income level of the industrialized nations by 2100, thereby providing ability to cope with hot temperatures. The only explanation for the IPCC's position appears to be to add support to the misguided dogma that man-made emissions will ruin the planet.

The IPCC is more balanced regarding the spread of vector-borne diseases, such as malaria, in their 2007 Fourth Assessment Report. They forecast a net negative effect on global health from vector-borne diseases due to global warming, but state that:

> ...there is still much uncertainty about the potential impact of climate change on malaria at local and global scales.[70]

The World Health Organization (WHO) has been more negative. Dr. Diarmid Campbell-Lendrum of WHO stated in 2006:

> Climate affects some of the most important diseases afflicting the world. The impacts may already be significant.[71]

Campbell-Lendrum went on to demand action by the "industrial north" to alleviate the disease burden on the south: "It's a global issue and a global justice issue."[72]

The issue of vector-borne disease is complicated. Temperature is only one element in the spread of disease. Human factors, host factors (the mosquito), and vector factors (the bacteria) are all important. Malaria was not uncommon in the Southern U.S. during the first half of the 20th century, but was eliminated primarily through use of window screens and pesticides. Yakutsk, Siberia in the former Soviet Union had 2,000 to 3,000 cases of malaria per year during the 1940s, even though winter temperatures periodically dropped below -50°C.[73]

Dr. Paul Reiter, professor at the Institut Pasteur in Paris, studied malaria in England hundreds of years ago. He found that malaria was an important cause of death from 1564 to 1730 during the coldest period of the Little Ice Age. According to Dr. Reiter, the first true cure for the "ague" (as malaria was called) was a mixture of cinchona powder and white wine, used by Robert Brady in England in 1660.[74] Dr. Reiter concludes:

> Claims the malaria resurgence is due to climate change ignore these realities and disregard history...Public concern should focus on ways to deal with the realities of malaria transmission, rather than on the weather.[75]

THE POLAR BEAR AND SPECIES EXTINCTION

The polar bear has become the poster child for Climatism. Al Gore has pictures of "poor polar bears" on top of melting ice bergs in most of his presentations. *The Hot Topic*, an alarmist book by Gabrielle Walker and Sir David King, has swimming polar bears on the cover.[76] In May, 2008, while under political and legal pressure, the U.S. Department of the Interior announced a decision to protect the polar bear under the endangered species act.[77]

The loss of the polar bear, and the larger issue of species extinction, is the *fifth* projected calamity of global warming. The IPCC warns:

> Globally about 20 to 30%...of species will be at increasingly high risk of extinction, possibly by 2100, as global mean temperatures exceed 2 to 3 °C above pre-industrial levels.[78]

This is a huge number of extinctions. Scientists estimate that there are between 10 million and 80 million species on Earth, of which 1.6 million have been identified.[79] So the IPCC is telling us that a *minimum* of 2-3 million species will become extinct due to a 2–3 °C rise in temperatures.

It's clear that human activities have been responsible for species extinction. The primary causes have been over-hunting or habitat destruction. Well-known examples are the dodo, the passenger pigeon, and the Japanese wolf. But since 1600, only about 1,000 cases of extinction have been documented, although this number is probably somewhat on the low side.[80]

Now it's time for the common-sense test. According to United Nations statistics, human population has grown from about 600 million in the year 1600 to over six billion today, a ten-fold increase.[81] Despite this increase and accompanying human effects of over-hunting, land clearance, and other forms of habitat destruction, we've lost somewhat over 1,000 species over the last 400 years. Now the IPCC tells us that we're going to lose 2–3 *million* species in 100 years with a 2–3° temperature rise?

Since global temperatures have increased 0.7°C over the last 100 years, we should ask: How many species have we lost from global warming to date? The answer is maybe one. Not one thousand, but *one* animal species. The golden toad of Costa Rica, which disappeared in 1989, is the only extinction the Climatists officially claim as a casualty of global warming. However, even this one case is in doubt. Scientists now believe that the extinction of the golden toad was due to clearing of forests on mountainsides, which altered the moisture received by the toad's mountain-top home.[82] If the IPCC claims are correct, we should have had hundreds or thousands of extinctions due to warming during the last century. It hasn't happened.

So how did the IPCC come up with their catastrophic extinction forecast? As they've done before, they used computer model projections without experimental basis. As Ian Murray, Senior Fellow at the Competitive Enterprise Institute has commented:

> The researchers created a model that dictated global warming will cause extinctions. Surprise, surprise! When they ran the model, that's exactly the result they got.[83]

What has become extinct here is common sense. Real experimental data tells us that increased atmospheric CO_2 fertilizes plants. Greater plant growth boosts the ecosystem, including the animal life that feeds on the plants. Biodiversity is also greater in the tropics than the polar regions, in part because of the warmer temperatures. Warming global temperatures would trade some polar lands for tropical lands, which would strengthen biodiversity, not reduce it. In their mad rush to create catastrophic scenarios, the IPCC has ignored these facts and has left common sense behind. In any case, we know from our previous discussion that if the Earth warms 2–3°C, it will be due to natural causes, not man-made greenhouse gas emissions.

So what about the polar bear? Winnipeg, Canada was the scene of a January, 2009 conference on the bear. Canadian Environment Minister Jim Prentice received starkly different inputs from scientists and the Inuit people who lived in the region. While alarmist scientists warned that vanishing sea ice and over-hunting threatened polar bear populations, the Inuit people said the threat was exaggerated. Inuit leaders spoke of "bear populations doubling over the last 50 years, proliferating to the point of becoming a pest in many northern communities..."[84] Gabriel Nirlungayuk, Director of Wildlife for Nunavut Tunngavik, Inc., an organization which cares for Inuit lands, provided this input on the polar bear:

> The current population is stable. It is not constructive to exaggerate the situation.[85]

Indeed, the polar bear population is stable. According to Dr. Mitch Taylor, polar bear biologist for the government of Nunavut, the population has *increased* from 8,000-10,000 fifty years ago to 22,000-25,000 today.[86] Dr. Taylor estimates that the increase is primarily due to greater restrictions on hunting.

It is true that the reduced arctic ice of recent years is affecting polar bear hunting and migration patterns. But this is due to natural causes and not man-made CO_2 emissions. A NASA study in late 2007 found that the rapid reduction in arctic sea ice in 2007 was due to unusual wind patterns that moved ice out of the Arctic into warmer waters. Dr. Son Nghiem, who led the study team, concluded:

> Unusual atmospheric conditions set up wind patterns that compressed the sea ice, loaded it into the Transpolar Drift Stream and then sped its flow out of the Arctic.[87]

There is also evidence that polar bears are a species that can adapt to warmer temperatures. Professor Olafur Ingolfsson from the University of Iceland has found the jawbone of a polar bear dated to be 100,000 to 130,000 years old.[88] This means the polar bear survived warmer climatic conditions 4,000 and 8,000 years ago, as well as the warm interglacial period before the last ice age. Dr. Ingolfsson remarks:

> This is telling us that despite the ongoing warming in the Arctic today, maybe we don't have to be quite so worried about the polar bear.[89]

MELTING GLACIERS AND WATER SHORTAGES

Melting glaciers leading to water shortages is the *sixth* disaster projected by the Climatists. A headline from *Guardian* declares: "Climate change lays waste to Spain's glaciers." The article states:

Spain has lost 90% of its glacial ice in the last century…scientists warn of potentially dramatic effects to agriculture."[90]

Al Gore uses 18 pages of pictures of shrinking glaciers in *An Inconvenient Truth* to proclaim the shrinkage an "abnormal event" due to global warming. He laments the loss of the snows of Kilimanjaro in Tanzania, East Africa, and the glaciers at Glacier National Park in the United States.[91] It's true that glaciers have been shrinking worldwide. But this loss of ice has been underway for more than a century due to natural warming of the Earth.

The shrinking snowcap at Mt. Kilimanjaro in East Africa has been an iconic image of global warming for the Climatists. However, scientists who have studied the mountain have concluded that the snow loss is due to other factors. Dr. Jeorg Kaser, of the University of Innsbruck, found a shift in the climate of East Africa around 1880. Kaser concludes that "the ensuing dryer climatic conditions are likely forcing glacier retreat on Kilimanjaro."[92] Most of the snows of Kilimanjaro were lost before 1940. In fact, the retreat of Kilimanjaro's glaciers has slowed after 1950. Figure 61 shows satellite data of temperatures near Kilimanjaro for the last 30 years, provided by Dr. John Christy, professor of Atmospheric Science at the University of Alabama, Huntsville. Temperatures near the

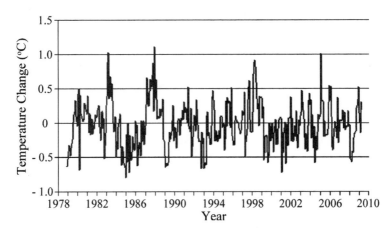

Figure 61. Temperatures Near Mt. Kilimanjaro. Temperature readings in the lower troposphere near Kilimanjaro from 1979 to 2008 from UAH satellite measurements. No temperature trend is observed. (Christy, 2009)[93]

mountain have been flat for the last three decades.

Mr. Gore is also concerned about the Himalayan Glaciers. From his book, *An Inconvenient Truth*:

> The Himalayan Glaciers on the Tibetan Plateau have been among the most affected by global warming. The Himalayas…provide more than half of the drinking water for 40% of the world's population…Within the next half-century, that 40% of the world's people may well face a very serious drinking water shortage, unless the world acts boldly and quickly to mitigate global warming. [94]

In fact, there is nothing the world can do to stop the shrinkage of glaciers in Tibet, since their ice loss is due to natural causes, not carbon dioxide emissions. Figure 62 shows the retreat of the Gangotri Glacier, which is the source of the Ganges River and one of the largest snow fields in the Himalayas. The glacier has been receding since 1780 with much of the ice loss before 1935.

Dr. Johannes Oerlemans from the University of Utrecht, Netherlands gathered historical data on the length of 169 glaciers

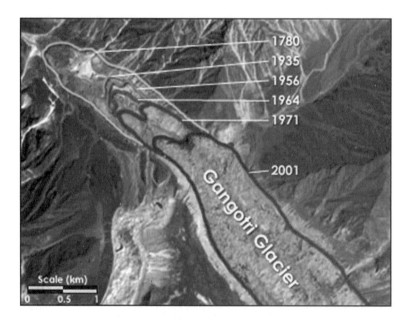

Figure 62. The Retreat of the Gangotri Glacier 1780–2001. The Gangotri Glacier in the Himalayas has been shrinking since 1780. (NASA, 2004)[95]

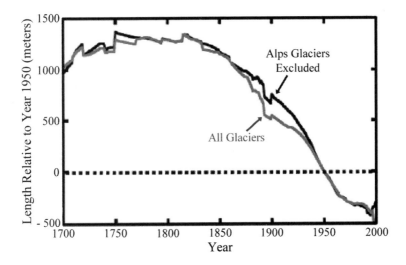

Figure 63. Trend in the Length of Global Glaciers. The chart plots average length of 169 glaciers relative to 1950. The gray curve is for all glaciers studied. The black curve excludes Alpine glaciers. Glacier length loss began about 1850, well before significant greenhouse gas emissions. (Oerlemans, 2005)[96]

from around the world over the last 300 years. He found glacier shrinkage began about 1850. His results are shown in Figure 63.

So, why have the world's glaciers been shrinking since the mid-1800s? The most logical explanation is ice loss due to warming of Earth as it recovers from the Little Ice Age of 1300 to 1850. Glacier retreat began 100 years before significant increases in atmospheric carbon dioxide.

CORAL AND ACIDIFICATION OF THE OCEANS

The *seventh* alarmist catastrophe is the death of coral reefs due to the acidification of the oceans. James Hansen summarizes the Climatist fears:

Coral reefs, the rainforest of the ocean, are home for one-third of the species in the sea. Coral reefs are under stress for several reasons, including warming of the ocean, but especially because of ocean acidification, a direct effect of added carbon dioxide. Ocean life dependent on carbonate shells and skeletons is threatened by dissolution as the ocean becomes more acid.[97]

The strong El Niño event in 1982–1983 and again in 1998 caused coral bleaching across a wide area. Coral bleaching occurs when coral expels algae (called zooxanthellae) that live within the coral in a symbiotic relationship, resulting in a whitening of the coral. Several papers have been written on this subject, including one by Dr. Ove Hough-Guldberg in 1998. Dr. Hough-Guldberg concluded:

> Reef-building corals...become stressed if exposed to small slight increases (1–2°C) in water temperature and experience coral bleaching...the rapidity of this change spells catastrophe for tropical marine ecosystems everywhere and suggests that unrestrained warming cannot occur without the complete loss of coral reefs on a global scale.[98]

Dr. Craig Idso, founder and Chairman for the Center for the Study of Carbon Dioxide and Global Change, has written CO_2, Global Warming and Coral Reefs, a book that thoroughly discusses the topic. Dr. Idso talks about increasing scientific evidence that coral is a resilient and adaptive species. He points out that bleaching is actually a coral adaptation mechanism. Bleaching is used in response to a rapid change in water temperature (warmer or cooler), changes in sunlight, sea water saltiness, bacterial infection, and other conditions. The coral bleaches, expelling its current zooxanthellae, and then adopts a different mix of zooxanthellae, which is more tolerant to ocean conditions. Dr. Idso points out that coral has been on Earth longer than man and has successfully adapted to higher levels of dissolved CO_2 and ocean temperatures, so alarms of coral extinction are unfounded.[99]

Dr. Glenn De'ath and others from the Australian Institute for Marine Science published a study in 2009 showing that calcification rates (growth rates) for giant coral at the Great Barrier Reef declined by 14% since 1990. They stated the "causes remain unknown," but the observed declines were "most likely" due to global warming.[100] Dr. Idso has a different interpretation of the data. He points out that the same data from De'ath shows calcification rates for the coral were 23% lower in the 1500s. Furthermore, calcification rates increased over the last 400 years up to 1990, as shown in Figure 64, during a period of rising ocean temperatures and rising atmospheric carbon

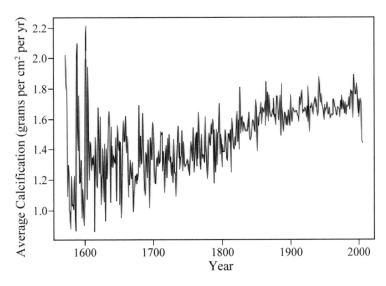

Figure 64. Coral Calcification at Great Barrier Reef. Coral calcification rates for Porites corals on the Great Barrier Reef from 1572 to 2001. Calcification rates have risen steadily over the period except for a recent downtrend after 1990. (Adapted from McIntyre, 2009)[101]

dioxide.[102] If warmer ocean temperatures and higher CO_2 concentrations reduced coral growth, then why would calcification rates be increasing during the modern warming period?

While there is still much debate on the issue, coral has prospered through many dramatic climate changes over thousands of years of Earth's history. Coral bleaching is a mechanism allowing coral to adapt successfully to warmer climatic conditions with higher levels of atmospheric carbon dioxide. Therefore, it's unlikely current global warming poses significant danger to ocean coral.

GULF STREAM SHUT DOWN AND NEW ICE AGE

The *eighth* Climatist calamity is a shut down of the Gulf Stream, triggering a new Ice Age. In the disaster movie *The Day After Tomorrow*, global warming causes the Greenland ice sheet to melt, causing the Gulf Stream current to shut down, resulting in abrupt global cooling and the next ice age.[103] Al Gore also worries about a similar event in *An Inconvenient Truth*.[104]

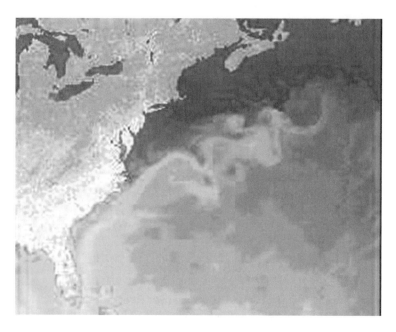

Figure 65. The Gulf Stream. A satellite image showing the Gulf Stream current as it flows along the Eastern United States and then across the North Atlantic Ocean. (NASA, 2009)[105]

The Gulf Stream is part of the Meridional Overturning Circulation (MOC), the great global ocean conveyor belt that flows from the North Atlantic Ocean to the Antarctic Ocean to the North Pacific Ocean and back again (see Figure 20 in Chapter 4). The Gulf Stream is the warm Atlantic Ocean portion of the current shown in the satellite photo in Figure 65. It gathers heat in the tropics of the Caribbean Sea, flows north along the Eastern United States, and then across the North Atlantic Ocean. This current brings heat from the tropics to warm Iceland, Scandinavia, and Northern Europe.

Since about 1998, alarmists have been warning that the Gulf Stream could shut down due to man-made global warming, causing colder temperatures in Europe and even a premature ice age (see text box). Alarm reached a peak in 2005 in a published paper from Dr. Harry Bryden of the U.K. National Oceanography Centre, who found "the Atlantic Meridional overturning circulation has slowed by about 30 per cent between 1957 and 2004."[106] Dr. Bryden had gathered data from a newly-installed system of ocean buoys. His

---

Gulf Stream Shutdown?

Along its northward journey, the Gulf Stream loses heat through evaporation, becoming cooler and saltier. As it reaches locations along Greenland, Iceland, and Norway, ice begins to form along coastlines. The remaining water becomes still colder and saltier, eventually sinking into the deep ocean to form the North Atlantic Deep Water current. It then flows south to the Southern Hemisphere, forming the deep water portion of the Meridional Overturning Circulation.

Scientists believe that at the end of the last ice age, about 12,000 years ago, the North Atlantic Deep Water current may have shut down. The theory is that huge quantities of fresh water from melting ice masses in Canada, Greenland, and Siberia poured into the North Atlantic Ocean. It is believed that Lake Agassiz, an enormous lake covering part of Canada at the time, burst its ice dam as part of this event, called the "Younger Dryas." This fresh water is believed to have mixed with Gulf Stream water, preventing the increase in water saltiness and density that drives the North Atlantic Deep Water Current. Global warming alarmists have predicted that the melting of glacial ice from Greenland could reduce the flow of the Gulf Stream, thereby triggering colder weather in Europe and even a premature ice age.[107] However, no such vast ice sheets exist today. Freshwater flows from Greenland are much smaller, so a similar event is unlikely.

---

results were trumpeted by Al Gore, *Guardian*, *BBC News*, and other media.

However, subsequent studies by different teams of researchers showed that Bryden's conclusions were premature. Dr. Friedrich Schott, oceanographer at Kiel University in Germany, led a team that looked at data from ship and buoy measuring stations in the Grand Banks area, east of Newfoundland. Dr. Schott found:

> Overall, the observations of Deep Water currents east of the Grand Banks reported here do not support suggestions of a basin-wide "slowdown" of the Atlantic Meridional Overturning Circulation.[108]

Dr. Bryden subsequently *reduced* his claimed variation in Atlantic currents from 30% over 50 years to just 10% over 25 years.[109] Note that most of the media that reported Bryden's first alarmist assertions did not find the new conservative findings newsworthy. *There is no experimental evidence of a slowing or halting of the Gulf Stream due to global warming, man-made or otherwise.*

SUMMARY: NO BASIS FOR CLIMATE DISASTERS

There is meager science for the eight catastrophes predicted by the alarmists, nor can such catastrophes be caused by man-made emissions. Predicted disasters are driven by *ifs*, *coulds*, and a high degree of imagination, along with model projections. The alarming predictions are at sharp odds with experimental data and geological history.

Sea levels continue to rise at a historical rate of seven inches (18 centimeters) per century. Earth's icecaps are stable, with the Antarctica Icecap growing, the Greenland Icecap stable, and the Arctic Icecap shrinking, but all within historical variation. Reported natural disasters have increased during the 20th century, but this is due to improved reporting and higher vulnerability of human populations due to demographic factors. Incidents of tropical storms, drought, and floods are not increasing and are all within the range of historical patterns. If our world warms, temperature-related deaths will decline overall, and there is no scientific evidence that vector-borne disease will increase. The claim that two million species will disappear is absurd and the polar bear is in no significant danger. Glacier shrinkage is due to natural causes, which began well before atmospheric carbon dioxide increases. Coral bleaching is a natural adaptation mechanism and coral calcification rates have increased over the last 400 years. The Gulf Stream is alive and well, and changes to it will be due to natural, not man-made forces.

Man-made greenhouse gas emissions are *not* the primary cause of global warming. As we have shown, today's warming and weather events are due to natural climatic cycles of Earth. But as we discussed in Chapter 1, governments of the world, major scientific organizations, most Fortune 500 companies, and the news media have accepted the false science that man-made emissions are causing catastrophic global warming. How did we get to this state of madness? Let's look at this in Part II.

Part II

Climatism—
The Ideology behind Global
Warming Alarmism
and
"Snake Oil" Remedies

"The only hope for the world is to make sure there is not another United States. We cannot let other countries have the same number of cars, the amount of industrialization we have in the U.S. We must stop third world countries right where they are."
Michael Oppenheimer, Environmental Defense Fund

CHAPTER 7

THE IPCC AND
THE ROAD TO WORLD DELUSION

"Controlling carbon is a bureaucrat's dream. If you control carbon, you control life." Dr. Richard Lindzen (1992)

The International Panel on Climate Control, an organization of the United Nations, engineered the current worldwide delusion that greenhouse gas emissions are causing global warming. The IPCC elevated a scientific theory to dogma and then required science to conform to the Global Warming Dogma. The IPCC has been remarkably effective at adopting the mantle of science, rallying worldwide support, and persuading all nations to accept its point of view. Let's look at the history of how this persuasion was accomplished.

UNITED NATIONS AND ENVIRONMENTAL ACTIVISM

The 1970s saw a rapid growth in the world environmental movement and the establishment of the first environmental ministries in government. The U.S. Environmental Protection Agency (EPA) was established in 1970. The international environmental organization Greenpeace was founded in Vancouver, Canada, in1971. Against this background, the UN moved to become the major player in global environmental legislation.

UN Secretary-General U Thant asked Maurice Strong (see text box) to organize the Stockholm Conference on Human Environment in 1972, which came to be known as the first "Earth Summit." The conference became a watershed event. It was the first time nations discussed environmental issues in terms of global governance.[1]

Among 69 recommendations adopted by the Stockholm Earth Summit, Recommendation 57 advised the Secretary-General to:

> ...ensure proper collection, measurement and analysis of data relating to the environmental effects of energy use and production...in particular, monitoring the environmental levels resulting from emission of carbon dioxide, sulphur dioxide, oxidants, nitrogen oxides...the objective is to learn more about the relationships between such levels and the effects on weather, human health, plant and animal life, and amenity values.[2]

It's clear the developing environmental policies of the UN were fertile soil for the claims of James Hansen 16 years before his Senate testimony in 1988. As a result of the Stockholm Conference, the United Nations Environmental Program (UNEP) was established in 1973 as a permanent UN institutional commitment to the environment. Maurice Strong was UNEP's first director general.[3]

In February 1979, the first World Climate Conference was held in Geneva, sponsored by UNEP and two other UN organizations, the World Meteorological Organization (WMO) and the International

---

Maurice Strong and the United Nations

Maurice Strong, a Canadian, is sometimes called the father of the 1997 Kyoto Protocol on global warming. Mr. Strong was born in 1929 and grew up in poverty during the Great Depression. During the 1950s and 1960s he became successful in the Canadian oil and utility industries. He describes himself as a "socialist in ideology" and a "capitalist in methodology."[4]

Mr. Strong became intimately involved in the world environmental movement and held a number of posts in the United Nations after 1970, developing close ties to five former UN Secretaries-General, from U Thant to Kofi Annan.[5] From positions of influence in the United Nations, he was a major driver in the formation of the International Panel on Climate Change.

Mr. Strong resigned from the UN in 2005 during the height of the UN's Oil-for-Food Scandal, when it was learned that he endorsed a check for $988,885 issued to him from a Jordanian bank. He currently lives in China.[6]

Council for Science (ICSU). The conference called on all nations to:

a) take full advantage of man's present knowledge of climate;
b) take steps to improve significantly that knowledge;
c) foresee and to prevent potential man-made changes in climate that might be adverse to the well-being of humanity.[7]

Dr. Bert Bolin, professor of meteorology at the University of Sweden, played a major role in promoting man-made global warming theory in the 1980s. Recall from Chapter 3 that Dr. Bolin proposed the "buffer factor" to defeat Henry's Law and provide a basis for false IPCC claims that carbon dioxide remains in the atmosphere for hundreds of years. Dr. Bolin was chosen director of the SCOPE project (Scientific Committee on Problems of the Environment), which was initiated in 1982 by the ICSU.[8]

A joint UNEP/WMO/ICSU conference in 1985 in Villach, Austria was heavily influenced by the SCOPE report. With Dr. Bolin as one of the editors, the Villach conference concluded:

As a result of the increasing concentrations of greenhouse gases, it is now believed that in the first half of the next century a rise of global mean temperature could occur which is greater than any in man's history.[9]

On what experimental data did the conference base this conclusion? The conference made the same great leap of faith that we discussed in Chapter 2. These conclusions are based on:

- The amounts of some trace gases in the troposphere, notably carbon dioxide (CO_2)...are increasing...

- The most advanced experiments with general circulation models of the climatic system show...[10]

In other words, the Villach conference based their conclusions on rising carbon dioxide and model projections, not experimental observations. Nevertheless, the Villach findings were important. National governments, including the United States, took note. At the recommendation of the Villach conference, the UN created the Advisory Group on Greenhouse Gases (AGGG) to further science and policy.

THE OZONE HOLE AND MONTREAL PROTOCOL

In September 1987, 29 nations and the European Community agreed to the Montreal Protocol on Substances that Deplete the Ozone Layer. The Montreal Protocol was the culmination of 13 years of research and scientific investigations regarding the "ozone layer" in Earth's stratosphere (see text box). Scientists concluded that chlorofluorocarbons (CFCs) generated by industry were destroying the ozone layer, increasing the risks to humans and environment due to higher intensity of ultraviolet rays. The signatories agreed to phase out and eliminate CFCs to prevent harm to the ozone layer.[11]

The Montreal Protocol was hailed as an international success. Nations of the world jointly agreed to solve a major environmental issue. The protocol helped support calls for a new organization to promote international action to address global warming. Climate change activists today point to the Montreal Protocol as an example of what can be accomplished by coordinated international effort. The table was now set for the hot summer of 1988.

THEATRICS IN THE SUMMER OF 1988

It's June, 1988 in the middle of a hot summer and temperatures approach the century mark in Washington D.C. Al Gore is chair of the Senate Committee on Energy and Natural Resources, which is holding its first hearing on the science of climate change. Witnesses and spectators fidget in the sweltering heat of the hearing room with an apparent lack of air conditioning. Former senator Tim Wirth disclosed to PBS *Frontline* how the hearings were staged:

> We called the Weather Bureau and found out what historically was the hottest day of the summer...So we scheduled the hearing that day, and bingo, it was the hottest day on record in Washington, or close to it...we went in the night before and opened all the windows...so that the air conditioning wasn't working inside the room.[12]

In his landmark testimony, a sweating James Hansen told the Senate that he was "99 percent confident that the world really was getting

The Montreal Protocol—Did We Save the Ozone Layer?

In 1974, Dr. Mario Molina and Dr. Sherwood Roland of the University of California put forth the theory that chlorofluorocarbons (CFCs) were destroying the ozone layer in our stratosphere. CFCs are Freon® gases used in spray bottles, refrigerators, and insulating foams. Molina and Roland won a Nobel Prize in Chemistry in 1995 for their work.[13]

The Ozone Layer is located between 10 and 40 kilometers (6 and 25 miles) above the Earth's surface, containing 90% of the ozone in the atmosphere. The ozone layer is known to block ultraviolet rays. Scientists warn that a thinning of the layer can cause increased rates of skin cancer and environmental problems.[14]

Three British researchers at the British Antarctic Survey, Farman, Gardiner, and Shanklin, announced in 1985 that they had found a hole in the ozone layer over Antarctica.[15] The scientific community mounted a number of expeditions to Antarctica to investigate the hole. They found a high level of ozone-destroying compounds and concluded that CFC emissions from industry were indeed destroying the layer. The nations of the world moved quickly to sign the Montreal Protocol in 1987, banning production of CFCs.

Did the Montreal Protocol save the ozone layer? Recent research challenges those scientific conclusions. A study from a team led by Dr. Francis Pope at the California Institute of Technology challenges the current chemical models postulated to destroy ozone.[16] Another study by Dr. Lu of the University of Ontario, Canada, attributes ozone variation to changes in cosmic ray intensity, or in other words, to natural causes.[17]

warmer and that there was a high degree of probability that it was due to human-made greenhouse gases."[18]

Hansen's testimony boosted media attention given to global warming. It coincided with the UN-sponsored Toronto Conference on the Changing Atmosphere in June 1988. Attending were 341 delegates from 46 countries, including scientists, government officials, industry representatives, and environmental activists.[19] Dr. Stephen Schneider reported the conference "attracted so many reporters that extra press rooms had to be added to handle the hordes of descending journalists."[20] At the Toronto Conference, the industrialized nations pledged to "voluntarily cut CO_2 emissions by 20% by the year 2005."[21] Just eight years after the World Climate Conference in 1979, many were now calling for cuts in emissions.

THE IPCC GATHERS A HOST OF SUPPORT

Finally in 1988, the Intergovernmental Panel on Climate Change was founded by the WMO and UNEP as an organization of the UN, headquartered in Geneva, Switzerland. The first IPCC session was held November 1988, and Bert Bolin was elected chairman. The establishment of the IPCC was the culmination of a number of trends and events that we have discussed. These include the newly adopted environmental activism of the UN, the claimed success of the Montreal Protocol, efforts of activist scientists such as James Hansen and Bert Bolin, efforts of political actors such as Al Gore and Maurice Strong, and a news media awakened by the hot summer events of 1988.

In many ways, the establishment of the IPCC was a master stroke for the groups that would embrace Climatism. The IPCC provided a platform of international legitimacy for alarmist claims of man-made global warming. Nations that reviewed IPCC reports and participated in the international conferences became supporters in fact.

The issue of man-made global warming provided a convenient means to an end for a diverse group of international players. The IPCC quickly gathered support from these organizations. These included the UN itself, environmentalists, the European Community (EC) and selected nations, the news media, and the activist scientific community.

The global warming issue appealed strongly to the United Nations in three major ways. First, the UN intended to take a leading global role in environmental issues. The UN-chartered Commission on Environment and Development published its report in 1987, *Our Common Future*. The commission became known as the Brundtland Commission, chaired by Gro Harlem Brundtland, then Vice President of the World Socialist Party. The Commission introduced the environmental concept of "sustainable growth," defined as "development that meets the needs of the present without compromising the ability of future generations to meet their own needs."[22] Maurice Strong was a contributor to the Brundtland Commission and later established and directed the UN Business Council on Sustainable Development.[23]

Second, the UN was interested in building a position of global governance. The idea of global control of greenhouse gases fit well with this concept. The UN established its Commission on Global Governance in 1992 with Mr. Strong as a commission member.[24]

Third, IPCC efforts fit with UN goals to reduce income differences between nations. For example, the UN Human

Development Report of 1996 stated, "Widening disparities in economic performance are creating two worlds—ever more polarized."[25] Reducing hydrocarbon emissions might constrain the growth of developed nations and allow developing nations a chance to catch up.

Control of greenhouse gas emissions also resonated strongly with environmental groups. Here was a global save-the-planet issue they could adopt. Control of hydrocarbon fuels was a natural extension for organizations such as the Club of Rome, which advocated limits on human economic development.

The EC also quickly supported the objectives of the IPCC. It viewed the control of greenhouse gases as a means to extend its reach into taxation and energy policy for Europe. The United Kingdom became a strong advocate to use the issue to achieve a measure of international environmental leadership.[26]

The activist scientific community also supported the IPCC. The IPCC and the news media became excellent channels for alarming projections from the climate models. Supercomputers are expensive. The media and IPCC served to report the latest alarming forecasts from the models, resulting in increased levels of funding from governments and environmental groups.

So from its birth in 1988, the IPCC found itself with a wide array of international support. Without wasting any time, it set out to define and warn about the dangers of greenhouse gas emissions.

SCORE DECIDED BEFORE THE GAME WAS PLAYED

From the first plenary meeting in November, 1988, the IPCC moved quickly to develop its first assessment report. Simultaneously with the IPCC plenary, the UN General Assembly adopted a resolution requesting the IPCC provide a "comprehensive review and recommendations" on climate change in 1990.[27] The IPCC succeed and issued its First Assessment Report in August 1990, concluding:

- emissions resulting from human activities are substantially increasing the atmospheric concentrations of the greenhouse gases...resulting on average in an additional warming of the Earth's surface

- based on current models...the global mean temperature will increase at a rate of about 0.3°C per decade during the next century, at a rate at least 10 times higher than any seen over the past 10,000 years[28]

- to stabilise concentrations of greenhouse gases at 1990 levels, emissions of CO_2 and other long-lived greenhouse gases must be reduced by more than 60% immediately

- without substantial reduction, global mean temperatures could rise...3°C by 2100[29]

How could the IPCC come to such a conclusion in less than two years of study? The mandate of the IPCC calls for an *objective* assessment of the risk of human-induced climate change (see text box). Most scientists agreed in 1990 that many aspects of the climate system were only poorly understood. But the IPCC chose to make a great leap of faith and issue alarming forecasts. As Dr. Richard Lindzen said years ago "the consensus was reached before the research had even begun."[30]

As evidence, note that the structure established for the IPCC in the 1988 plenary meeting consisted of three working groups: WGI (Science), WGII (Impacts), and WGIII (Response Strategies). Dr. John Zillman, principal delegate from Australia to the IPCC, states that it was necessary for each of the working groups to"work in parallel rather than in sequence."[31] For Working Groups II and III to assess the impacts and response strategies to climate change, a decision must have already been made to accept the catastrophic

---

Role of the Intergovernmental Panel on Climate Change

According to the web site of the IPCC, its role is:

...to assess on a comprehensive, objective, open, and transparent basis the latest scientific, technical and socio-economic literature produced worldwide relevant to the understanding of the risk of human-induced climate change, its observed and projected impacts and options for adaptation and mitigation.[32]

"Adaptation" is action taken by nations to adjust to temperature change as global warming occurs. "Mitigation" is action taken by nations to reduce greenhouse gas emissions to prevent global warming.

temperature rise projections of the climate models. In effect, *the score was decided before the game was played.*

From the outset, the IPCC authors decided to set a declarative tone, ignoring the uncertainties of climate science. Dr. Zillman describes discussions at the February 1990 meeting of Chapter Lead Authors in Edinburgh:

> After long debate, the meeting abandoned a more cautiously written narrative Executive Summary in favour of the more punchy and much quoted headlines:
>
> - We are certain of the following...
> - We calculate with confidence that...
> - Based on current model results, we predict...
> - There are many uncertainties...due to our incomplete understanding of...
> - Our judgement is that...
> - To improve our prediction capability, we need...[33]

And so the Global Warming Dogma was born. Future IPCC efforts were aimed at finding climate signals attributed to human activity and developing the impacts and response strategies to man-made global warming. In other words, efforts used to defend the dogma.

Why didn't conservative climate scientists rein in the assertions of the IPCC? The reality of the greenhouse effect provided a theoretical basis for the IPCC position. Nevertheless, many researchers criticized the IPCC claims. But the 99%-certain position of Hansen and others carried the day against the "we-just-don't-know" position of moderate scientists. As we'll discuss, Climatism used techniques to drive the "consensus" of Global Warming Dogma at the expense of alternative points of view.

1992 EARTH SUMMIT AND THE FCCC

With the IPCC's First Assessment Report in hand, the United Nations sought international agreement to reduce greenhouse gas emissions. The UN held five intergovernmental negotiating meetings during 1991 to 1992 to establish a Framework Convention on Climate Change (FCCC). The FCCC is essentially a treaty between all the nations of the world to agree to reduce greenhouse gas emissions. At the same time, planning was underway for the Second Earth Summit, to be held in Rio De Janeiro, Brazil in June 1992. Maurice Strong was chosen to be conference Secretary General.

The Second Earth Summit, also called the United Nations Conference on Environment and Development (UNCED), is viewed as a landmark convention by the environmental movement. The summit was attended by 172 nations, 108 at the level of heads of State. Twenty-four hundred representatives from environmental groups also attended.[34] The summit produced the Agenda 21, a 2,500-page radical-environmental document that called for a "profound reorientation of all human society."[35] In his keynote speech, Mr. Strong stated that industrialized nations will need to change their way of life:

> It is clear that current lifestyles and consumption patterns of the affluent middle class–involving high meat intake, consumption of large amounts of frozen and convenience foods, use of fossil fuels, appliances, home and work-place air-conditioning, and suburban housing–are not sustainable. A shift is necessary toward lifestyles less geared to environmentally damaging consumption patterns.[36]

The FCCC treaty was presented at the summit, signed by 41 nations and the EC.[37] It's significant that, while the IPCC had defined "climate change" to include natural causes, the FCCC redefined it to mean "a change of climate which is attributed directly or indirectly to human activity." The signing parties of the FCCC were committed to:

> ...adopt national policies and take corresponding measures on the mitigation of climate change, by limiting its anthropogenic emissions of green house gases...to...1990 levels.[38]

Largely unreported by the press was a statement issued by 47 atmospheric scientists in February of 1992, stating:

> ...we are concerned by the agenda for UNCED, scheduled to convene in Brazil in June 1992...to impose a system of global environmental regulations, including onerous taxes on energy fuels, on the population of the United States and other industrialized nations. Such policy initiatives derive from highly uncertain

scientific theories…based on the unsupported assumption that catastrophic global warming follows from the burning of fossil fuels and requires immediate action. We do not agree.[39]

In spite of this statement, the governments of the world signed the FCCC treaty and stamped the Global Warming Dogma with international government approval. *In only three and a half years since the formation of the IPCC, the world had rushed to agree to a dogma based on a false scientific theory.* The world had accepted the IPCC assertions as scientific proof. After 1992, most of the arguments on the world scene have involved the size and timing of emissions cuts, not whether they are needed.

INCREASING "CERTAINTY" OF IPCC REPORTS

Since 1992, the IPCC has worked hard to strengthen the accepted position of man-made climate change. A large number of meteorologists and climatologists were gathered to publish the Second Assessment Report in 1995, the Third Assessment Report in 2001, and the latest version, the Fourth Assessment Report in 2007. The Fourth Assessment approached an impressive 3,000 pages long. Each of these reports comes with a Summary for Policy Makers (SPM) that tends to be more assertive and alarmist.

The levels of warning and "certainty" increase with each report. The Summary of the Second Assessment Report states: "The balance of evidence suggests a discernible human influence on global climate change." Like all IPCC reports, its conclusions are heavily influenced by model results.[40]

The Third Assessment Report of 2001 introduced a subjective system of "confidence and likelihood statements" to attempt to qualify the validity of its conclusions. For example, "very likely" meant a 90%–99% chance "that a result is true" and "likely" meant a 66–99% chance. The Third Assessment found that:

Most of the observed warming over the last 50 years is *likely* [our emphasis] due to increases in greenhouse gas concentrations due to human activities.[41]

The Third Assessment Report also introduced what came to be

known as the infamous "Mann Hockey Stick Curve" to claim that 20th century temperatures were the warmest in over 1,000 years.

The Fourth Assessment Report of 2007 stated:

> Most of the observed increase in global average temperatures since the mid-20th century is *very likely* due to the observed increase in anthropogenic greenhouse gas concentrations.[42]

This increased the IPCC's certainty from 66–90% to 90–99%. Yet, as we showed earlier, science was moving away from this conclusion.

RE-WRITING HISTORY WITH THE MANN CURVE

Recall that in Chapter 4, we spent much time presenting the historical and proxy evidence for existence of the Medieval Warm Period, an era of warm global temperatures from about 950 to 1300 AD. The Medieval Warm period is important to determine the cause of global warming. If natural cycles caused Earth to warm 1,000 years ago, these same cycles could have caused the 20th century warming, rather than man-made CO_2 emissions. Surprisingly, *the IPCC admitted the historical fact of the Medieval Warm Period in their 1995 Second Assessment Report (SAR).* The graph shown in our Figure 66 is reproduced from the IPCC's Second Assessment Report Figure 22.

Dr. David Deming, of the University of Oklahoma, published an article in *Science* in June of 1995 on climatic warming in North America. In his testimony before the United States Senate in 2006, Dr. Deming made some amazing comments about events that occurred after his publication:

> I received an astonishing email from a major researcher in the area of climate change. He said, "We have to get rid of the Medieval Warm Period"…In 1999, Michael Mann and his colleagues published a reconstruction of past temperature in which the MWP simply vanished…This unique estimate became known as the "hockey stick,"…Normally in science, when you have a novel result that appears to overturn previous work, you have to demonstrate why the earlier work was wrong. But the work of Mann…was initially accepted uncritically, even though it contradicted the results of more than 100 previous studies.[43]

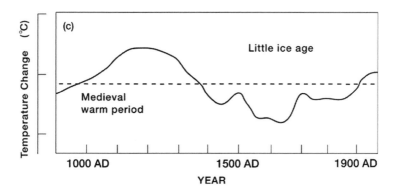

Figure 66. 1,000 Years of Temperatures from the IPCC. Reproduced from the Second Assessment Report, Figure 22. (IPCC, 1995)[44]

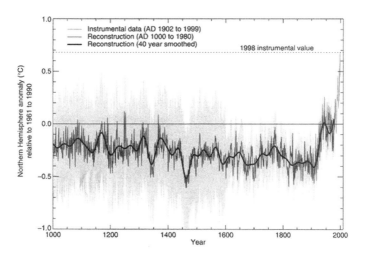

Figure 67. Mann's Infamous "Hockey Stick Curve." Rewriting history without the Medieval Warm Period or the Little Ice Age. (IPCC, 2001)[45]

Dr. Deming refers to a paper published in *Nature* 1998 by Dr. Michael Mann of the University of Massachusetts, along other authors.[46] The IPCC adopted Mann's graph of global temperatures, which became known as the "Hockey Stick Curve," in its Third Assessment Report (TAR) in 2001 (Figure 67). This "evidence" soon became the focus of the TAR. On page 2 of the summary, the IPCC states:

New analyses of proxy data for the Northern Hemisphere indicate that the increase in temperature in the 20th century is likely to have been the largest of any century during the past 1,000 years. It is also likely that, in the Northern Hemisphere, the 1990's was the warmest decade and 1998 the warmest year.[47]

The Mann curve is shown on page two of the assessment. The assessment later spends two pages dismissing historical and proxy evidence for both the Medieval Warm Period and the Little Ice Age. But the real story is that the IPCC put its stamp of approval on faulty science and adopted it in an attempt to change history.

After publication of the Mann curve in the 2001 TAR, two Canadians, Dr. Ross McKitrick of the University of Gulph and Stephen McIntyre, the primary author of the web site Climate Audit, requested Dr. Mann's data in order to try to reproduce the Mann curve. Reproducing the results of a study is a standard test in scientific circles. The data was provided to McKitrick and McIntyre only reluctantly and then incompletely. Apparently none of the peers who reviewed the 1998 publication of Mann's work had bothered to check his data. After finally extracting all the data from Dr. Mann, McIntyre and McKitrick published a critique and corrections to Mann's work in 2003, finding numerous errors:

> The data set of proxies of past climate used in Mann...for the estimation of temperature from 1400 to 1980 contains collation errors, unjustifiable truncation or extrapolation of source data, obsolete data, geographical location errors, incorrect calculation of principal components and other quality control defects.[48]

After corrections were made to the proxy data, McIntyre and McKitrick found the hockey stick shape disappeared, the Medieval Warm Period reappeared, and temperature values "in the early 15th century exceed any values in the 20th century."[49]

In 2006, a hearing on the Mann Curve was held in the U.S. House of Representatives. Dr. Edward Wegman, an expert in statistics, was chosen by the National Academy of Sciences to lead a team to provide an independent critique of Dr. Mann's work for the House committee. Regarding Mann's assertions, Wegman found:

> ...that the decade of the 1990's was likely the hottest decade of the

millennium and that 1998 was likely the hottest year of the millennium cannot be supported by their analysis.[50]

But the damage was done. Immediately after publication of the Third Assessment Report in 2001, the IPCC placed heavy emphasis on the "new analysis" provided by the Mann curve. Climatists rushed to show the world that temperatures in the late 20th century were warmer than any time in the last 1,000 years. John Houghton, Chairman of IPCC Working Group I, used the curve in many of his presentations (see Figure 68). Al Gore used a form of the Mann curve in *An Inconvenient Truth* and repeatedly in presentations. The Canadian Government mailed a copy of the curve to every household.[51] Britain featured the curve prominently in their 2003 Energy White Paper, showing it as evidence of man-made global warming.[52]

After the Congressional hearings on the Mann Curve, the IPCC quietly removed it from the Fourth Assessment Report of 2007, without any explanation or retraction. They replaced it with a "spaghetti graph" of proxy curves, but still held to their "warmest 20th century" dogma. The graph consists of 12 proxy reconstructions specially selected by the IPCC from a much broader background of evidence (see Figure 69 next page). Recall from Chapter 4 that more than 200 peer-reviewed studies are available that show a historical Medieval Warm Period, with most showing temperatures warmer than those of today. But the IPCC ignored these studies.

Figure 68. John Houghton at 2005 Climate Change Conference. Note the Mann Hockey Stick Curve in the background. (Tai, 2005)[53]

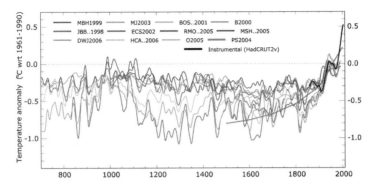

Figure 69. Temperature Reconstructions from the IPCC. Graph TS.20 from 2007 Fourth Assessment Report. (IPCC, 2007)[54]

CONSENSUS SCIENCE FROM THE IPCC

The IPCC is a political organization masquerading as a scientific organization. From the beginning, the IPCC focused on proving climate change is caused by man. As a political organization, it works to achieve consensus, which is a political objective, not a scientific objective. Science involves observing nature, developing a hypothesis that describes nature, and testing the hypothesis with experimentation and observation. Hypotheses that survive enough testing become theories and eventually laws, but even laws are never fully proven. Consensus does not have a role. Nevertheless, the IPCC created a process for "science consensus" in favor of man-made global warming, by effective use of six techniques.

The IPCC's first technique is to assemble a *large quantity of scientists* to agree on a consensus. About 2,500 scientists were involved in the last assessment report. However, only about 600 scientists drafted Working Group I on the science of climate change. The other 1,900 contributed to Working Group II (Impacts) and Working Group III (Response Strategies) blindly supporting the catastrophic temperature increases and sea level rises projected by WGI. But the number of scientists in Working Group I addressing the whole man-made climate change claim was small, according to Dr. Timothy Ball:

Of the 600 in WGI, 308 were independent reviewers, but only 32 reviewers commented on more than three chapters and only five reviewers commented on all 11 chapters of the report.[55]

Most scientists were not questioned about anything other than the one or two pages on which they worked. Yet the IPCC and the news media always point to 2,500 scientists, so how can the IPCC be wrong?

The second technique is to *select authors of critical chapters that support the man-made global warming dogma*. Dr. John McLean, climate scientist from Australia writing for the Science and Public Policy Institute, analyzed the IPCC author list for Chapter 9 of the Fourth Assessment Report. Chapter 9 is the critical chapter discussing whether global warming is man-made. Dr. McLean concludes:

> More than two-third of all authors of chapter 9 of the IPCC's 2007 climate-science assessment are part of a clique whose members have co-authored papers with each other...the majority of scientists who are skeptical of a human influence on climate significant enough to be damaging to the planet were unrepresented in the authorship of chapter 9.[56]

McLean further noted that many Chapter 9 authors are climate modelers. In particular, more Chapter 9 authors were from the U.K.'s Hadley Centre for Forecasting than any other organization.[57] Hadley Centre is home of the leading climate-modeling team in the U.K.

The third technique to achieve science consensus is to *provide chapter lead authors with authoritarian editorial power*. Dr. John Zillman, delegate from Australia to the IPCC, comments:

> ...the IPCC Chairman and the WG Co-chairmen were meticulous in insisting that the final decision on whether to accept particular review comments should reside with chapter Lead Authors. This was at variance with the normal role of journal editorial boards and led to suggestions that some Lead Authors ignored valid critical comments or failed to adequately reflect dissenting views when revising their text.[58]

Dr. Richard Lindzen of MIT, writing for the Heartland Institute, states that the IPCC process is not like a standard scientific peer-reviewed process:

> Under true peer-review...a panel of reviewers must accept a study
> before it can be published in a scientific journal. If the reviewers have
> objections the author must answer them or change the article to take
> reviewers' objections into account. Under the IPCC review process,
> the authors are at liberty to ignore criticisms.[59]

Over the last 20 years, the IPCC process resulted in a continuous
winnowing of dissenting scientists from the list of reviewers. Dr.
Lindzen and other reviewers of the first and second assessments had
their comments ignored and refused to participate in later
assessments. Some, like hurricane-expert Dr. Christopher Landsea,
were forced to threaten the IPCC with legal action to remove their
name from the list of expert reviewers.[60] As a result, few scientists
remain in the IPCC process who are not in lockstep with the dogma.

The fourth IPCC technique is *pressure for consensus* from lobby
groups and the IPCC itself. Dr. Zillman, who generally favors the
IPCC process, straightforwardly states:

> ...lobbyists were not averse to reporting stances taken by delegates or
> experts to home country media and governments in ways that would
> embarrass, left scientific members of some IPCC delegations feeling
> distinctly uncomfortable in presenting their views openly and
> honestly...dissenting individuals were subject to considerable peer
> pressure to agree in order to avoid the stigma of being seen to have
> prevented the IPCC from achieving a consensus report.[61]

With the lobbyists, the press, and the IPCC itself pushing for man-
made global warming consensus, is this an open and honest process?

The fifth technique is to *write the conclusions before the science is
agreed.* For the last three assessment reports, the Summaries for Policy
Makers were written and published *several months prior* to the full
report. After publically issuing the SPM, pressure was put on
scientists to align their work with the already announced summary
conclusions. There's nothing like writing the conclusions first and
then selecting the science to support the conclusions. That's politics,
not science.

Dr. Frederick Seitz, former president of the U.S. National
Academy of Sciences, specifically criticized the process to produce the
SPM for the 1995 Second Assessment Report. Dr. Seitz charged that
lead authors made key changes to the report after scientists had

accepted the text:

> But more than 15 sections in Chapter 8 of the report—the key
> chapter setting out the scientific evidence for and against a human
> influence over climate—were changed or deleted after the scientist
> charged with examining this question had accepted the supposedly
> final text… The following passages are examples of those included in
> the approved report but deleted from the supposedly peer-reviewed
> published version:
>
> - "None of the studies cited above has shown clear evidence that we
>   can attribute the observed [climate] changes to the specific cause
>   of increases in greenhouse gases."
>
> - "No study to date has positively attributed all or part [of the
>   climate change observed to date] to anthropogenic [man-made]
>   causes."
>
> - "Any claims of positive detection of significant climate change
>   are likely to remain controversial until uncertainties in the total
>   natural variability of the climate system are reduced."[62]

**In other words, the lead authors changed the text to remove any
statements that disagreed with their conclusion that "the balance of
evidence suggests a discernible human influence on climate."**

The sixth technique is *selective choice of science* to support the
dogma and minimize natural explanations for climate change. We
have discussed how the Mann curve attempted to re-write history and
eliminate the Medieval Warm Period. Another example is the IPCC's
conclusion that the sun is an insignificant factor in climate change.
The IPCC discounts all the evidence we discussed in Chapter 5 on
long-term trends in solar irradiance, sunspot links to cloudiness, and
solar activity proxies. As shown in Figure 70 (next page), they set the
magnitude of solar effects less than four different anthropogenic
(man-made) effects. Even classifying aerosols as man-made is
incorrect, since aerosol effects are primarily from natural sources, as
we discussed in Chapter 3.

Note the tiny radiative forcing effect of the "solar irradiance" line
at the bottom of Figure 70, showing IPCC disdain for any solar
influence on global temperatures. Scientists agree that radiation from
the sun indirectly causes *all weather* on Earth. It's true that direct

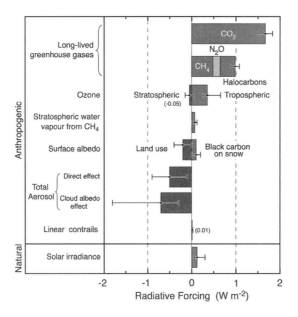

Figure 70. Radiative Forcings According to the IPCC. Radiative forcings to global temperature in watts per square meter at the top of the troposphere are shown. The shaded bars represent the magnitude of the forcing and the grey lines the uncertainty. The IPCC concludes that CO_2 and other greenhouse gases have a much larger effect than the sun. (Adapted from IPCC, 2007)[63]

links between the sun and global temperatures are not yet well understood. But does it make sense that the four CO_2 molecules of every 10,000 in the atmosphere have ten times the impact of the sun? The IPCC conclusion that the sun has little radiative forcing effect is certainly an exercise in selective science to support their predetermined conclusions.

THE KYOTO TREATY: LESSON IN FUTILITY

According to the United Nations, the 1992 Framework Convention on Climate Change "encouraged industrialized countries to stabilize greenhouse gas emissions," and the Kyoto Protocol "commits them to do so."[64] During the mid-1990s, the UN, supported by the European Commission and the United Kingdom, led international lobbying efforts to establish a follow-on agreement to the FCCC with binding emissions cuts. The signing of the Kyoto Protocol achieved this in December 1997 in Kyoto, Japan.

Thirty-eight nations and the European Community agreed to numerical limits for greenhouse gas emissions relative to a 1990 base year. The average was a reduction of about 5% from 1990 for the nations so limited. Developing nations such as China, India, and Brazil were signatories to the agreement, but were not required to agree to emissions limits. It was recognized that developed nations were "principally responsible" for current greenhouse gases in the atmosphere, so their burden is heavier under the Protocol. The Protocol also set up a system for trading "emission reduction units" to allow nations that were over emissions limits to buy reduction units from nations that were under the limits. The treaty was to cover emissions up to 2010, when a follow-on agreement would take over.[65]

The choice of the 1990 base year was particularly advantageous to the United Kingdom, which was in the process of converting power plants from coal to gas, and to Germany, which had the opportunity to reduce emissions from modernization of unified East German territory. The treaty would "enter into force" when nations with "55 per cent of the total emissions for 1990" had ratified the treaty.[66]

Negotiating the signing of the Kyoto Protocol was difficult, but achieving the needed level of ratifying nations was more difficult. Vice President Gore was part of the U.S. negotiating team at Kyoto, but the Clinton administration was unable to win in the U.S. Senate. The Senate voted 95–0 to reject the Protocol, primarily because China and developing nations did not have emissions limits. Australia and Russia were the other major holdouts. The Russian Science Academy did not agree with the dogma of man-made global warming. However, the projected sale of billions of dollars in emissions trading units, along with European support for Russian membership in the World Trade Organization, bought Russian ratification.[67] The treaty finally went into effect on February of 2005. To date, 184 nations have ratified the Kyoto Protocol, including Australia. The United States is the only major exception.

What has the Kyoto Protocol accomplished? For all practical purposes–nothing. It's as useful as a screen door in the Space Shuttle. Analysis shows that even *if* all nations met their emissions target, including the U.S., the global-temperature reduction would be only about seven *hundredths* of a degree Celsius by the year 2050.[68] This temperature difference is not measureable by any method. But as we shall see in Part III, implementation of Kyoto has already caused large negative economic dislocations where misguided "solutions" have been implemented.

ALARMING IPCC TEMPERATURE FORECASTS

According to the IPCC, the disasters that we debunked in Chapter 6 will be driven by rising global temperatures. Temperatures are projected to the year 2100 by the General Circulation Models based on several scenarios, as shown in Figure 71. These scenarios are

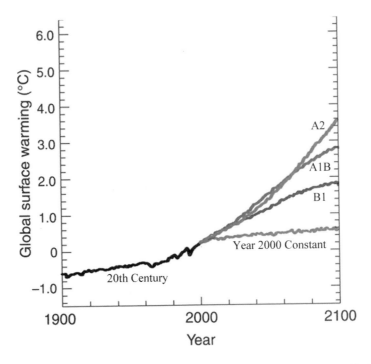

Figure 71. IPCC Projected Global Temperature Increases. Projected temperature increases from models for the IPCC scenarios. A2 is high population growth, slow economic development, and slow technological change, A1B is rapid economic growth and technological advance, with a balance between fossil and renewable energy, and B1 is rapid economic growth with a move to service economies. The bottom curve shows temperature rise with emissions held to year 2000 atmospheric concentrations. Since global temperatures have cooled since 2002, all IPCC projections are already high. (Adapted from IPCC, 2007)[69]

alternative global futures, covering "a wide range of demographic, economic and technological driving forces and resulting in greenhouse gas emissions" according to the IPCC.[70] Since the world has cooled over the last eight years, all IPCC projections are already high. Nevertheless, acceptance of these projections and the resultant forecasted disasters is forcing radical re-alignment of world economies.

SUMMARY

The creation of the IPCC was a master stroke for Climatism. **Only three and a half years after the establishment of the IPCC, the majority of the world signed the treaty of the Framework Convention on Climate Change, validating the false science of man-made global warming with international acceptance.** The consensus science of the IPCC, along with fear-generating catastrophic projections by Al Gore, James Hansen, and the news media, moved the world down the road of delusion.

Global warming alarmism began with science. It started with the theory of greenhouse gases and the role played in heating Earth's atmosphere. But along the way, it left the path of science and chose the road of politics and fear.

CHAPTER 8

CLIMATISM: THE IDEOLOGY BEHIND GLOBAL WARMING ALARMISM

"The Earth has cancer and the cancer is man." Club of Rome, 1974

Climatism is an ideology promoting the belief that man-made greenhouse gas emissions are destroying Earth's climate. It's a form of extreme environmentalism, an outgrowth of the population-control theories of the 20th century. Climatism promotes a false belief system that, after twenty years of repetition in the news media, has been accepted by most people. Climatism uses fear of global catastrophe, the mantle of science authority, time urgency, repetitive propaganda, and other tactics to further their agenda. Advocates include a wide array of governmental, environmental, scientific, and business persons and organizations.

Climatism opposes the free development of human society and seeks to substitute autocratic control from centralized government bureaucracy. It calls for the radical transformation of our way of life, regardless of cost. Climatism demands adherence from *all* nations. Indeed, each and every citizen *must* accept such doctrine or be forced to do so. Václav Klaus, President of the Czech Republic and 2009 President of the European Union warns:

The climate alarmists believe in their own omnipotence, in knowing

161

better than millions of rationally behaving men and women what is right or wrong, in the possibility to give adequate instructions to hundreds of millions of individuals and institutions and the resulting compliance or non-compliance of those who are supposed to follow these instructions.[1]

Let's examine the roots of Climatism in this chapter and its beliefs, objectives, and methods in Chapter 9.

MALTHUS AND THE ROOTS OF CLIMATISM

The roots of current Climatism stretch back to a 19th-century Anglican minister named Thomas Malthus. Malthus published *An Essay on the Principle of Population* in 1798, which discussed the relationship between population growth and resources needed to support the population. He warned that, while the food supply tends to grow arithmetically (i.e. 1, 2, 3, 4...), population tends to grow geometrically (i.e. 1, 2, 4, 8...). As a result, Malthus projected economic disaster with population outstripping resource availability.[2]

Malthus argued that "the period when the number of men surpass their means of subsistence has long since arrived." He recommended population control by non-marriage, late marriage, and "vice." He opposed urbanization, industrialization, and vaccination for smallpox, and favored housing shortages to curb population growth.[3]

However, Malthus was wrong. The ingenuity of mankind delivered an agricultural revolution during the 1800s and 1900s, supporting rapid global population growth (Figure 72). Advances in mechanization and energy were applied to agriculture, producing inventions such as plows, seed drills, reapers, tractors, and harvesters. Science added new fertilizers and genetically engineered seeds. Even though global population increased from one billion in 1800 to 6.5 billion by 2005, food production increased faster, contrary to the warnings of Malthus. Figure 73 shows the ramp in world grain production for the last half of the 20th century.

Although Malthus was one of the first, many subsequent philosophers, economists, and environmentalists warned of the dangers of overpopulation and advocated various forms of population control. Sir Francis Galton, cousin of Charles Darwin, developed the concept of eugenics in 1883. Proponents of eugenics presume that

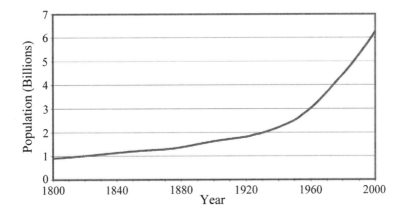

Figure 72. World Population Growth 1800–2000. (Adapted from Kruse, 2008, original source Delong, 1998)[4]

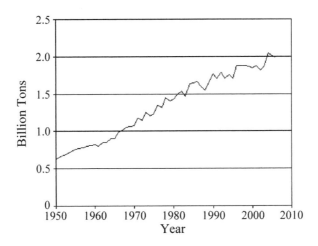

Figure 73. World Grain Production 1950–2006. (Adapted from Earth Policy Institute, 2006, data from USDA, 2006)[5]

the qualities of humanity can be improved by selective human reproduction. Galton believed human civilization prevented Darwin's natural selection from working in the human species. He advocated "promoting early marriage in the classes to be favoured" in order to promote "the speedier evolution of a more perfect humanity."[6] Support for eugenics grew in Europe and the United States in the early 1900s, often in concert with the conservation movement, the forerunner of today's environmental movement.

The wealthy Rockefeller and Osborn families played a leading role in funding both conservation and eugenics. John D. Rockefeller, Jr. contributed millions to help establish Great Smoky Mountain and Grand Teton National Parks.[7] He also founded the Bureau of Social Hygiene in 1911, which promoted eugenics.[8] Henry Fairfield Osborn presided at the 2nd International Congress on Eugenics in 1911. His keynote speech at the 3rd International Congress on Eugenics in 1933 was strongly anti-population and anti-industrialization, complaining about "six overs":

> ...over-destruction of natural resources...over-mechanization...over-construction of means of transport, over-production of food...over-confidence in future demand and supply...[and] over-population.[9]

After World War II, Nazi eugenic activities were exposed, discrediting eugenics as acceptable social policy. But population control and conservation continued to be connected. In 1945, John Rockefeller, Jr. donated land in New York City on which the United Nations stands today.[10] His son, John Rockefeller III, founded the Population Council in 1952, an organization dedicated today to "improve the well-being and reproductive health of current and future generations around the world and to help achieve a humane, equitable, and sustainable balance between people and resources."[11]

Henry Osborn's son, Henry Fairfield Osborn, Jr., was a conservationist and a long-time president of the New York Zoological Society. He wrote *Our Plundered Planet* in 1948, which called for "world-wide planning" to control population growth. He followed up with *The Limits of the Earth* in 1953, arguing that Earth could never feed a population of four billion.[12]

The Rockefellers and the Osborns are only a part of a consistent effort by first conservationists and later environmentalists. These are documented in the book *Environmentalism: Ideology and Power*, by sociologist Donald Gibson.[13] Consistent themes have been 1) world population growth is unsustainable, 2) industrialization and technological development are ruining the planet, and 3) governments must take measures to solve these problems. As we entered the last of the 20th century, these themes became the ideological basis for Climatism.

THE NEW MALTHUSIANS ADOPT CLIMATISM

As the environmental movement expanded in the 1960s and 1970s, some amplified the theme that human population growth was pushing mankind toward catastrophe. One of these "new Malthusians" was Dr. Paul Ehrlich. In his 1968 book, *The Population Bomb*, Ehrlich stated:

> The battle to feed all of humanity is over. In the 1970s the world will undergo famines—hundreds of millions of people are going to starve to death in spite of any crash programs embarked upon now.[14]

Dr. Ehrlich was certain Earth would not be able to feed itself in the short term. He warned of looming resource shortages. Only famine and disease would stop the growth of human population. Ehrlich's remedy was severe and coercive:

> We must have population control at home, hopefully through a system of incentives and penalties, but by compulsion if voluntary methods fail. We must use our political power to push other countries into programs which combine agricultural development and population control. And while this is being done we must take action to reverse the deterioration of our environment before population pressure permanently ruins our planet. The birth rate must be brought into balance with the death rate or mankind will breed itself into oblivion.[15]

Ehrlich's ideas in 1968 are similar to today's claims of Climatism. He warned of short-term disasters, advocating compulsive solutions that required compliance by all countries. This closely mirrors the claims and proposed remedies of today's global warming alarmists.

In 1974, Ehrlich wrote *The End of Affluence* with his wife Anne, upping his forecast of doom. The book warned of a nutritional disaster by the 1980s "likely to overtake humanity":

> Due to a combination of ignorance, greed and callousness, a situation has been created that could lead to a billion or more people starving to death.[16]

However, the mass famines of the 1970s and 1980s did not occur.

In October 1980, the late economist Julian Simon challenged Ehrlich's assertion that severe-resource scarcity was just ahead. Ehrlich, Dr. John Holdren, and others wagered Simon $1,000 that the price of five metals—chrome, copper, nickel, tin, and tungsten—would increase within 10 years. Although 800 million people were added to the world's population during the 1980s, the most in any decade, Simon won the bet as prices of all five metals declined in both actual and inflation-adjusted terms by 1990.

The reasons for the price declines are a lesson in economics and the ingenuity of man. Technological improvements as well as substitution of alternative materials brought the price of all five metals down. More efficient refining techniques and new sources of supply reduced the price of chrome, nickel, and the other metals. Aluminum was substituted for tin in cans, ceramics for tungsten in tools, and copper wires were replaced with fiber optics.[17]

Despite his forecasting failures, Ehrlich continues to be revered in environmental circles. Beginning in the early 1970s, Dr. Ehrlich, John Holdren, and others developed the "population impact equation" (see text box). They estimated the impact of human population on Earth's environment to be a function of three factors: 1) size of the population, 2) affluence or per capita consumption of the population, and 3) environmental damage "inflicted" by the population's technology. Ehrlich and Holdren concluded that *both*

Ehrlich, Holdren and the Population Impact Equation

Ehrlich and Holdren described the impact of global population on the environment (I) by the equation:

$$I = P \cdot A \cdot T$$

Where P is the size of the population, A is affluence or per capita consumption, and T is the environmental damage inflicted by technology to supply each unit of consumption. According to Ehrlich and Holdren, increases in population, affluence, and technology multiply together to increase the environmental impact on the Earth. It's difficult to measure affluence and technology separately, but energy usage can be a substitute for the product of A and T. Ehrlich and Holdren therefore argue that growth in both population and energy usage must be limited.[18]

Note that John Holdren, a strong proponent of "zero growth" in population and energy usage, has reached the highest levels of the United States Government. In February, 2009 Holdren was appointed assistant to the president for science and technology and Director of the Office of Science and Technology Policy.[19]

population and energy usage growth must be severely limited.[20]

The Club of Rome, an organization favoring radical limits to population growth and resource consumption, was organized in 1968. The Club issued its first report, *The Limits to Growth*, in 1972. The report was prepared by Donella Meadows and others at the Massachusetts Institute of Technology based on analysis from models of population and resource growth. The report found:

> If the present growth trends in world population, industrialization, pollution, food production, and resource depletion continue unchanged, the limits to growth on this planet will be reached sometime within the next one hundred years. The most probable result will be a rather sudden and uncontrollable decline in both population and industrial capacity.[21]

The report viewed an ideal situation as "stopping population growth in 1975 and industrial capital growth in 1985" and warned:

> We cannot say with certainty how much longer mankind can postpone initiating deliberate control of its growth before it will have lost the chance for control.[22]

In 1991, Alexander King and Bertrand Schneider authored another report for the Club of Rome titled *The First Global Revolution*. They detailed how global warming would be the instrument to drive global revolution:

> In searching for the new enemy to unite us, we came up with the idea that pollution, the threat of global warming, water shortages, famine and the like would fit the bill. In their totality and in their interactions these phenomena do constitute a common threat which demands the solidarity of all peoples. But in designating them as the enemy, we fall into the trap about which we have already warned, namely mistaking symptoms for cause. All these dangers are caused by human intervention and it is only through changing attitudes and behaviors that they can be overcome. The real enemy, then, is humanity itself.[23]

The anti-population, anti-affluence, and anti-technology philosophies of radical environmentalism had discovered global warming as the perfect vehicle for revolution of world society. Climate could be framed as a threat of global proportion. It could generate a wide variety of looming catastrophes, including floods, heat waves, storms, and droughts to capture the attention of the world. It could be portrayed as a time-sensitive crisis (i.e., "We must act now!"). It required establishment of permanent United Nations organizations, such as the IPCC. It called for leveling of industrialized nations relative to developing nations. It would allow world leaders to reduce standards of living, restrict freedom, and increase governmental control in the name of solving the crisis. By controlling carbon, alarmists could retard technological growth, regulate affluence, and reduce population growth, without a direct attack on any of the three.

Thus, Climatism was born. President Klaus warns about the real objective of Climatism:

> I am afraid there are people who want to stop the economic growth, the rise in the standard of living (though not their own) and the ability of man to use the expanding wealth, science and technology for solving the actual pressing problems of mankind, especially of the developing countries.[24]

The Club of Rome is today an influential advisor to the UN. A search of the UN website displays over 400 references to "Club of Rome," many for publications. Club membership includes Al Gore,[25] Maurice Strong,[26] Anne Ehrlich, activist scientist Stephen Schneider,[27] and Mikhail Gorbachev, founder of Green Cross International.[28]

What if the population-explosion alarmists-turned-Climatists are correct? They forecast famine and economic disaster from overpopulation and over-industrialization. They forecast increasing environmental damage as population, affluence, and technology increase. But what does the evidence show? Let's take a closer look at the real state of the world.

THE IMPROVING STATE OF THE WORLD

Advances in science, technology, agriculture, industry, energy, and medicine during the last two centuries have provided vast gains to the health and welfare of mankind. Contrary to the warnings of the alarmists, the improvements in human society have been little short of miraculous (Figure 74). Since 1600, Gross National Product per

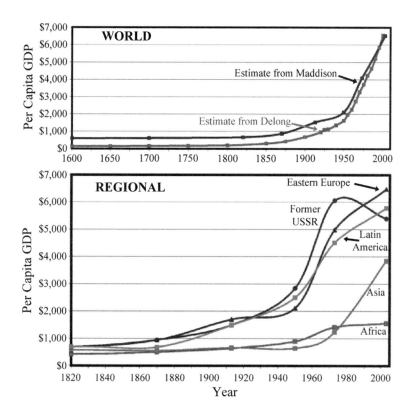

Figure 74. World and Regional GDP Growth. The top chart shows two estimates of world gross domestic product per person from 1600 to 2003. (Adapted from Kruse 2008, data from Delong 1998, Maddison, 2007)[29] The bottom chart shows GDP growth from 1820 to 2003 for selected regions. (Adapted from Kruse 2008, data from Maddison, 2007)[30]

person has increased from under $700 to over $5,000 for most of the world. Major gains have also been made in each of the world's regions. Gross Domestic Product has more than tripled since 1950 for Eastern Europe, the former USSR, Asia, and Latin America. The one exception is Africa, with only a 50% increase in GDP.

These improvements are best reflected in gains in human life expectancy. For most of history, human life expectancy was 20 to 30 years. This is because early childhood disease and infant mortality has plagued mankind for most of existence. An examination of gravestones, mummies, and skeletons shows that the average citizen of Rome 2,000 years ago lived only 22 years.[31]

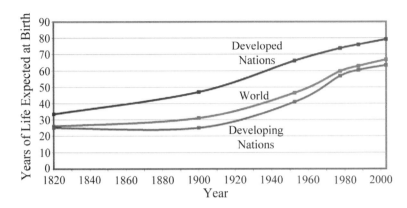

Figure 75. Improving Life Expectancy 1820–2003. Trends in life expectancy at birth for developed nations, developing nations and the world from 1820–2003. (Adapted from Kruse, 2008, from Goklany, 2007)[32]

Life expectancy has tripled since 1800. As shown in Figure 75, a child born today will live 80 years in developed nations and over 60 years in developing nations. A child born into the United States in 1901 could expect to live 49 years. By the year 2000, expected life for a U.S. baby was more than 77 years. In 1950, life expectancy in China and India was about 40 years. By the year 2000, this had grown to 63 years.[33]

Increases in life expectancy have been primarily driven by declines in worldwide death rates. Improvements in medicine, better nutrition, and control of infectious diseases have significantly reduced infant mortality. These improvements have also shifted the cause of most deaths in much of the world from diseases early in life to diseases of aging. Figure 76 shows the decline in both birth rates and and death rates since 1950. Note the steady decline in death rates, without any catastrophic interruptions in the trend. The "billion deaths" predicted by Dr. Ehrlich for the 1980s did not occur.

Because the global birth rate is declining faster than the global death rate, population growth rates are also declining, not exploding as some have warned. The United Nations predicts world population growth will reach zero between 2050 and 2100. Total global population is projected to peak at nine to ten billion people.[34]

Major gains have also been made in reducing the number of people living in poverty. Figure 77 shows the percentage of global population and number of people living on less than one U.S. dollar per day. While 20% of the population of the world still lives in

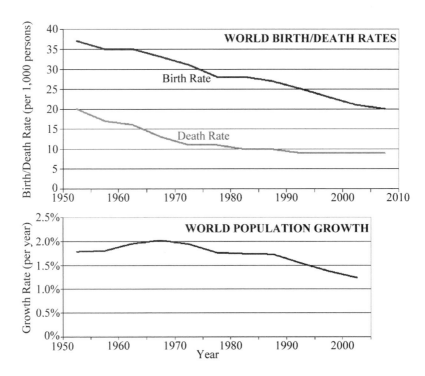

Figure 76. Birth Rate, Death Rate, and Population Growth. Global death rates and birth rates per 1,000 persons (top chart). Global population growth rate per year (bottom chart). Birth rates have declined faster than death rates, reducing world population growth rates. (EarthTrends, 2006)[35]

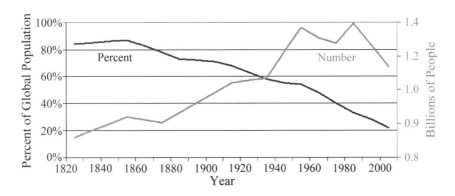

Figure 77. World Population Living on $1.00 per Day. The left axis measures the percent of world population living on one U.S. dollar per day from 1820 to 2000. The right axis measures the total number of people. (Adapted from Wolf, original data from World Bank and US Bureau of the Census, 2002)[36]

extreme poverty, this is down from 60% of the population in 1930. For many years, the number of persons living on less than one U.S. dollar per day grew as global population grew at a rapid pace. But since 1980, the number of persons in extreme poverty has declined at a faster rate than population growth.

The quality of life for people has also improved. One way to measure this is by years of education per person. Dr. Indur Goklany finds the average number of years of education per person in France, Japan, the U.S., and the U.K. now exceeds 16, up from fewer than four years in 1820. Years of education for China and India has quadrupled in the last 50 years.[37] The UN reports enrollment in primary education reached 88% worldwide in 2007.[38]

Certainly, all is not well in many parts of the world today. More than one billion people are trying to survive on less than one U.S. dollar per day. This includes almost one-half of the people living in sub-Saharan Africa.[39] About 25,000 people die every day of hunger or hunger-related causes.[40]

Disease continues to be a major cause of death in the developing nations. AIDS is now the second largest epidemic in history, after the Black Death of the Middle Ages. AIDS kills almost 2 million people each year, with the largest toll in Africa. Malaria, tuberculosis, pneumonia, and diarrheal diseases, such as cholera and dysentery, *each* kill one million or more persons every year.[41]

An estimated 32% of people in the developing nations, 1.6 billion persons, do not have access to electricity. This includes 75% of people in sub-Saharan Africa.[42] Many still rely on burning firewood and dung for fuel, resulting in lung infections from smoke and four million deaths every year.[43] The UN estimates that 2.5 billion people do not have adequate sanitation and 900 million do not have access to sources of clean water.[44]

But in many of these areas, continued improvements are being steadily made, decade-by-decade. The UN reports that from 1990 to 2007, the number of persons living in extreme poverty declined 23%, deaths of children under five declined 29%, and over one billion people gained access to improved sanitation and water supplies. Even the number of AIDS deaths appears to have peaked in 2005.[45]

Mankind's pursuit of economic growth, affluence, and technological advance is extending and improving the quality of the lives of people all over the globe. For further information, read *The*

Skeptical Environmentalist, Measuring the Real State of the World by Dr. Bjorn Lomborg.[46] Also read *The Improving State of Our World* by Dr. Indur Goklany.[47]

But what about pollution? The environmentalists-become-Climatists say mankind's gains are at the expense of the environment. Are we "wrecking the planet?"

THE ENVIRONMENT AND KUZNETS CURVES

Ask almost any American high school student, "Is pollution better or worse today than 20 years ago?" and you'll hear "worse." But, in fact, air and water pollution levels are considerably improved. Figure 78 shows steady declines in U.S. airborne levels of carbon monoxide, nitrogen dioxide, lead, sulfur dioxide, ozone, and particulates over the last 25 years. Nevertheless, the education system and the media have convinced students that air pollution is increasing.

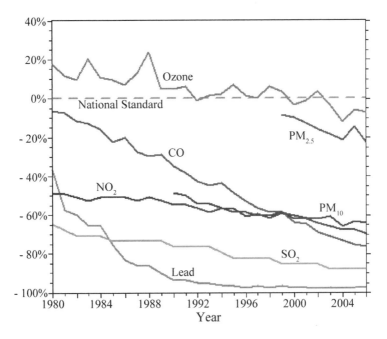

Figure 78. Air Pollution in the U.S. 1980–2006. Declining levels of carbon monoxide (CO), nitrogen dioxide (NO₂), sulfur dioxide (SO₂), ozone, lead, and particulates (PM₁₀ and PM₂₅) are shown. (EPA, 2006)[48]

I visited South Korea on business in the 1992. My lungs took a beating. The outdoor air in Seoul was visibly polluted. The indoor air was not much better, since most Korean businessmen smoked. But if you visit Korea today, the situation is much improved, both indoors and out. South Korea's population, affluence, and technology have all grown rapidly since 1992, *yet pollution has declined.*

Much of the misconception about air pollution arises because Climatism has labeled carbon dioxide a pollutant and added it to the mix. Traditional air pollutants are declining in all developed nations and even many of the newly industrialized countries. But because CO_2, a harmless, odorless gas, has been reclassified as a pollutant, alarmists can mislead the public about how the air is getting dirtier.

We see similar improvement patterns in water pollution. Figure 79 shows measured water pollution for France, New Zealand, the U.K., and the U.S. from 1980 to 2000. During this period, Gross Domestic Product for each of these nations has grown (France up 41%, New Zealand up 29%, the U.K. up 55%, and the U.S. up 53%),[49] but water pollution has declined. According to the over-population alarmists, increasing GNP should increase environmental impact, including water pollution, but this is not the case.

In 1954, Simon Kuznets delivered an address at the American Economic Association. In his lecture, he theorized that as per capita

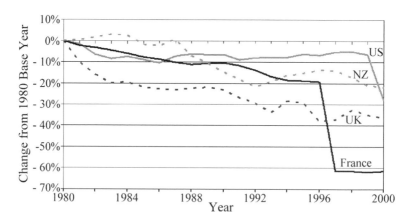

Figure 79. Water Pollution in Selected Countries 1980–2000. Declining levels of water pollution measured in terms of biochemical oxygen demand (BOD) for France, New Zealand, United Kingdom, and United States. Not for comparison between nations. (Data from Earth Trends, 2009, and World Bank, 2006)[50]

income of nations rises, income inequality increases at first, but then at some point, begins declining. Dr. Kuznets was awarded the Nobel Prize in 1971 for his study of the economic growth of nations. Economists from Princeton University, Dr. Gene Grossman and Dr. Alan Krueger, expanded the work of Kuznets by reporting a relationship between the income of a nation and environmental quality, characterized by an Environmental Kuznets Curve (EKC) in 1991.[51]

A simple graph of an EKC is shown in Figure 80. Many economists believe environmental damage is related to increasing income by the inverted U-curve shown. In the early economy of a developing nation, high priority is given to material gain at the expense of the environment. However, as life expectancy and wealth increase, greater priority is given to clean air and water. Societies also gain a growing ability to reduce environmental pollution. Eventually a "turning point income" is reached for a developed nation where improvements in the environment are more important than material gain. From this point on, the environment of a nation improves as per capita income increases.

The existence of Environmental Kuznets Curves has been verified by more than 100 peer-reviewed papers since 1991. The turning point income has been found to vary, depending upon the pollutant and the state of property rights and rule of law in a nation. Grossman and Krueger place turning point incomes between $9,000 and

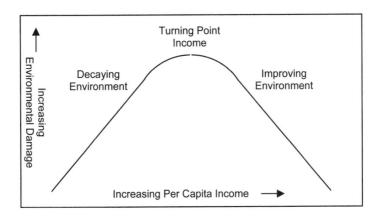

Figure 80. The Environmental Kuznets Curve. At some turning point level of income, increasing per capita income leads to environmental improvement.

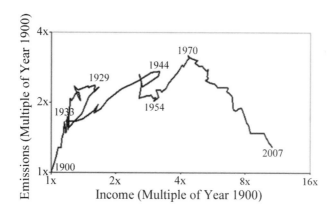

Figure 81. U.S. Sulfur Dioxide Emissions 1900–2007. Sulfur dioxide emissions peak in the U.S. in 1970, following an Environmental Kuznets Curve shape. (Ausubel, 2009)[52]

$41,000 in 2003 U.S. dollars for five different air pollutants.[53]

Figure 81 shows sulfur dioxide (SO_2) airborne emissions for the United States from 1900 to 2007. Note the rough Kuznets Curve-shape peaking in 1970. Back in Figure 79, we showed water pollution declining for France, New Zealand, United Kingdom, and United States. These developed nations can be considered on the downward slope of their Kuznets Curve.

Environmental Kuznets Curve characteristics can be seen today as some of the world's nations move to a more developed status. Figure 82 shows water pollution levels for China, Egypt, and South Korea from 1980. Each nation has increased their Gross Domestic Product during the 20-year period (China up 510%, Egypt up 71%, South Korea up 338%).[54] A Kuznets shape can be seen for all three nations, with pollution peaking and then declining.

Deforestation is considered a major environmental problem. In 1995, a United Nations press release warned:

> …despite substantial progress in the formulation and implementation of national forest policies, deforestation and forest degradation continue at an alarming rate.[55]

The IPCC estimates carbon dioxide emissions due to "land-use change" (deforestation) to be about 1.6 billion tons per year, or about 20% of total greenhouse gas emissions.[56]

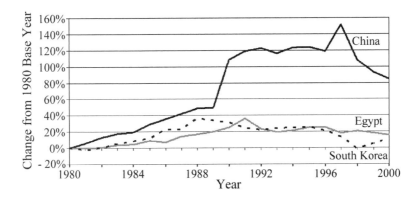

Figure 82. Water Pollution and Kuznets Curves. Levels of water pollution measured in terms of biochemical oxygen demand (BOD) for China, Egypt, and South Korea from 1980 to 2000, showing a Kuznets Curve characteristic. Not for comparison between nations. (Data from Earth Trends, 2009, and World Bank, 2006)[57]

However, it appears forest growth for nations also follows a Kuznets Curve. Dr. Pekka Kauppi of the University of Helsinki led a team that analyzed the growth of forests for 50 nations, using data from the UN Global Forest Resources Assessment Report of 2005. Dr. Kauppi's team found *net forest expansion* from 1990 to 2005 in more than one-third of the 50 nations, including China, France, India, Italy, Japan, Poland, Spain, Sweden, Ukraine, and the U.S. Kauppi also found forests to be increasing for all nations with per capita Gross National Product numbers above $4,600. Brazil, Indonesia, and other developing nations were losing forests, so total global forests continued to decline.[58] But it appears that once national incomes reach a certain level, nations go through a "forest transition" from depleting forests to adding forests. If environmental impact is larger with affluence, then developed nations should be losing forests, but *actual data does not show this.*

At the first Earth Summit in Stockholm in 1972, Indira Gandhi said "Poverty is the worst form of pollution."[59] Indeed, Environmental Kuznets Curves show that **environmental damage can best be reduced by increasing the standard of living for people.**

It may seem we have gotten off-track, but it's important to establish some facts about global economic prosperity and environmental impacts. The population-control alarmists-turned-

Climatists want to restrict population growth, affluence, and technology. They tell us we already have too many people on Earth; we need to use less energy; and we're ruining the planet. But both income and life expectancy continue to rise for developed and developing nations. Population growth is declining as birth rates fall faster than death rates. Pollution is falling for industrialized nations and developing nations are seeking to reduce their environmental impact as income rises.

THE ROOTS OF CLIMATISM

The roots of Climatism are radical environmental forces seeking to control the growth of population, affluence, and the advance of technology. Suppression of greenhouse emissions to combat global warming is the perfect vehicle to achieve these aims.

The last 200 years have seen spectacular improvements in agricultural output, economic growth, and human well being. Kuznets Curves show declining pollution for industrial nations and environmental turning points for developing nations as economic prosperity increases. The disaster predictions of Malthus and Ehrlich have been spectacularly wrong. The need for coercive growth-control policies is not supported by the evidence. Nevertheless, the juggernaut of Climatism rolls on. Let's examine the beliefs, objectives, and tactics of Climatism in Chapter 9.

CHAPTER 9

THE BELIEFS, OBJECTIVES AND TACTICS OF CLIMATISM

"The whole aim of practical politics is to keep the populace alarmed—and hence clamorous to be led to safety—by menacing it with an endless series of hobgoblins, all of them imaginary." H. L. Mencken

Like communism and socialism, Climatism is about global societal revolution. Along the way, the advocates of Climatism are sure to prosper. As energy industry professional Vinod Dar states:

> Climatism is the exploitation of the fear of nature to gain power, wealth, and social esteem.[1]

Let's examine the beliefs, objectives, and tactics of Climatism.

EIGHT FALSE BELIEFS OF CLIMATISM

In the twenty years since the global warming movement was legitimized by the formation of the IPCC, the doctrine of Climatism has developed its own belief system. These ideas are false, unsupported by scientific or economic observations. Nevertheless, the Climatist-driven media bombards us with a steady diet of these beliefs

---

Eight False Beliefs of Climatism

#1: Greenhouse gases emitted from man's industrial activities are causing global warming.

#2: The climate of Earth was optimum prior to the growth and industrialization of human population.

#3: Human activities are destroying the climate of the earth.

#4: The climate is the top priority. It's more important than human lives, freedom, western industrial civilization, and prosperity in the developing nations.

#5: We can save the planet if we all work together.

#6: Carbon dioxide is a pollutant.

#7: Fossil fuels are dirty and should not be used.

#8: Wind and solar power are free and should become our energy sources.

---

in a massive propaganda campaign. See the text box for the eight beliefs of Climatism.

Belief #1: Man-made greenhouse gases are causing global warming.

If you have followed our discussion, it's clear that scientific evidence shows global warming is due to natural, not man-made causes. Even the decline in global temperatures since 2002 has not shaken Climatism of this belief, but has resulted in a change in their message. During the last several years, "climate change" has replaced "global warming" as the leading phrase to describe the crisis.

If you think global warming mania is science rather than belief, reflect on this nugget from U.S. Senator Debbie Stabenow. When told Earth had cooled since 2001 she replied:

> But climate change is not just about temperatures going up. It's also about volatility.[2]

So, increased atmospheric carbon dioxide is increasing climate volatility? Not much science here.

Belief #2: The climate of Earth was optimum prior to man.

This belief is the underlying assumption for many Climatist alarms. The "temperature is the warmest," "the icecaps are the smallest," "the hurricanes are the most violent," "the glaciers are the shortest," and on and on. We saw in Chapter 4 that Earth's climate is continuously changing as part of long-term, medium-term, and short-term cycles. This Climatist belief shows a profound disregard for the geological history of Earth. Dr. Tom Segalstad of Oslo University states:

> The IPCC needs a lesson in geology to avoid making fundamental mistakes…Most leading geologists, throughout the world, know that the IPCC's view of Earth processes is implausible if not impossible.[3]

Belief #3: Human activities are destroying the climate.

There is no question human activity does influence the Earth's climate on a local scale. The Urban Heat Island Effect (see Chapter 2) raises local temperatures. Land clearing impacts local wildlife. Over-irrigation reduces river flows and has even caused the Aral Sea to dry up. But on a global scale, Earth's climate is dominated by natural cycles driven by the sun. Nevertheless, Climatists fervently believe in climate catastrophe from human activities. Note this comment from Harry Reid, majority leader in the U.S. Senate:

> Coal makes us sick. Oil makes us sick. It's global warming. It's ruining our country. It's ruining our world.[4]

Belief #4: The climate is the top priority.

According to Sir David King, formerly chief scientific advisor for Prime Minister Tony Blair:

> In my view, climate change is the most severe problem that we are facing today, more serious even than the threat of terrorism.[5]

Climatists believe climate change is the biggest problem. *All else is less important and must be sacrificed to this belief.* David Shearman and Joseph Smith argue in *Climate Change and the Failure of Democracy* that democracy is inadequate to meet the climate change crisis.[6] They advocate replacement of democracy with autocratic government to cope with global warming. According to Judi Bari, a principle organizer of Earth First, capitalism must be sacrificed:

> If we don't overthrow capitalism, we don't have a chance of saving the world ecologically.[7]

Of course, freedom is forfeit, according to Rowan Williams, Archbishop of Canterbury:

> We must support government coercion over enforcing international protocols and speed limits on motorways if we want the global economy not to collapse and millions, billions of people to die.[8]

The Western way of life, particularly as experienced by most residents of the United States, must be sacrificed, according to Dr. Ehrlich:

> A massive campaign must be launched to de-develop the United States. De-development means bringing our economic system into line with the realities of ecology and the world resource situation.[9]

Even people must be sacrificed to the beliefs of Climatism. Jonathon Porritt, chairman of the U.K. Sustainable Development Commission and former chairman of the Green Party states:

> I think we will work our way towards a position that says that having more than two children is irresponsible.[10]

Why allow even two children? Many Climatists regard people only as "carbon footprints," rather than special beings with heart, mind, and soul. Billionaire Ted Turner summarizes for PBS:

> We're too many people; that's why we have global warming...on a voluntary basis, everybody in the world's got to pledge to themselves that one child is it.[11]

Ironically, Mr. Turner has five children.

Alarmists even wish for disasters so more people will believe in the climate crisis. Economist Thomas Schelling states:

> I sometimes wish we could have, over the next five or ten years, a lot of horrid things happening—you know, like tornadoes in the

Midwest and so forth—that would get people very concerned about climate change.[12]

Belief #5: Together we can save the planet.

The web site wecansolveit.org, founded by Al Gore, states its objective to:

> ...build a movement that creates the political will to solve the climate crisis—in part through repowering America with 100 percent of its electricity from clean energy sources within 10 years.[13]

As we'll discuss in Part III, the idea that we can power America from "clean" energy sources (meaning wind and solar) is also a belief and based on faulty economics. But, this inspirational message sounds good. According to the site, over two million people have joined in the effort. Mr. Gore also appeals to patriotism, self-sacrifice, and idealism in his speeches, exhorting his audience to "rise to the challenge" needed to solve global warming.

I'm amazed at the arrogance of Climatism. Mankind is unable to control the weather in a single location on Earth. Countries are unable to consistently control their own national economies, as the 2008 financial meltdown shows. Yet, some think they can control the climate of our planet, a system far more complex than any economy.

The crusade to control Earth's climate is pointless. As Dr. Willie Soon says: "a magical CO_2 knob for controlling weather and climate simply does not exist!"[14]

Belief #6: Carbon dioxide is a pollutant.

This belief is the propaganda driver for Climatism. Al Gore, James Hansen and many others have successfully established carbon dioxide as a pollutant in the minds of the public. However, as we discussed earlier, CO_2 is odorless, colorless, non-toxic, exhaled by man and animals, food for plants, and 96% of it is generated by nature.

News media headlines talk about worsening air pollution, but it's no longer clear whether they mean traditional pollutants such as carbon monoxide, sulfur dioxide, and ozone, which cause real human health problems, or CO_2, which is harmless. Data shows concentrations of traditional pollutants are declining. Yet, Climatists have successfully confused the issue by branding CO_2 a pollutant.

The height of folly was reached in recent actions by the United States Supreme Court and the Environmental Protection Agency. In the April 2007 case of Massachusetts vs. EPA, the Supreme Court ruled that carbon dioxide was a pollutant under the Clean Air Act. The court ordered the EPA to determine whether CO_2 endangered human health and to consider the regulation of CO_2. In December 2009, the EPA found that emissions of greenhouse gases "threaten the public health and the welfare of future generations" under the Clean Air Act.[15] Senator Barrasso of Wyoming comments on the negative impacts for the U.S. economy:

> This misuse of the Clean Air Act will be a trigger for overwhelming regulation and lawsuits based on gases emitted from cars, schools, hospitals and small business…This will affect any number of other sources, including lawn mowers, snowmobiles and farms. This will be a disaster for the small businesses that drive America.[16]

Why has the EPA issued the endangerment finding? The EPA has decided to the use of the Clean Air Act to address climate change. But let's examine this for a minute. Every eighth-grade science student knows the simple chemical formula for the combustion of a hydrocarbon in oxygen, which is:

Fuel + Oxygen → Heat + Carbon Dioxide + Water

Note that water vapor is produced when we burn hydrocarbon fuel. In fact, the combustion of each of propane, methane, and octane produces more molecules of water vapor than carbon dioxide. Recall from Chapter 3 that water vapor is the most abundant greenhouse gas in our atmosphere. By the EPA's own logic, since water is produced from hydrocarbon combustion and adds to the greenhouse effect, *the EPA should now declare water to be a pollutant.* Clearly, logic is what is endangered here.

Belief #7: Fossil fuels are dirty and should not be used.
Belief #8: Wind and solar power are free and should become our energy sources.

Propaganda rages on both sides of these statements. Al Gore labels fossil fuels "dirty" and the coal industry proclaims "clean-coal" technology. The "dirty" label is purely subjective and designed to elicit negative feelings within the public.

Wind and solar power are anything but free. In fact, they require massive subsidies by governments to sustain their industries. Comparisons of one energy source to another should be based on utility, reliability, deliverability, safety, cost, and *real* environmental impact, not propagandistic labels.

THE OBJECTIVES OF CLIMATISM

Ultimately Climatism is not about the climate. *Climatism at its roots is about global social revolution.* It's about population and affluence control, global economic equality, political power, and profits. The United Nations, environmental groups, politicians, businesses, scientists and the news media are all active participants in Climatism.

The United Nations is the central international player and focal point for Climatism. Environmental organizations have been closely connected to the UN since before Maurice Strong paid for travel for green representatives to the 1972 Stockholm Earth Summit.[17] The UN provides a platform for the philosophies of environmental groups and organizations such as the Club of Rome. Scientists and climate modelers also use the IPCC of the UN to legitimize their findings.

The United Nations has two main objectives served by climate alarmism: 1) global population and affluence control and 2) global economic equality. These objectives were captured in the concept of "sustainable development" in Agenda 21 of the 1992 Earth Summit. Sustainable development was reinforced in Principle 6 of the UN's International Conference on Population and Development in 1994:

> To achieve sustainable development and a higher quality of life for all people, States should reduce and eliminate unsustainable patterns of production and consumption and promote appropriate policies, including population-related policies, in order to meet the need of current generations with out compromising the ability of future generations to meet their own needs.[18]

We've seen that levels of traditional air and water pollutants are declining for the developed nations, but the classification of carbon dioxide as a "pollutant" has allowed the United Nations to raise the level of alarm. The "Programme for the Further Implementation of Agenda 21," adopted by the UN General Assembly in 1997, talks about pollution and the need for sustainable development:

> ...greenhouse gases and waste volumes are continuing to increase... Unsustainable patterns of production and consumption, particularly in the industrialized countries, are identified in Agenda 21 as the major cause of continued deterioration of the global environment... All countries should strive to promote sustainable consumption patterns;...while avoiding those unsustainable patterns, particularly in industrialized countries, generally recognized as unduly hazardous to the environments, inefficient and wasteful, in their development processes.[19]

The same UN document also reports on global inequality:

> ...the gap between the least developed countries and other countries has grown rapidly in recent years.[20]

The concern about economic inequality between industrial and developing nations is easy to understand, but what exactly does the phrase "unsustainable patterns of production and consumption" mean? A look at today's website for the UN's Division for Sustainable Development tells the story. The Division has developed the concepts of "sustainable lifestyles" and "sustainable consumption." The Division laments the spread of Western consumerism:

> These Western lifestyles of consumerism are spreading all around the world through products and services, media and trade policies. Western type restaurants and coffee shops are as common on the streets of Beijing, as international brands of clothing and other products...Goods and services previously seen as luxuries—TVs, mobile phones and cars—have now become necessities...However, the price is paid in the form of degradation of many ecosystem services and the exacerbation of inequities and disparities between people.[21]

According to this, the UN wants to curtail or eliminate Western consumerism, including restaurants, coffee shops, TVs, mobile phones, and cars. Imagine what they think about aircraft, private boats, personal computers, and your Sport Utility Vehicle. They also want to prevent this consumer "disease" from spreading to the developing nations. *This is no less than an attack on our way of life.*

Many others warn about population and affluence and want to bring about equality between nations. Sir James Lovelock warns "The big threat to the planet is people: there are too many, doing too well economically and burning too much oil."[22] Christine Stewart, former Canadian Minister of the Environment has said:

> No matter if the science is all phony, there are collateral environmental benefits...Climate change [provides] the greatest chance to bring about justice and equality in the world.[23]

Control of carbon dioxide provides the perfect vehicle for the UN and others to achieve the objectives of population and affluence control and global economic equality. Industrial nations can be placed under emissions restrictions, slowing energy usage and economic growth. "Wasteful" consumer lifestyles can be curtailed. Since each person has a carbon footprint, population growth can be limited to avoid global warming catastrophe.

Global economic equality can also be promoted. The U.S. and other industrial nations are blamed for most of the CO_2 in the atmosphere. Under "polluter pays," another UN principle, the burden for mitigation falls on these industrial nations. Payments to developing nations from industrial nations will be required, to help developing nations cope with the impacts of global warming. This is no idle comment. As part of ongoing Kyoto II negotiations, the developing nations are asking for annual transfer payments of 0.5% to 2% of the Gross National Product of the industrial nations—hundreds of billions of dollars!

An additional hidden objective of the United Nations is global governance. The UN is trying to achieve a measure of this through Climatism and other initiatives. The UN Convention on Climate Change, the Convention on Biological Diversity, the Convention on the Rights of the Child, the International Criminal Court, the Convention on the Law of the Sea, and many other conventions have become global governance in fact. Climatism provides a method to further the global power of the UN.[24] At the November, 2000 UN meeting in Netherlands, Jacques Chirac, then-President of France, described the Kyoto Protocol as the "first component of an authentic global governance."[25]

The objective of environmental groups appears to be greater influence in society through Climatism. Global warming is labeled a global crisis by most environmental groups, and a global crisis is always good for fund raising, membership, and boosting political power. But the environmental message is mixed when it comes to supporting measures to actually reduce carbon dioxide emissions. The Greens continue to take positions against nuclear power, the one major energy source that does not emit CO_2. The 2001 Marrakech amendments to the Kyoto Protocol do not allow use of Clean Development Mechanism (CDM) credits from nuclear power.[26] Clean Development Mechanism is a method set up by the Kyoto Protocol for industrial countries to invest in projects in developing nations to offset emissions in their own countries. By disallowing the use of CDM credits, the UN and supporting environmental groups have created a bias against using nuclear power to reduce emissions.

Many environmental groups continue to oppose dam building and renewable hydroelectric power. David Foreman, co-founder of Earth First!, is praised by Paul Ehrlich as setting "an example for us all." Here's an example Mr. Foreman would like to set:

> We must make this an insecure and inhospitable place for capitalists and their projects. We must reclaim the roads and plowed land, halt dam constructions, tear down existing dams, free shackled rivers and return to wilderness millions of acres of presently settled land.[27]

It's clear environmental groups want greenhouse gas emissions reduced, but not enough to use nuclear or hydroelectric power.

Our political leaders are always open to pursuing solutions to a crisis. U.S. Secretary of State Hillory Clinton told an audience in Brussels in March, 2009:

> Never waste a good crisis...Don't waste it when it can have a very positive impact on climate change and energy security.[28]

Crises have been frequent in the United States during the last year. In September, 2008 we were told by the Bush administration that the $700 billion Troubled Asset Relief Program (TARP) must be approved "within the next few days" to avoid total collapse of the U.S. financial system.[29] In December, U.S. citizens were told to lend General Motors and Chrysler $17.4 billion dollars.[30] We were further told approval was necessary "by the end of the month" to prevent bankruptcy of the two automobile companies (which eventually filed for bankruptcy anyway). In February, the U.S. Congress rushed through a $787 billion stimulus plan to battle "the worst economic crisis since the Great Depression."[31] The urgency of the crisis required the plan to be rammed through Congress so fast that most legislators had not even read the bill.

Global warming is an excellent lever to increase political power. Carbon emissions control measures, such as Cap & Trade, require tens of thousands of government employees to monitor emissions and enforce statutes. Carbon credits are awarded to businesses, increasing political power, as they have been in Europe. It's a great opportunity to build government control and strengthen your constituency. In addition, it's an excuse for tax increases, in the name of environmental responsibility. President Václav Klaus, author of the book *Blue Planet in Green Shackles*, comments on the motivation of political leaders:

> Being often with many leading politicians, I feel frustrated that they do not listen. They already know. They fully subscribed to the idea that talking about "saving the planet" is an effective way to show their "caring" for humanity and that it is the easiest way to maximize votes irrespective of any relevant activity which would aim at the real needs of people. The global warming dogma has become a very easy form of escapism from the current reality.[32]

Business leaders have also joined the Climatist bandwagon. Corporations are pragmatic and their primary objective is always profits. Billions of dollars in profits can be made in the artificial market of carbon credits trading and in state-subsidized industries such as wind and solar power. The energy company Enron was a major investor in "climate capitalism" until its demise. Millions more are available to consultants who advise on climate change. Yet more millions are available to the legal industry from lawsuits against "carbon polluters."

The U.S. Climate Action Partnership is an association of more than 20 companies and organizations, including Caterpillar, Conoco Philips, DuPont, Ford Motor, General Electric, General Motors, PepsiCo, Shell, and Siemens, who have joined to:

...call on the federal government to quickly enact strong national legislation to require significant reductions of greenhouse gas emissions.[33]

The Partnership wrote the framework for the Waxman-Markey Cap & Trade bill introduced to Congress in May 2009. By writing the framework for the bill, U.S. Climate Action Partnership companies seek billions of dollars in free carbon credits to gain a competitive advantage in the marketplace.

General Electric is a special player in the pursuit of profits from global warming. GE is a leading supplier of wind turbines, the preferred technology for generating renewable electricity. Coincidentally, GE owns broadcaster NBC. Of course, NBC is a leading advocate for solving the global warming crisis. Effective persuasion from NBC is great for GE's wind turbine business.

Business leaders can read the writing on the wall. Since they believe carbon-control madness will be enacted, their objective is to determine how to make profits under such a regime. The key is to be well positioned by the time legislation is enacted. All of the Fortune 500 Companies have either enthusiastically jumped on the Climatist bandwagon or given politically correct support. Climate change lobbyists in Washington, D. C. now exceed 2,300, up 300% from five years ago, more than four lobbyists for each member of Congress.[34] Many businesses have decided to support Climatism to avoid confrontations with environmental groups and boost their public relations image. This brings to mind another quote from Sir Winston Churchill: "An appeaser is one who feeds the crocodile, hoping it will eat him last."[35]

Scientists and the news media have formed a symbiotic Climatist relationship to further their own careers and prosperity. We'll discuss this in the Chapter 10.

THE TACTICS OF CLIMATISM

The tactics of Climatism have been effective in creating worldwide acceptance of global warming catastrophe. These tactics include: 1) exploitation of fear, 2) donning the mantle of science and authority, 3) application of propaganda and pressure, and 4) ruthless attack on opponents. Iain Murray at the Competitive Enterprise Institute, describes similar tactics of the environmental movement in his book *The Really Inconvenient Truths*, many now adopted by Climatism.[36]

The average citizen is unable to understand the science of climate change without a major time investment. His or her experience from the weather is also of little help. Temperatures in many locations vary by 100°F over a year, while climatic temperatures may change by 1°F over a century. Many people don't believe the UN, parts of the scientific community, and our political leaders could advocate a position contrary to actual scientific evidence. These citizens are easy prey for the tactics of Climatism.

Exploitation of fear is the primary tool of Climatism. We have already provided many catastrophic quotes by Mr. Gore, Dr. Hansen and others. It's important to understand that Climatists are engaging in *deliberate distortion* of natural effects and using scientific theories to create a climate of fear. See the text box for statements from alarmists about permissible exaggeration of man-made global warming.

Jim Hansen's "tipping points" are an example of fear generation. Alarmists regard Earth's climate as an unstable system that can be pushed to a tipping point, after which climate would become more extreme, regardless of any possible human actions. A high level of atmospheric CO_2, a melting ice cap, and deforestation of the Amazon Basin are claimed tipping points where we would "step off a cliff." But scientists have no historical evidence of a runaway climate.

A short time to disaster is always part of the fear scenario. We have

---

Climatists Approve of Deliberate Exaggeration

"…it is appropriate to have an over representation of factual presentations on how dangerous it is, as a predicate for opening up the audience to listen to what the solutions are, and how hopeful it is that we are going to solve this crisis." —Al Gore[37]

"Unless we announce disasters, no one will listen." —John Houghton, 1st Chairman of IPCC Working Group 1[38]

"…we are not just scientists but human beings as well. And like most people we'd like to see the world a better place, which in this context translates into our working to reduce the risk of potentially disastrous climatic change. To do that we need to get some broad based support, to capture the public's imagination. That, of course, entails getting loads of media coverage. So we have to offer up scary scenarios, make simplified, dramatic statements, and make little mention of any doubts we might have." —Dr. Stephen Schneider[39]

30 years, 15 years, 7 years, *fill in the blank* years, until the disaster occurs, so "we must act now." A short time, often arbitrarily chosen, heightens the sense of urgency to the public.

Consider a fact sheet released by environmental activist group Greenpeace in May 2006 in reaction to President George Bush's efforts to promote nuclear energy during a visit to Pennsylvania. Although not about the climate, the memo is indicative of systematic tactics to generate a climate of fear. The memo said:

> In the twenty years since the Chernobyl tragedy, the world's worst nuclear accident, there have been nearly [*fill in alarmist and armageddonist factoid here*].[40]

The memo was mistakenly released to the public by Greenpeace prior to "filling in the factoid."

Having generated a fear of future disaster, Climatism offers to solve the problem. But the cost is very high. The solution is carbon suppression by control of human society, including restriction of freedoms, changing our way of life, and putting chains on progress in developing nations.

The second tactic of Climatism is to don the mantle of science and authority. In Chapter 7 we discussed how the IPCC legitimized the dogma of man-made global warming in the scientific community and on the international stage. It's informative to watch climate change hearings in the U.S. Senate.[41] The scientists representing Climatism typically follow the same line, which is 1) Earth's temperatures are increasing, 2) carbon dioxide is increasing, and 3) "the models show..." The opposing skeptical scientists point to geologic history showing natural warming and cooling of Earth, weather impacts, and the absence of positive feedback. At this point, the debate usually stops. Rather than address the points raised by the skeptics, the Climatist scientists point out that the IPCC, the National Academy of Sciences, the American Meteorological Society, and other scientific organizations support man-made global warming. In other words, they showcase authority, rather than discuss science.

Climatist scientists are almost always certain. In 1988, James Hansen had no quantitative idea how precipitation, cloud formation, and aerosols affected global climate, but he told the U.S. Senate he was 99% certain the world was warming up and it was likely due to man-made greenhouse gases. An alarmist scientist can usually be found to pronounce any particular flood, storm, or drought "caused or worsened" by global warming. In contrast, skeptical climatologists are often uncertain, and rightly so. Skeptics often disagree with each other, which the news media denounces as weakness. However, such disagreement is actually a characteristic of healthy science.

The General Circulation Models provide another element of authority. Climate models are the tools for generating apocalyptic scenarios. Multi-million dollar supercomputers are impressive, so the public assumes their results must be correct. But recent global cooling indicates that projections by *all* of the models were wrong. Dr. Timothy Ball remarks:

> Perhaps the greatest scientific deception of the IPCC is the abuse and misuse of computer climate models. They allow them to make their reports and deliberations appear credible. They allow them to bamboozle the public because computer models are a complete mystery to most people.[42]

The third tactic of Climatism is effective use of propaganda and pressure. Climatist propaganda created terms such as "dirty fuels" for coal and oil, and "free energy" for wind and solar energy. "Carbon footprint," "tipping point," "negative ice balance," and "radiative forcing" were developed to serve the cause. Environmental groups and politicians expanded our vocabulary with "green cars," "green houses," "green energy," and "green jobs" referring to processes with low carbon emissions. The United Nations invented "sustainable development," "sustainable consumption," and "sustainable lifestyles" and also their "unsustainable" counterparts. These propaganda terms are effectively used by Climatists, in partnership with the media, to build acceptance of man-made global warming dogma.

Environmental groups are the sword of Climatism. Greenpeace has been the leading global actor in efforts to halt man-made greenhouse gas emissions. Greenpeace is a powerful organization, highly effective in the use of propaganda and pressure.

Greenpeace has opposed coal plant construction in U.K., Poland, Indonesia, and around the world. They have pressured nations to ban incandescent light bulbs, with success in Australia, Argentina, Belgium, Canada, China, France, Ireland, Italy, and the U.S. They have opposed Heathrow Airport runway expansions. They have pressured major European automobile makers, in one case labeling

Mercedes Benz cars "climate pigs." Polar bear costumes are often worn by Greenpeace organizers to call attention to the "plight" of the bear. Greenpeace members travel to the Great Barrier Reef, the Amazon Rainforest, the Congo River Basin, and palm oil plantations in Indonesia to advocate protection and criticize human activities.[43]

Greenpeace demonstrations are often spectacular. Examples are video projection of images on the Rome Coliseum and the Washington Monument. Members climb cooling towers of coal and nuclear plants. They hang huge banners from visible locations such as the bridge to Nitiroi in Rio de Janiero or the escarpment of Mt. Rushmore. Figure 83 shows a Greenpeace banner hung on Esso headquarters in Leatherhead, England.

A notable example of Greenpeace work is the story of the "Kingsnorth Six." In October 2007, six members of Greenpeace scaled the 200-meter high smokestack of the Kingsnorth power station in Kent, United Kingdom to protest government plans to build a new coal-fired plant in Kingsnorth. The six painted "GORDON" on the smokestack (for Prime Minister Gordon Brown), were arrested, and were charged with causing £30,000 in

Figure 83. Greenpeace Banner in England. This protest banner was hung on Esso headquarters in Leatherhead, U.K. (Happ, 2003)[44]

criminal damage. At trial in 2008, the defendants admitted that they caused the damage, but argued it was lawful for them to damage the chimney to "protect other property…from climate change."[45]

Dr. James Hansen testified on behalf of the defendants, warning that if the world continues with business-as-usual, our descendants will be "left with a much more desolate planet and much less biodiversity." Hansen projected several hundred species deaths directly from the Kingsnorth power station. Jurors acquitted the Kingsnorth Six in September 2008.[46] This ruling means, at least in the U.K., that destruction of property in the name of preventing climate change is now lawful.

The environmental groups who drive Climatism are very well funded. Greenpeace has 45 organizations around the world, about three million members, and an annual budget of over 200 million euros.[47] U.S. environmental groups have annual revenues in excess of $2 billion, led by the Nature Conservancy with 2008 revenues of $1.1 billion.[48] U.S. foundations, such as the Ford Foundation, the Rockefeller Foundation, and the Hewlett and Packard Foundation, provide significant funding to the cause of climate change. In March, 2009 the McKnight Foundation announced it was joining the Hewlett Packard Foundation to commit $1 billion over five years to combat climate change.[49] Such a pool of funds allowed Al Gore to conduct a $300 million advertising campaign in favor of solving global warming, during the 2008 U.S. presidential elections.[50]

Pressure on corporations is enormous to accept Climatism and reduce their "carbon footprint." Climatist groups target major firms, such as Apple Computer, British Petroleum, Coca-Cola, and McDonald's, using letter writing, boycotts, shareholder pressure, and negative publicity campaigns. The Carbon Disclosure Project is a non-profit organization funded by foundations and the UN, directed by Rockefeller Philanthropy Advisors. It annually publishes its Carbon Disclosure Project Report for the Global 500 companies. The report consists of questionnaire replies, analysis, and data on company emissions levels. According to the CDP, the purpose of the report is to encourage "private and public sector organizations to measure, manage, and reduce emissions and climate change impacts."[51] But the report provides ammunition for environmental groups and investors to pressure companies.

In the best spirit of totalitarian regimes, Climatists have active programs to indoctrinate students with man-made global warming dogma. Despite its court-declared fallacies, Mr. Gore's movie *An Inconvenient Truth* is shown in many schools. One Canadian high school student publically complained he was forced to watch it four times, including once in gym class. No opposing opinions were offered.[52] Many global warming alarmist books are now available for children. Among these is the *Down-to-Earth Guide to Global Warming*, by Laurie David, producer of Gore's movie, and environmentalist Cambria Gordon. The book is rife with mistakes, including telling children that carbon dioxide and temperature go together like "peanut butter and jelly."[53] Our children are being taught Climatism based on false science.

The fourth tactic of Climatism is ruthless attack on opponents. Scientists who voice skepticism that global warming is man-made are viciously attacked in the news media and internet blogs. Federal government scientists and state climatologists are fired or muzzled. Opponents are labeled "deniers," as in Holocaust deniers, and said to be in the pay of oil and gas companies. University faculty members who don't embrace man-made global warming are chastised or worse. *This ruthless attack effectively suppresses views opposed to global warming dogma.* The Climatist attack on opponents is a big topic, which we'll discuss in Chapter 10.

SUMMARY

Climatism is a powerful global movement embraced by most political, environmental, commercial, scientific, and media organizations. The movement is based on a false belief system which is broadcast to the citizens of the world on a daily basis.

The objectives of Climatism are nothing short of global social revolution. The desired results of such a revolution are centralized government control over affluence and population, curtailment of the "wasteful" consumerism, and global redistribution of wealth from the industrial nations to the developing nations. The tactics of Climatism include the creation and exploitation of fear, donning the mantle of false science, effective use of propaganda and pressure, and ruthless attack on opponents.

CHAPTER 10

THE SCIENCE IS NOT SETTLED

"The truth is found when men are free to pursue it."
Franklin Delano Roosevelt (1936)

In the wake of release of his movie *An Inconvenient Truth* in May, 2006, Al Gore appeared many times as a guest on U.S. television stations. An example was the CBS *Early Show* with Harry Smith on May 31. On the show, Mr. Gore claimed "the science is settled," which has become a propaganda line for Climatism. He also went on to ridicule the skeptics of man-made global warming:

> ...the debate among the scientists is over. There is no more debate. We face a planetary emergency. There is no more scientific debate among serious people who've looked at the science...Well, I guess in some quarters, there's still a debate over whether the moon landing was staged in a movie lot in Arizona, or whether the Earth is flat instead of round."[1]

The assertion that "the science is settled" is repeated over and over in the media by man-made warming alarmists.

Climatism downplays and suppresses the opinions of its opponents. The news media willingly partners in the corruption of science and building the delusion of man-made global warming. Most scientific organizations have succumbed to environmental, political, and media pressure and endorsed the dogma. The scientific method itself has been bypassed. But a growing number of courageous climate realists are challenging this false science.

THE ATTACK ON DENIERS

George Monbiot, U.K. journalist and columnist for the *Guardian*, uses the term "denier" to describe those who disagree with global warming dogma. He describes deniers as people who "deny that climate change is happening."[2] Denier is used as a label to equate opponents to Holocaust deniers. Monbiot published a list of "Top 10 Climate Change Deniers," which includes: President Václav Klaus, Pat Michaels of the Cato Institute, Christopher Monkton (former advisor to Margaret Thatcher), Sarah Palin, U.K. botanist David Bellamy, and U.S. Senator James Inhofe.[3] British economist Sir Nicholas Stern also called opponents deniers, stating:

> If you look at all the serious scientists in the world, there is no big disagreement on the basics of this…it would be absolute lunacy to act as if climate change is not occurring."[4]

Both Monbiot and Stern distort the position of their opponents. I've talked with dozens of climate skeptics and *none* take the position that climate change is *not* happening. None of the six persons listed above in Monbiot's top-10 list takes such a position. They all agree Earth warmed over the last 25 years of the 20th century, but that the warming was due to natural cycles of our planet, not greenhouse gas emissions. Mr. Monbiot and Mr. Stern are both intelligent men, yet they still say skeptics "deny climate change is occurring." Why would they misstate the position of those who disagree with their viewpoint?

Since they believe the science is settled, the environmental and media arms of Climatism assume that scientists opposing climate change efforts are in the pay of big "carbon-polluting" corporations or "right-wing think tanks." According to Gore:

> The misconception that there is serious disagreement among scientists about global warming is actually an illusion that has been deliberately fostered by a relatively small but extremely well-funded cadre of special interests, including Exxon Mobil and a few other oil, coal, and utilities companies. These companies want to prevent any

new policies that would interfere with their current business plans that rely on the massive unrestrained dumping of global warming pollution into the Earth's atmosphere every hour of every day.[5]

Many realist scientists of prominence have been attacked on internet blogs and in the media as being in the pay of the energy companies. An example is "Global Warming Skeptics: A Primer," a pamphlet published by Environmental Defense. The Primer lists 24 skeptics and their supposed connections to energy companies.[6]

As part of a massive media campaign, the oil, coal, and natural gas industries have been demonized. Anyone at a conservative organization, such as The Heartland Institute or the Competitive Enterprise Institute, who writes in favor of a rational climate or energy policy, must be stooges for big energy. Columnists such as George Monbiot and Andy Revkin at *The New York Times* beat the "in-the-pay-of-the-oil-companies" drum, ignoring millions channeled from foundations and environmental groups for Climatism.

Well folks, I've met the realist scientists, and regarding being in the pay of the energy companies—*it just isn't so*. These courageous men and women are true believers in science integrity, not in any retainer from energy companies. In fact, those who oppose the politically-correct dogma of Climatism suffer repeated attacks and negative career consequences.

More than just economic or scientific disagreement, Climatists have gone further and made climate change a moral issue. Dr. Gro Harlem Brundtland, former chairperson of the Norwegian Labor Party and the Brundtland Commission, states:

> The diagnosis is clear, the science is unequivocal—it's completely immoral, even, to question now, on the basis of what we know, the reports that are out, to question the issue and to question whether we need to move forward at a much stronger pace as humankind to address the issues.[7]

According to Dr. Brundtland, it's now *immoral even to question* global warming dogma. Once skepticism becomes immoral, then skeptics become criminals (see text box next page). A Google search of "climate criminals" finds over 83,000 links. Greenpeace provided a "Field Guide to Climate Criminals," listing 16 persons, to delegates at the 2005 UN Climate Change Conference in Montreal.[8]

To Disagree Is Criminal

"When we've finally gotten serious about global warming, when the impacts are really hitting us and we're in a full worldwide scramble to minimize the damage, we should have war crimes trials for these bastards—some sort of climate Nuremberg."[9] —David Roberts, staff writer for *Grist*

"We can no longer tolerate what's going on in Ottawa and Edmonton. What I would challenge you to do is to put a lot of effort into trying to see whether there's a legal way of throwing our so-called leaders into jail because what they're doing is a criminal act."[10] —Canadian environmentalist David Suzuki

"...every time someone dies as a result of floods in Bangladesh, an airline executive should be dragged out of his office and drowned."[11]
  —George Monbiot, *Guardian*

"I wonder what sentences judges might hand down at future international criminal tribunals on those who will be partially but directly responsible for millions of deaths from starvation, famine, and disease in the decades ahead."[12] —Mark Lynas, U.K., journalist and environmentalist

---

Climate skeptics are under frequent attack by the environmental arm of Climatism. Christopher Horner, author of *Red Hot Lies*, reports that Greenpeace was "stealing his trash."[13] Outspoken journalist Alexander Cockburn, reports on his experience:

> Since I started writing essays challenging the global warming consensus, and seeking to put forward critical alternative arguments, I have felt almost witch-hunted. There has been a hysterical reaction. One individual, who was once on the board of the Sierra Club, has suggested I should be criminally prosecuted...There was a shocking intensity to their self-righteous fury, as if I had transgressed a moral as well as an intellectual boundary and committed blasphemy. I sometimes think to myself, "Boy, I'm glad I didn't live in the 1450's, because I would be out in the main square with a pile of wood around my ankles."[14]

The use of the term "denier" is a direct attack on our right to disagree. It's a highly pejorative term for labeling man-made climate change skeptics as evil or criminal. It demands that certain speech or points of view are forbidden. The label "denier" and the phrase "the science is settled" are just two of the tactics used by Climatists to suppress debate on climate science and excoriate those who hold the realist point of view. Let's examine two particular cases.

REVELLE, GORE, AND SUPPRESSION OF SCIENCE

Al Gore writes in his books *Earth in Balance* (1992) and *An Inconvenient Truth* (2006) that Dr. Roger Revelle, his college professor, introduced him to the idea that carbon dioxide accumulation in the atmosphere posed an environmental threat. Recall from Chapter 2 that Revelle began measuring atmospheric CO_2 concentration in Hawaii in 1957. But over the next 25 years, Mr. Gore's and Dr. Revelle's views diverged on the magnitude of the global-warming threat. This lead to Gore's effort to change history and suppress science in 1992, as documented by Dr. Fred Singer in the book *Politicizing Science*, edited by Michael Gough.[15]

Singer, Revelle, and Dr. Chauncey Starr published the article "What to Do about Greenhouse Warming: Look before You Leap" in *Cosmos* in April 1991, although this was unknown to Gore at the time. The article concluded, "the scientific base for a greenhouse warming is too uncertain to justify drastic action at this time."[16] Revelle had also sent letters to members of the U.S. Congress with similar points of view in 1988. Dr. Revelle died in July 1991, three months after the article was published.

Gore published *Earth in Balance* in 1992, in which he advocated a "Global Marshall Plan" that would include "the stabilizing of world population" and a "Strategic Environment Initiative" to "deal comprehensively with the global environmental crisis."[17] After Gore's book was published, journalist Greg Easterbrook, writing for *Newsweek*, pointed out that Mr. Gore did not mention the *Cosmos* article co-authored by Mr. Revelle. Other media picked up on the *Newsweek* story. Then, as a senator and vice-presidential candidate, Gore received a question regarding Revelle's cautious views during a 1992 vice-presidential debate, which he deflected, saying Dr. Revelle's views had been taken "completely out of context."[18]

Soon after, Al Gore began an effort to remove Revelle as a co-author of the *Cosmos* article. Dr. Lancaster, an associate of Gore, called Singer and demanded that Revelle's name be removed. When Singer refused, Lancaster and Dr. Anthony Socci, a member of Gore's staff, claimed through different channels that Singer had added Revelle's name to the paper "over his objections" and that Revelle's mental capacities were failing at the time, neither of which was true.

Singer filed a libel suit in April 1993. A public-relations campaign mounted by Lancaster to build support for his side of the case failed. Singer won, and Lancaster published a full retraction in April 1994.[19]

In February 1994, while the suit was in progress, Gore contacted Ted Koppel of *ABC News* in an effort to discredit Dr. Singer and other scientists skeptical of global warming.[20] But the effort backfired. In the February 24, 1994 televised addition of *Nightline* titled "Is Environmental Science for Sale?," Mr. Koppel stated:

> There is some irony in the fact that Vice President Gore, one of the most scientifically literate men to sit in the White House in this century,...is resorting to political means to achieve what should ultimately be resolved on a purely scientific basis...The issues of global warming and ozone depletion are undeniably important. The future of mankind may depend on how this generation deals with them. But the issues have to be debated and settled on scientific grounds, not politics.[21]

DANISH COMMITTEE ON SCIENTIFIC DISHONESTY

On the other side of the world, scientific suppression was also in full swing. In 2001, Dr. Bjorn Lomborg, Associate Professor of Statistics at the University of Aarhus in Denmark, published *The Skeptical Environmentalist, Measuring the Real State of the World*,[22] the English-version text of the original book from 1998. Lomborg was a self-described "left-wing Greenpeace activist,"[23] who read an article by Dr. Julian Simon in 1997 in which Simon professed that "our doomsday conceptions of the environment are not correct."[24] Lomborg at first disagreed and set out to disprove Simon, but found that data on global trends supported Simon's position. As a result, Lomborg wrote his book to describe the real state of the world.

The book contains statistics on almost every global measure of human society and the environment, including human welfare, life expectancy and health, energy, pollution, biodiversity, and global warming. It cites over 2,900 references, most of these from recognized international sources, such as the United Nations, the World Bank, and the U.S. Environmental Protection Agency. After four years of work, Dr. Lomborg optimistically concluded:

...children born today—in both the industrialized world and developing countries—will live longer and be healthier, they will get more food, a better education, a higher standard of living, more leisure time and far more possibilities—without the global environment being destroyed.[25]

Regarding global warming, Lomborg states: "global warming is not anywhere near the most important problem facing the world."[26] He believes warming is man-made, but that adaptation, rather than mitigation (emission-reduction efforts), is the best economic policy.

In writing his book, Lomborg crossed swords with the Climatists, as well as population control and environmental organizations. He was viciously attacked by alarmist groups around the world, including World Resources Institute, Union of Concerned Scientists, World Wildlife Fund, and the Green Alliance. A website, *anti-lomborg.com*, was even established. Environmentalists followed him to book-signing promotions to disrupt his efforts. He received a well-documented pie in the face from author Mark Lynas in a bookstore in Oxford, England in 2003.[27]

The most notable assault was sponsored by *Scientific American* in January 2002. The magazine chose four scientists and devoted 12 pages to an editorial attack on Lomborg's book. The authors were Dr. Stephen Schneider, former member of the Club of Rome and famous for his "we-must-offer-up-scary-scenarios" quote, Dr. John Holdren, an advocate of population control and loser of the bet with Julian Simon, Dr. John Bongaarts, Vice President of Population Council, a radical population control organization, and Dr. Thomas Lovejoy, former director of the World Wildlife Fund. The editorial was subtitled "science defends itself," but should have been subtitled "*radical activist science defends itself*," based on the scientists chosen.[28]

Another serious attack came from the Danish Committee on Scientific Dishonesty (DCSD). This sounds like something from George Orwell's book *1984*, doesn't it? In January 2003 the DCSD ruled Lomborg's book was scientifically fraudulent and seriously misleading.[29] After the ruling, Lomborg's position at his university was considered in jeopardy. But in December of 2003, the Danish Ministry of Science overturned the DCSD's ruling and removed the stigma on Lomborg from the committee ruling.[30]

THE CLIMATIST INQUISITION

The cases of Revelle and Lomborg are only two of many efforts to sacrifice science to Climatism. A "Climatist Inquisition" continues to be applied to skeptics by man-made global warming advocates in the U.S., Europe, and across the world. The combination of political pressure and campaigns to discredit, along with increased funding for those who "join the team," lined up most major global scientific organizations behind Climatism.

Al Gore ran his own inquisition. During 1991 and 1992, then-Senator Gore held congressional hearings, calling in noted skeptics to challenge and attempt to discredit them. Two of these were Dr. Sherwood Idso and Dr. Lindzen. Holman Jenkins Jr., writing for *The Wall Street Journal*, reported on Idso's experience:

> Two years ago, he was dragged before Mr. Gore's subcommittee and accused, in effect, of being a scientific shill for earth-raping coal companies. "A Gore staffer told me that the hearing was going to be an 'exploration of views,'" says another scientist who testified that day. "But actually the whole purpose of the hearing as far as I could see was to hammer Idso." Adds a career scientist from DOE who was also present: "It was a setup."[31]

At a similar hearing, Gore pressed Dr. Lindzen to reject a hypothesis on water vapor in the upper troposphere, which Lindzen calls a "rather arcane discussion." Lindzen recalls: "Gore then called for the recording secretary to note that I had retracted my objections to 'global warming.'"[32] The hearing transcript was sent to *The New York Times,* which published an article later in the month stating:

> Senator Albert Gore of Tennessee, who convened the scientific group, said that while Dr. Lindzen remained skeptical on general grounds, he had abandoned the specific theory he previously advanced.[33]

Mr. Gore afterward claimed in *Earth in Balance* that Lindzen had retracted his objections to catastrophic global warming.[34]

Senator Gore became Vice President Gore, taking office in 1993 with President Bill Clinton. Gore set the tone for what he considered acceptable science early in the term with physicist William Happer. Dr. Happer was Director of Energy Research for the U.S. Department of Energy, in the previous administration. He was initially asked to stay on in the new administration, but Happer's skeptical science ran afoul of Mr. Gore's views on the ozone layer.

In *Earth in Balance*, Mr. Gore warns about thinning of the ozone layer from man-made chlorofluorocarbons:

> A thinner ozone layer allows more ultraviolet radiation to strike the Earth's surface. Many life forms are vulnerable to large increases in this radiation, including many plants…We too are affected by extra ultraviolet radiation. The best known consequences include skin cancer and cataracts…[35]

Happer challenged some of the ozone assumptions with data showing measured levels of ultraviolet radiation were flat to slightly decreased. Other studies showed damage to plants from ultraviolet radiation was much less than expected. Apparently, the thinning ozone layer was not producing the expected effects on Earth's surface.

Happer's scientific skepticism was not appreciated. He was fired by Mr. Gore and returned to a position at Princeton University. "I was told that science was not going to intrude on policy," he says.[36]

The firing of Happer set the tone for science in the U.S. government during the 1990s. It was clear to the community that questioning the politically accepted version of science, including global warming, was fraught with risk. This "Climatist Inquisition" extended to state governments and universities in the United States, and at the same time was underway in many nations of Europe.

U.S. science organizations including the National Academy of Sciences, the American Meteorological Society (AMS), the American Geophysical Union (AGU), and the National Aeronautic and Space Administration (NASA) have all embraced Climatism. Dr. William Gray, the famous hurricane forecaster, leveled withering criticism at the AMS. Gray says the AMS leadership has capitulated to:

> …the lobby of the climate modelers and to the outside environmental and political pressure groups who wish to use the now AMS position on AGW [anthropogenic (man-made) global warming] to help justify the promotion of their own special interests …Instead of organizing meetings with free and open debates on the

basic physics and the likelihood of AGW induced climate changes, the leaders of the society…have chosen to fully trust the climate models and deliberately avoid open debate on this issue.[37]

Dr. Gray himself has been under attack for his skeptical views. Unable to maintain funding for hurricane research, he provided $100,000 from his own funds to continue his most important projects.[38] Dr. Peter Friedman of the University of Massachusetts is a member of the American Geophysical Union. He reports pressure on skeptics from high levels of the AGU:

> Several respected climate scientists have told me that there would be even more vocal skeptics if they were not afraid of losing funding, much of which is controlled by politically correct organizations.[39]

Many different forms of pressure are stifling debate. Scientific papers challenging man-made warming dogma are difficult to publish. The magazines *Science* and *Nature* have taken a particularly hard line, often dismissing climate-skeptical research for limited reasons, such as "no interest to the readership."[40] Pressures on young scientists at universities are also strong. Some are told directly that their realist position on global warming is out of step with the position of the university. Others are concerned about becoming a tenured member of the staff. Most outspoken skeptical scientists are older or already established in the scientific community.

The experiences of U.S. state meteorologists are notable examples of the gagging of skeptical scientists. Dr. Patrick Michaels, formerly State Climatologist for Virginia and former President of the American Association of State Climatologists, documents this in his book *Climate of Extremes*. Dr. Michaels was told by his University of Virginia that he could no longer discuss climate change using the State Climatologist title. Dr. David Legates at the University of Delaware was told by Governor Ruth Ann Minner that he could no longer speak on global warming as the State Climatologist of Delaware. Oregon State Climatologist George Taylor challenged Climatist dogma showing a decline in Pacific Northwest snowpack from 1950 to 1995. He pointed out that if one looks at a longer history from 1900 to 2000, the recent snowpack decline is not unusual. Taylor was ordered not to show the full data. He resigned in 2008. Finally, Mark Albright, Assistant State Climatologist for the state of Washington, was fired for providing a complete snowfall record for the Cascade Mountains, rather than a subset of the data that supported Climatist dogma.[41]

Pressure on the European scientific community to conform to Climatism is equally strong. The Royal Society, Britain's premier scientific organization with a history dating back to the 1600s, has succumbed to Climatist dogma and now seeks to suppress skeptical opinions. The Society sent a letter to Exxon in September 2006 stating Exxon funded 39 organizations in the U.S. that:

> …misrepresented the science of climate change, by outright denial of the evidence that greenhouse gases are driving climate change, or by overstating the amount and significance in uncertainty and knowledge…[42]

The letter further wants to know which organizations in Europe are receiving funding from Exxon.

For those of you who know science, we trust you are as shocked about the actions of the Royal Society as we are. That a prestigious scientific organization would stoop to sending a pressure letter to a corporation is astonishing. The logo of the Royal Society says "dedicated to promoting excellence in science,"[43] but this letter to Exxon seems more in character with a religious inquisition.

There are many incidents in both the U.S. and Europe where scientists have publically questioned the dogma of man-made global warming, and paid the price. These are well-documented in the book *The Deniers* by Lawrence Solomon, a Canadian journalist.[44] Dr. Zbigniew Jaworowski, now chairman of the Scientific Council of the Central Laboratory for Radiological Protection in Warsaw, published an article at the Norwegian Polar Institute criticizing methods for measuring CO_2 levels from ice cores in 1990. The IPCC relies on these results to claim that atmospheric CO_2 levels today are the highest in history. The Norsk Polarinstitutt came under attack from the global-warming establishment and subsequently fired Dr. Jaworowski.[45] Dr. Henk Tennekes was dismissed from his position as research director at the Royal Dutch Meteorological Institute, for questioning the scientific basis of man-made global warming.[46] Dr. Henrik Svensmark, who developed the theory that cosmic-ray intensity, modulated by sunspot activity, drives global temperature change, has been heavily criticized for over 10 years. As we discussed in Chapter 5, his theory provides a viable alternative to the IPCC theory of radiative forcing from greenhouse gases. After Svensmark's first article about his theory was published in 1996, Bert Bolin, former chairman of the IPCC, publically pronounced Svensmark's theory "completely naive and irresponsible." Svensmark's book *The Chilling Stars* also attracted much criticism in Europe.[47]

THE PURCHASING OF CLIMATE SCIENCE

The political views of a funding organization can have a powerful influence on research results. A number of years ago, while working at a Fortune 100 corporation, my boss asked me to write an analysis to generate a specific answer regarding Asian electronics competition. When I dragged my feet due to the pre-ordered conclusion, my boss gave the job to another employee, who produced the requested answer. Someone can always be found to produce an analysis showing a requested conclusion, in science as well as business.

Government funding has been used to purchase modern-day climate science. In the wake of Al Gore's 1988 senate hearings and

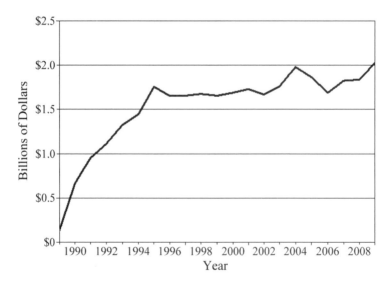

Figure 84. U.S. Climate Change Science Program Funding. Funding for years 1989–2009. (Data from GAO, OMB, USGCRP, 2009)[48]

the forming of the IPCC in 1989, United States funding for the Climate Change Science Program soared. Figure 84 shows a 10-fold increase from 1989 to 1996, and funding stands today at $2 billion. Total U.S. climate-change-related funding from all sources is more than double this amount, but the curve is representative of the growth in funding. Much of this funding ramp paid for computer-model generated futures, rather than real experimental science.

Dr. Roger Pielke Sr., senior scientist at the Cooperative Institute for Research in Environmental Sciences, says recent climate science is "short circuiting the scientific process."[49] A science fair summary of the scientific method lists the following six steps:

1. Ask a question.
2. Do background research.
3. Construct a hypothesis.
4. Test your hypothesis by doing an experiment.
5. Analyze your data and draw a conclusion.
6. Communicate your results.[50]

However, much of today's climate science consists of generating scenarios from computer models, including future predictions of temperatures, rainfall, droughts, melting ice caps, species extinction, storm intensities, and other climate events with a potential for catastrophe. These projections are step three (construct a hypothesis) of the scientific process above. Since the projections are usually made for decades into the future, the key experimental step, step four, cannot be completed. Conclusions are drawn directly from the computer simulations without experiment. The results are published and then adopted by the IPCC and the alarmists. Pielke laments:

> What the current publication process has evolved into, at the detriment of proper scientific investigation, is the publication of untested (and often untestable) hypotheses…This is the main reason that the policy community is being significantly misinformed about the actual status of our understanding of the climate system and the role of humans within it.[51]

What Dr. Pielke is saying, in other words, is that no experimental testing is used. Climatist scientists have been running models and publishing forecasts of the future. *This process seems to belong to a new profession of high-tech fortune telling—not science.*

The greatest exaggerations seem to get the most attention. Monika Kopacz, Ph.D. candidate at Harvard, made this frank assessment:

> It is no secret that a lot of climate-change research is subject to opinion, that climate models sometimes disagree...The problem is, only sensational exaggeration makes the kind of story that will get politicians'—and readers'—attention. So, yes, climate scientists might exaggerate, but in today's world, this is the only way to assure any political action and thus more federal financing to reduce the scientific uncertainty.[52]

Dr. Lindzen of the Massachusetts Institute of Technology names the funding quest the "triangle of alarmism." He describes the process as: 1) scientists make meaningless or ambiguous statements about the climate, 2) advocates and media translate statements into alarmist declarations, and 3) politicians respond to alarm by feeding scientists more money.[53] It's instructive to watch Congressional testimonies by officials from the National Oceanic and Atmospheric Administration, the Center for Disease Control, or other government organizations supporting man-made global warming. No testimony is complete without a plea for "more funding to further study climate change."

Dr. Roy Spencer, research scientist at the University of Alabama in Huntsville, was with the National Aeronautic and Space Administration for many years, and was awarded NASA's Medal for Exceptional Scientific Achievement. Dr. Spencer retired early from NASA more than seven years ago "to have more freedom to speak my mind on global warming."[54] Spencer mentions Dr. John Theon and Dr. Joanne Simpson as notable team members who publically announced skepticism regarding the role of mankind in global warming *after* leaving NASA. According to Spencer, NASA scientists curbed skeptical views to get required project funding:

> Manmade global warming was a potentially serious threat, and NASA wanted Congress to fund new satellites to study the problem. It was a team effort to get that accomplished.[55]

NASA administrator Michael Griffin was one who made the mistake of publically airing his views on climate change. In March, 2007, he told National Public Radio:

I have no doubt that…a trend of global warming exists. I am not sure that it is fair to say that it is a problem we must wrestle with. To assume that it is a problem is to assume that the state of Earth's climate today is the optimal climate, the best climate that we could have or ever have had and that we need to take steps to make sure that it doesn't change…I think that's a rather arrogant position for people to take.[56]

The public assault on Griffin was quick and venomous. Jerry Mahlman of the National Center for Atmospheric Research said Griffin was either "totally clueless" or "a deep anti-global warming ideologue." James Hansen added his criticism of Griffin. "Clarifying" announcements were also made by the White House and NASA.[57]

Dr. Ferenc Miskolczi, working as a NASA contractor, was involved in another case of "protect the funding." In 2004 he developed a mathematical theory that placed an upper limit on Earth's greenhouse effect. In short, Miskolczi hypothesized that once the limit was reached, each increase in carbon-based greenhouse gas would be naturally offset by a decrease in the atmosphere's largest greenhouse gas—water vapor. His supervisors at NASA would not allow him to publish his work. Miskolczi believes his supervisors were concerned with a potential loss of funding. He resigned in 2005 and had his hypothesis peer-reviewed and published in Europe.[58]

In fact, *the vast majority of today's climate research is paid for by world governments that have already made up their mind that greenhouse gases are warming the globe.* The answers requested from the government grant providers are overwhelmingly "to determine the impacts of man-made climate change," not to objectively determine the cause. If Albert Einstein was funded by a directed government grant, would he have developed his Theory of Relativity?

THREE NOTABLE DISSENTERS

It often seems that more scientists move from believers in man-made warming to the realist camp, rather than the other way around. Here are three notable persons who have courageously joined the realists.

Dr. David Bellamy was one of the best known faces on British television. As a botanist and author of 35 books, he presented 400-plus television programs over more than a decade. He was president of several environmental organizations. But all this was before 1996 when, according to Bellamy, he criticized wind farms on the British children's television show *Blue Peter* and also published an article describing man-made global warming as "poppycock."[59]

After that year, Bellamy was shunned by both the BBC and environmental groups. He has not appeared on British television since. He was president of Plantlife International for 15 years and was also president of The Royal Society of Wildlife Trusts. Both organizations asked him to step down.[60] Says Dr. Bellamy:

> Global warming is part of natural cycle and there's nothing we can actually do to stop these cycles. The world is now facing spending a vast amount of money in tax to try to solve a problem that doesn't actually exist.[61]

Atmospheric scientist Dr. John Theon was the high-level supervisor for James Hansen, in charge of all weather and climate research at NASA from 1983 to 1994.[62] Theon retired from NASA in 1995 and came out as a public opponent of man-made global warming in January 2009, stating "the climate models are worthless." He warns "there is no rational justification for using climate model forecasts to determine public policy."[63]

Geochemist Dr. Claude Allègre is one of the most decorated scientists of France. He headed the Institut de Physique de Globe in Paris from 1976 and was president of the French geological survey from 1992 to 1997.[64] He authored more than 100 scientific articles and 11 books and received numerous scientific awards. In 1992 he was one of 1,500 scientists who signed the "World Scientists' Warning to Humanity" letter, warning that global warming's "potential risks are very great."[65] But he shocked the world in 1996 with the article "The Snows of Mount Kilimanjaro" for *L'Express*, in which he stated "the cause of this climate change is unknown."[66] Since then, Allègre has become France's most outspoken skeptic.

THE CULPABLE NEWS MEDIA

The news media has been a major contributor to climate delusion. Citizens are continuously bombarded by alarmist messages directly fostering the creation and exploitation of fear. Most major magazines have put global warming on the cover, some more than once.

Two examples are noteworthy. *Business Week* put a picture of Earth in flames on the cover of their August 16, 2004 issue, with the title "Global Warming, Why Business is Taking It so Seriously."[67] Not to be outdone, the *Time* April 3, 2006 issue had polar bears on the cover with the title "Be Worried. Be Very Worried."[68]

Climate disaster stories are not new. Note the excerpt of the alarming climate story written by *Newsweek* in 1975 (text box). It sounds like many of global warming disaster stories of today, does it not? But it's actually about *global cooling*, a popular topic in 1975.

News media is a tough business. Forty years ago the U.S. media consisted mainly of network TV stations ABC, CBS, and NBC, the public network station PBS, radio, newspapers, and magazines. Today we have added a flood of internet information, along with new cable-TV channels and satellite and internet radio. Breaking news in one location is re-transmitted around the world in real time. Competition for readers and viewers is intense and global.

The media's objective has always been to increase its market, whether readers or viewers. With increased competition, this objective has become a matter of survival. As Julian Cribb, one of Australia's leading science journalists, pointed out:

> For years I was a newspaper editor and I knew—as most editors know—that if you print a lot of good news, people stop buying your paper. Conversely, if you publish the correct mix of doom, gloom and disaster, you circulation swells. I have done the experiment! The publication of "bad news" is not a vice peculiar to editors…It's what people on average demand.[69]

Climate Change in 1975?

There are ominous signs that the earth's weather patterns have begun to change dramatically and that these changes may portend a drastic decline in food production—with serious political implications for just about every nation on earth. The drop in food output could begin quite soon, perhaps only ten years from now…The evidence in support of these predictions has now begun to accumulate so massively that meteorologists are hard pressed to keep up with it…farmers have seen their growing season decline…During the same time, the average temperature around the equator has risen by a fraction of a degree—a fraction that in some areas can mean drought and desolation. Last April, in the most devastating outbreak of tornadoes ever recorded…— "The Cooling World," *Newsweek*, April 28, 1975[70]

This tendency of the media to print gloom and disaster plays into the hands of the Climatists as they create their climate of fear. Exaggeration of scientific projections is assured. When a scientist announces projections of sea-level rise between four and 20 feet, the media will invariably lead with "sea rise of up to 20 feet projected."

Beyond the effort to promote disasters to build viewership, a large portion of the media appears to have an agenda to convince the public that global warming is man-made. Ann Curry of NBC's *Today*, while on her way to Antarctica on October 29, 2007, stated: "Our mission, of course, is to find evidence of climate change."[71]

The leading U.S. television networks, ABC, CBS, and NBC, have been among the most biased in promoting Climatist dogma. The Business and Media Institute analyzed 205 stories about global warming or climate change by the three networks during the last six months of 2007. They found 13 man-made global warming advocates featured for every skeptic. ABC was the most balanced with a 7-to-1 ratio, while CBS was the least, with 38 man-made warming advocates featured for every single skeptic. In the articles analyzed, only 15% of the persons interviewed by the networks were scientists, while 85% were politicians, celebrities, and journalists.[72] CBS journalist Scott Pelley worked hard to make CBS the most one-sided. In March 2006, when asked why his reports did not pause to acknowledge global warming skeptics, he responded:

> If I do an interview with [Holocaust survivor] Elie Wiesel, am I required as a journalist to find a Holocaust denier?

Mr. Pelley further went on to say that his team tried hard to find a respected skeptical scientist, but was unable to do so.[73]

Media bias is also strong in Europe. Peter Sissons, veteran newsreader, retired in June 2009 after a notable 20-year career with BBC News. Mr. Sissons describes the bias of BBC involving an interview in December 2008:

> On a wintry Saturday last December, there was what was billed as a major climate change rally in London. The leader of the Green Party, Caroline Lucas, went into the Westminster studio to be interviewed by me on BBC News channel. She clearly expected what I call a 'free hit'; to be allowed to voice her views without being challenged on them. I pointed out to her that the climate didn't

seem to be playing ball at the moment. We were having a particularly cold winter, even though carbon emissions were increasing. Indeed, there had been no warming for ten years, contradicting all the alarming computer predictions. Well she was outraged…Miss Lucas told me angrily that it was disgraceful that the BBC—the BBC!—should be giving any kind of publicity to those sort of views…But it is effectively BBC policy, enthusiastically carried out by the BBC environment correspondents, that those views should not be heard—witness the BBC statement last year that 'BBC News currently takes the view that their reporting needs to be calibrated to take into account the scientific consensus that global warming is man-made.'[74]

GROWING NUMBER OF CLIMATE REALISTS

In the first half of this book, we've cited many qualified scientists challenging the theory that greenhouse gas emissions are causing catastrophic global warming. This number of realist climate scientists is growing, despite the suppressive tactics of Climatism, as every year science learns more about Earth's climate. A trickle of skepticism in 1990 has become a raging river of voices today. Commendations to these courageous scientists, who stand up to ridicule, suppression, and worse from the forces of alarmist politics.

One of the first statements opposing alarmist environmental dogma was the Heidelberg Appeal. The Heidelberg Appeal was released during the 1992 Earth Summit in Rio de Janeiro. More than 425 scientists and intellectual leaders signed the Appeal by the end of 1992, and the Appeal has more than 4,000 signatories today. The Appeal is a call for reason and recognition that science and technology are the solution to the world's environmental problems, rather than the problem itself as claimed by the UN's Agenda 21 document of the 1992 Summit. The Appeal states:

The greatest evils which stalk our Earth are ignorance and oppression, and not Science, Technology, and Industry, whose instruments, when adequately managed, are indispensable tools of a future shaped by Humanity, by itself and for itself, overcoming major problems like over population, starvation and worldwide diseases.[75]

The full text of the Appeal and 103 of the most distinguished signers is found on the Science and Environmental Public Policy website.[76]

Senator James Inhofe from Oklahoma is the leading climate skeptic in the United States Senate. Marc Morano, while part of Senator Inhofe's staff, compiled a list of leading scientists from all over the world who disagree with Climatist global warming dogma. The list included 400 scientists when first published in 2007, but had grown to over 700 in 2009. The list contains statements by the scientists, as well as links to further information. This is probably the best list of realist scientists available. It can be found at the U.S. Senate Committee for Environment and Public Works website.[77]

The "science is settled" claims of the Climatists are now acting to boost the number of opposing scientists. Dr. William Briggs, meteorologist and statistical expert, and one of the 700, remarks: "After reading [UN IPCC chairman] Pachauri's asinine comment [comparing skeptics to] Flat Earthers, it's hard to remain quiet." Another of the 700, meteorologist Hajo Smit of Holland stated "Gore prompted me to start delving into the science again and I quickly found myself solidly in the skeptic camp."[78]

The Global Warming Petition Project provides a list of more than 31,000 U.S.-based scientists, engineers, and trained professionals, including over 9,000 PhDs, who disagree with the theory of man-made global warming. The Project is directed by Dr. Arthur Robinson, Professor of Chemistry of the Oregon Institute of Science and Medicine, an institution which he founded. The Petition states:

> We urge the United States government to reject the global warming agreement that was written in Kyoto, Japan in December, 1997, and any other similar proposals. The proposed limits on greenhouse gases would harm the environment, hinder the advance of science and technology, and damage the health and welfare of mankind.

> There is no convincing scientific evidence that human release of carbon dioxide, methane, or other greenhouse gases is causing or will, in the foreseeable future, cause catastrophic heating of the Earth's atmosphere and disruption of the Earth's climate. Moreover, there is substantial scientific evidence that increases in atmospheric carbon dioxide produce many beneficial effects upon the natural plant and animal environments of the Earth.[79]

The number of scientists on each side of a debate is not important to scientific validity. Albert Einstein reportedly said "no amount of experimentation could ever prove me right, and it takes only one experiment to prove me wrong." However, a large number of skeptical scientists are a good indication that *the science is not settled.*

THE NONGOVERNMENTAL INTERNATIONAL PANEL ON CLIMATE CHANGE

In 2003, Dr. Fred Singer, founder and president of the Science and Environmental Policy Project (SEPP), along with other realist scientists, began a project to create an independent second opinion to the assessment reports of the IPCC. The Nongovernmental International Panel on Climate Change (NIPCC) was established to develop an alternative report. Their work culminated in 2009 in *Climate Change Reconsidered,* authored by Singer and Dr. Craig Idso, an 880-page volume that is the most extensive critique of the position of the IPCC ever published.[80] The report cites thousands of peer-reviewed articles that were ignored or dismissed by the IPCC. The NIPCC report concludes that "natural causes are very likely to be the dominant cause" of global warming and that "the net effect of continued warming and rising carbon dioxide concentrations in the atmosphere will be beneficial to humans, plants, and wildlife."[81]

CLIMATEGATE AT EAST ANGLIA

The Climatic Research Unit (CRU) at the University of East Anglia is regarded as the world's leading source for global temperature information. The CRU works with the Hadley Centre, a branch of the U.K. Meteorological Office, to collect data from land temperature stations, ocean stations, radiosonde (balloons) and satellites, adjusts it, and produces the global temperature graph that we showed in Figure 6 of Chapter 2. The CRU collects data from "more than 4,000 weather stations distributed around the world."[82]

On November 19, 2009, an information "explosion" occurred that may shake the world of climate science for years to come. An unknown hacker downloaded more than 1,000 documents and emails from the CRU and posted them on a server in Russia.[83] Within hours, these documents were accessed by web sites around the world. The emails are a subset of candid conversations between top Climatist scientists in the U.K., the U.S., and other nations, between March, 1996 and November, 2009. They provide inside information about development of the IPCC reports, conflict with skeptical scientists, efforts by CRU personnel to avoid freedom of information requests, and strong bias within the alarmist scientific community to promote the theory of man-made global warming. The incident was soon named "Climategate" by the realist scientific community and the news media that reported the story.

The emails reveal a high level of bias toward man-made warming by scientists at not only the CRU, but also at NASA, NOAA, the University Corporation for Atmospheric Research (UCAR), and other major scientific organizations. In one email, Dr. Kevin Trenberth of UCAR laments the current global cooling and failure to forecast the cooling:

> Well I have my own article on where the heck is global warming? We are asking that here in Boulder where we have broken records the past two days for the coldest days on record...The fact is that we can't account for the lack of warming at the moment and it is a travesty that we can't...Our observing system is inadequate...

The email containing this statement was sent in October, 2009 to top alarmist scientists across the world, including James Hansen and Gavin Schmidt at NASA, Michael Mann at Pennsylvania State, Stephen Schneider at Stanford, Michael Oppenheimer at Princeton, Thomas Karl, director of NOAA, Ben Santer at Lawrence Livermore Laboratory, Tom Wigley at UCAR, Myles Allen at Oxford, Peter Stott of the Hadley Centre, and Phil Jones, director of the CRU.[84]

Dr. Keith Briffa of the CRU reveals strong bias during development of the IPCC Third Assessment Report (TAR), published in 2001. Recall that the TAR contained the Mann Curve and the claim that 20th Century temperatures were the warmest in over 1,000 years.[85] In an email sent in September, 1999, Briffa talks about pressure to come to the "warmest in thousand years" conclusion:

> ...I know there is pressure to present a nice tidy story as regards "apparent unprecedented warming in a thousand years or more in

the proxy data" but in reality the situation is not quite so simple...I believe that the recent warmth was probably matched about 1,000 years ago...[86]

Briffa was expressing concern to his fellow alarmist scientists two years prior to publication of the TAR, but eventually the IPCC did adopt the "warmest in 1,000 years" conclusion.

Dr. Jones, director of CRU, wrote emails clearly expressing a wish for global warming to validate CRU projections. In a mail to the Hadley Centre U.K. Meteorological Office in January, 2009, he states "I hope you're not right about the lack of warming lasting until about 2020." At another time in 2008 he says "I'd like the world to warm up quicker."[87]

This zeal for the cause of man-made global warming apparently resulted in selective use of scientific data, suppression of freedom of information requests, and efforts to suppress publication of skeptical scientific papers. In an email in 1999 from Phil Jones to Michael Mann, Malcolm Hughes of the University of Arizona, and Ray Bradley of the University of Massachusetts, Dr. Jones stated:

> ...I've just completed Mike's Nature trick of adding in the real temps to each series for the last 20 years (i.e. from 1981 onwards) and from 1961 for Keith's to hide the decline...[88]

The statement by Jones reveals how the Mann Curve was constructed. Michael Mann used tree-ring and other proxy data to reconstruct 1,000 years of past temperatures as shown in Figure 67 of Chapter 7. Note how the curve shows a slight downward slope from year 1000 until about 1950, nicely fitting Climatist dogma by eliminating the Medieval Warm Period and the Little Ice Age. But then Mann encountered a problem. The proxy data did not show the late 20th century temperature increase. At this point, many scientists would conclude that the proxies selected did not accurately portray global temperatures. Instead, Mann replaced the declining proxy data at the end of the 20th century with real temperature data, referred to by Jones as the "trick" to "hide the decline."

In another email, Dr. Tom Wigley of UCAR discusses with Phil Jones efforts to try to reduce the sea surface temperature measurements from the 1940s:

...If you look at the attached plot you will see that the land also shows the 1940s warming blip (as I'm sure you know). So, if we could reduce the ocean blip by, say 0.15 deg C, then this would be significant for the global mean—but we'd still have to explain the land blip...[89]

Note that by reducing 1940 temperatures, the 1975-1998 global temperature rise is emphasized, thereby supporting the alarmist case. In yet another example, Dr. Mick Kelly of the CRU tells Phil Jones he will remove some recent colder data from his temperature curve:

...Anyway, I'll maybe cut the last few points off the filtered curve before I give the talk again as that's trending down as a result of the end effects of the recent cold-ish years...[90]

The same alarmist scientists that showed bias in these Climategate emails wrote the IPCC reports, concluding that global warming was very likely due to man-made greenhouse gas emissions. The text box shows that many of the scientists mentioned in the Climategate emails were authors for the IPCC 2007 Fourth Assessment Report (FAR). Michael Mann is missing because the hockey stick scandal precluded his participation as an FAR author.

While the CRU collects and compiles temperature data, they've been less than cooperative in sharing the raw data with the scientific community. Warwick Hughes, a climate researcher based in Perth, Australia, requested temperature data by station from Phil Jones in 2004. Dr. Jones refused, stating:

...We have 25 or so years invested in the work. Why should I make the data available to you, when your aim is to try and find something wrong with it...[91]

Multiple Climategate emails show Jones in repeated efforts to avoid Freedom of Information Act (FOI) requests. For example, he states in February, 2005, in an email to Michael Mann:

...If they ever hear there is a Freedom of Information Act now in the UK, I'll delete the file rather than send to anyone...[92]

In a December, 2008 email to Ben Santer and Tom Wigley, and copied to others, he states:

Hacked CRU Emails Involve Many Authors of the IPCC 2007 4th Assessment Report[93]

Myles Allen, University of Oxford
    Contributing Author–Working Group 1, Chapters 1 and 9
    Review Editor–Working Group 1, Chapter 10

Keith Briffa, East Anglia Climate Research Unit
    Lead Author– Working Group 1, Chapter 6

Chris Folland, Hadley Centre, U.K. Meteorological Office
    Contributing Author–Working Group 1, Chapter 3

James Hansen, NASA
    Contributing Author–Working Group 1, Chapter 10

Thomas Karl, Director, NOAA
    Review Editor–Working Group 1, Chapter 3

Phil Jones, Director, East Anglia Climate Research Unit
    Draft Contributing Author–Summary for Policy Makers
    Contributing Author–Technical Summary
    Coordinating Lead Author–Working Group 1, Chapter 3

Michael Oppenheimer, Princeton University
    Contributing Author–Working Group 1, Chapter 10

Jonathan Overpeck, University of Arizona
    Drafting Author–Summary for Policymakers
    Lead Author–Technical Summary
    Contributing Author–Working Group 1, Chapter 10

Ben Santer, Lawrence Livermore Laboratory
    Contributing Author–Working Group 1, Chapters 1 and 9

Gavin Schmidt, NASA
    Contributing Author–Working Group 1, Chapter 10

Susan Solomon, NOAA
    Drafting Author–Summary for Policymakers
    Coordinating Lead Author–Technical Summary

Peter Stott, Hadley Centre, U.K. Meteorological Office
    Lead Author–Working Group 1, Chapter 9
    Contributing Author–Working Group 1, Chapter 10

Kevin Trenberth, UCAR
    Draft Contributing Author–Summary for Policy Makers
    Contributing Author–Technical Summary and WG 1, Chapters 1 and 7
    Coordinating Lead Author–Working Group 1, Chapter 3

...I am supposed to go through my emails and he can get anything I've written about him. About 2 months ago I deleted loads of emails, so have very little—if anything at all.[94]

Finally, the emails show that Climatist scientists have been actively engaged in suppressing the publication of skeptical articles. In reaction to the paper "Nature, not CO_2 rules the climate" by Dr. David Douglas and others in 2003,[95] Ben Santer stated that he contacted the editor, who agreed to try to slow down publication of the paper in the International Journal of Climatology:

> I just contacted the editor, Glenn McGregor, to see what he can do...he may be able to hold back the hardcopy (i.e. the print/paper version) appearance of Douglass et al., possibly so that any accepted Santer et al. comment could appear along side it.[96]

In 2004, Phil Jones discussed skeptical publications in an email to Michael Mann:

> ...I can't see either of these papers being in the next IPCC report. Kevin [Trenberth] and I will keep them out somehow, even if we have to redefine what the peer-review literature is![97]

The immediate result of Climategate was the resignation of director Phil Jones and the launch of an investigation of the CRU by the University of East Anglia in December 2009.[98] The global temperature data from Hadley Centre and CRU is also being questioned. The larger long-term impact may be to spur the world to take a second look at climate science, rather than relying on the flawed conclusions of the IPCC.

CONCLUSION

It's clear the science is not settled, at least not in favor of man-made global warming. Despite the best efforts of the Climatist propaganda machine to belittle, criminalize, and suppress their climate realist opponents, the list of courageous skeptics is growing. At the same time, more and more questions are being raised about the integrity of IPCC conclusions. The truth will rise again.

CHAPTER 11

SNAKE OIL REMEDIES TO "SAVE THE PLANET"

"Nothing is more terrible than ignorance in action."
Johann Wolfgang von Goethe

A favorite character in American books and movies is the doctor who travels from town to town, peddling elixirs and potions to cure "whatever ails you." "Why, this painkiller even includes snake oil," he'd pitch to his audience. An accomplice in the crowd would offer personal testimony of a "miraculous cure" to boost sales. After selling a few bottles, the quack would typically leave town, before his snake oil was found to be useless.

How do you solve a problem that doesn't exist? The Climatists are very creative. They're like the quack except they *believe* their snake oil remedies will actually stop global warming. Since greenhouse gases are emitted from *all human activity*, the remedy is actually hundreds of different solutions, from downsizing your car, to regulating your diet, to "greening" your house, to even limiting your family size. But these peddlers don't even need to "leave town by sundown." They claim we "must act now," but mankind won't be able to measure results from their extreme solutions for decades, if ever.

Nancy Pelosi, United States Speaker of the House of Representatives, recently spoke about climate change:

We have so much room for improvement. Every aspect of our lives must be subjected to an inventory...of how we are taking responsibility.[1]

Well get ready, because Climatists plan to inventory and regulate every aspect of our lives. This is a direct attack on our freedoms and our livelihood. We'll lay out their illusory solutions in the rest of this chapter, with deeper discussion of the impacts in Part III.

THE PLAN: REDUCE CO_2 EMISSIONS 50% TO 85%

The model-predicted impacts of catastrophic global warming are captured in the "burning embers" diagram in Figure 85. This diagram is from the IPCC Third Assessment Report of 2001 and is still used by many alarmist scientists.[2] The right scale shows global temperature increase above 1990 in degrees Celsius. Not shown is the "pre-industrial global temperature" which alarmist scientists assume to be 0.6°C below the 1990 temperature. This assumption ignores the temperature cycles of Earth discussed in Chapter 4. The five risk bars represent: I) impact on ecological systems, II) impact on weather, III) distribution of impacts on regions, IV) aggregate impacts on Earth, and V) risk of large-scale discontinuities when "tipping points" are passed. When in color, the chart goes from yellow at the bottom to red at the top (hence the burning embers name).

Climatism has established a target of a maximum of 2°C of temperature rise from the "pre-industrial level," or 1.4°C above 1990 global temperatures. We are told that if we keep the rise to just 2°C, then most of the climate disasters in Figure 85 can be avoided. Remember, this is all based on forecasts by General Circulation Models that can't be experimentally tested.

The IPCC has estimated global temperature rise from atmospheric carbon dioxide concentrations, which is summarized in Figure 86. According to the IPCC, to keep global temperatures from rising more than 2°C, we need to "stabilize" atmospheric carbon dioxide at about 450 ppm (parts per million). This is shown by the gray-shaded bar in Figure 86. Note that *this requires a global cut in projected greenhouse gas emissions of between 50% and 85% by 2050.* This would be a huge change to our hydrocarbon-based society. Let's see how big this is.

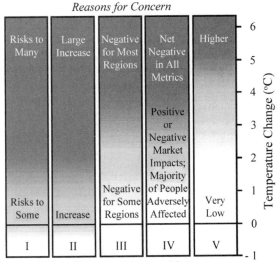

Reasons for Concern

I Risks to Unique and Threatened Systems
II Risks from Extreme Climate Events
III Distribution of Impacts
IV Aggregate Impacts
V Risks from Future Large-Scale Discontinuities

Figure 85. Burning Embers Diagram. Impacts shown from global temperature rise. Zero degrees corresponds to 1990 global temperatures. Pre-industrial temperatures are assumed to be -0.6°C. (Adapted from IPCC, 2001)[3]

CO_2-Equivalent Concentration at Stabilization	Peaking Year for CO_2 Emissions	Change in CO_2 Emissions in 2050 (% of 2000 Emissions)	Global Average Temperature Increase above Pre-Industrial Level
445–490 ppm	2000–2015	-85% to -50%	2.0–2.4°C
490–535 ppm	2000–2020	-60% to -30%	2.4–2.8°C
535–590 ppm	2010–2030	-30% to +5%	2.8–3.2°C
590–710 ppm	2020–2060	+10% to +60%	3.2–4.0°C
710–855 ppm	2050–2080	+25% to +85%	4.0–4.9°C
855–1130 ppm	2060–2090	+90% to +140%	4.9– 6.1°C

Figure 86. IPCC Prediction of Temperature Increase. Temperature increase for a given stabilization of CO_2-equivalent greenhouse gases (includes all human-emitted greenhouse gases). To meet the 2°C rise above industrial level (grey bar), global emissions must be cut 50% to 85% by 2050. (Reproduced from IPCC, 2007)[4]

THE STABILIZATION WEDGES

The Princeton Environmental Institute has a Carbon Mitigation Initiative project that shows how mankind can achieve emissions cuts to stabilize atmospheric levels of carbon dioxide. They define a "stabilization triangle" of reductions needed to provide a flat path of zero CO_2 emissions growth, as shown in Figure 87. They state that current annual emissions are about seven billion tons of carbon per year, expected to rise to 14 billion per year in 50 years. According to the Climatists, if we divide the stabilization triangle itself into seven "wedges" of emissions reduction of one gigaton (one billion tons) each, we can achieve flat carbon emissions growth. Fourteen gigatons of carbon may sound like a large amount, but recall from Chapter 3 that all human emissions in history have added only one molecule of CO_2 for each 10,000 molecules in Earth's atmosphere.

The emission reduction wedges were originally introduced by Dr. Stephen Pacala and Dr. Robert Socolow at Princeton University in 2004.[5] This concept has generally been adopted by the Climatist community. Fifteen possible wedges are shown in the text box, of which seven are needed to stabilize atmospheric CO_2.

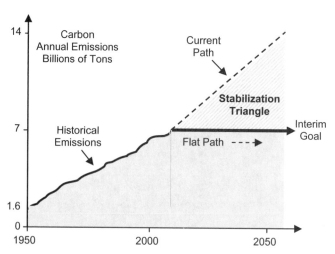

Figure 87. The Stabilization Triangle. Global carbon emissions today are estimated at about seven billion tons per year, with mankind on a path to 14 billion tons per year in 2056. The stabilization triangle concept calls for flat emissions from today forward via various methods. (Adapted from CMI, 2007)[6]

At first glance, many of these wedges seem reasonable. There are, however, at least five major problems with the wedges listed. First, none of these solutions will naturally occur through market forces, so governments must impose solutions upon us. Consumers favor cars that average 30 mpg, because they choose style, carrying capacity, roominess, acceleration, and other features over additional fuel economy. Consumers prefer cars to mass transit for reasons of convenience and lifestyle. Buildings are designed for a measure of energy efficiency in balance with other functional features demanded by residential and industrial customers. Utilities build coal plants, balancing efficiency with cost, serviceability, and other factors to maximize profit. To implement emissions-reduction wedges, heavy government intervention will be required.

Second, as we'll detail in Part III, most of these solutions will significantly boost the cost—and limit the availability—of energy. Electricity from wind power costs almost twice as much as power from coal. Solar-generated electrical power is even more expensive. Electricity from both wind and solar is less reliable than hydrocarbon

---

15 Proposed Global Emissions Reduction Wedges[7]

Efficiency
1. Double fuel efficiency of 2 billion cars from 30 to 60 mpg
2. Decrease the number of car miles traveled by half
3. Use best efficiency practices in all buildings
4. Double efficiency of today's coal-based electricity plants

Electricity Production
5. Replace 1,400 coal-fired power plants with natural gas plants
6. Double nuclear electricity capacity and replace coal plants
7. Use 2 million large windmills for 50x wind-capacity increase
8. Install 700x today's solar-electricity capacity

Carbon Capture and Storage
9. Capture and store CO_2 from 800 coal-fired power plants
10. Capture and store CO_2 from 180 coal-to-synfuels plants

Biomass and Hydrogen Fuels
11. 50x ethanol production using 1/6th of world cropland
12. Produce hydrogen from coal at 6x today's rate, capture and store CO_2
13. Use 4 million large windmills to produce fuel for hydrogen fuel cell cars

Natural Sinks
14. Eliminate tropical deforestation and double new forest planting
15. Adopt agricultural conservation tillage worldwide

or nuclear power. Implementation of these wedges will impose cost and reduce productivity of the world's economies.

Third, the wedges call for radical changes in agriculture and commerce. The most important of these is using a massive one-sixth of the world's agricultural land to produce biofuels for vehicles. This will cause major distortion in agricultural markets and also raise the price of vehicle fuel. Dr. Kunihiko Takeda of Chubu University of Japan says, "...making fuel out of corn makes sense only if you want to increase the price of corn and fuel at the same time."[8]

Fourth, some of these solutions are not yet technologically viable. Carbon Capture and Storage (CCS) is a technology that involves capturing the carbon dioxide ~~from~~ produced by electrical power plants, liquefying it, and then pumping it into an underground reservoir. This technology is far from being commercially ready. Even if technical problems can be solved, analysis shows CCS will boost the cost of electrical power. Hydrogen-fuel for cars is another technology not ready for prime time. Lack of viability eliminates four of the 15 wedges as not feasible in the near term.

Fifth, the magnitude of the effort of some of these wedges is staggering. Replacing 1,400 coal plants with natural gas plants would require replacement of 28 plants per year or more than two per month for the next 50 years. More than 100 large windmills must be installed *per day* over 50 years, to implement the wind power wedge. Hundreds of billions of dollars in subsidies will be needed to provide energy approaching competitive costs. Increasing solar cell electricity by 700 times will similarly require massive subsidies.

The alarmists believe they know better than free-market dynamics, so they'll impose mandates to force adoption of seven-wedge policies. This has already been underway for several years in Europe. The primary tool is to raise the price of hydrocarbon energy using a "cap-and-trade" taxation system. Other tools are subsidies for renewable energy systems and regulation to force "responsible behavior."

Evidence is mounting that the wedges framework is already inadequate. In 2007, China passed the U.S. as the world's largest greenhouse gas emitter. China reportedly intends to build over 500 coal-fired plants by 2012. This will keep Asian emissions growing at more than 10% per year in the near future, double IPCC estimates.[9] It's probable that many more than seven wedges will be demanded.

It's clear the price of this snake oil is very high. Lord Nicholas

Stern raised his global cost estimate in 2008 to 2% of GDP, or over one trillion dollars per year.[10] Yet, even this is probably an underestimate for the societal dislocation planned by Climatism.

EMISSIONS AFTER KYOTO PROTOCOL

Recall that the Kyoto Protocol was signed in December, 1997 by 38 nations, became active in 2005, and has now been ratified by 184 nations, with the United States the major exception. The U.S. is a favorite target for "climate responsible" nations. In 2005, Juergen Trittin, Environment Minister of Germany said:

> The American president [George Bush] closes his eyes to the economic and human damages that are inflicted on his country and the world economy by natural disasters, like Katrina, through neglected climate protection.[11]

Figure 88 shows per capita greenhouse gas emissions growth for the U.S. and seven most populous European nations, using data from the U.S. Department of Energy. Since the nations of Europe signed

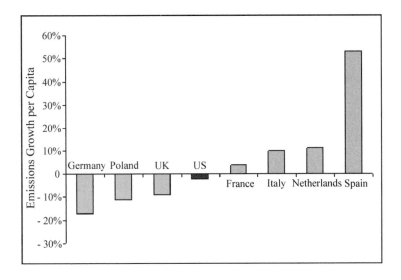

Figure 88. Per Capita Emissions Growth 1990–2005. Emissions growth per person for the U.S. and seven most populous nations of Europe. (EIA, 2006)[12]

Kyoto and have adopted policies to reduce carbon emissions for ten years, we would expect U.S. emissions to be growing faster than those of European countries. However, this is not the case. Growth in U.S. emissions is right in the middle of the pack over the last 15 years.

In addition, the base year of 1990 was successfully negotiated by the U.K. and Germany to provide them emissions advantages. When Kyoto was formulated in the mid-1990s, the U.K. was already well along the path of converting many of its coal-fired electricity plants to gas. Germany similarly gained advantages by the union of East and West Germany and the opportunity to replace inefficient carbon-emitting factories in the East. As a result, emissions reductions for both the UK and Germany are boosted by the choice of the 1990 base year. Note also the huge emissions growth in Spain over the 15-year period. Spain is often held high as a model, due to heavy recent investment in wind and solar power. It seems climate change talk is cheap for the change advocates, but real results are something else.

KYOTO II CONFERENCE IN COPENHAGEN

All 193 member nations of the UN met in Copenhagen in December, 2009 to negotiate "Kyoto II," the follow-on agreement to the Kyoto Protocol. The event was officially called the United Nations Climate Change Conference, part of the UN Framework Convention on Climate Change (UNFCCC). In all, about 20,000 delegates from governments, non-governmental organizations, businesses, and media attended the event, including more than 100 heads-of-state.[13] Carbon-emitting vehicles flooded the Danish capital, including more than 1,200 limousines and 140 private jets.[14]

Hopes were high for success, reinforced by expected climate activism from first-year U.S. President Barack Obama. Climatists sought: 1) legally binding emissions-reduction commitments from all countries, including the developing nations, and 2) agreed financial transfer payments from industrialized to developed nations. Kofi Annan, former UN Secretary General, called the Copenhagen Conference "a great opportunity which should not be squandered."[15]

The event was preceded by an intense lobbying effort from the UN, using the campaign slogan "Seal the Deal! Power green growth. Protect the Planet!" The Seal the Deal campaign spent tens of millions of dollars, using UN officials, environmental groups, Hollywood celebrities, youth groups, businessmen, scientific organizations, and news media to boost public pressure for a climate treaty.[16] The UN's Economic and Social Council sponsored "The First International Conference on Broadcast Media and Climate Change" in Paris in September, 2009.[17] More than 250 broadcasters attended, agreeing to "comprehensively report" on the Copenhagen Conference, and declaring that:

> "...an increased public understanding of the urgency of climate change is essential to mitigate its negative impacts and to avert human suffering."[18]

At Copenhagen, economic and political realities quickly scaled back treaty expectations. Developing nations demanded billions in climate transfer payments, but industrialized nations feared voters might not approve such payments. Emissions reduction commitments were sought from China and India, but these nations did not want to slow their economic growth. Ronen Sen, India's ambassador to the U.S., summarizes India's problem:

> We can't tell an electorate...when you have something like in excess of 300 million people without any access to electricity at all that you have to put a cap on this.[19]

In the end, no treaty was produced, but only the Copenhagen Accord. The five-page accord contains no legally binding commitments, or even agreed-to emissions targets. It states that "climate change is one of the greatest challenges of our time" and recognizes that "the increase in global temperature should be below 2 degrees Celsius." Participating nations are to set their own reduction targets for 2020. It proposes a "collective commitment" of $30 billion in payments from the developed to the developing countries for the period 2010-2012 and a goal of $100 billion per year by 2020 to flow through a new entity, the Copenhagen Green Climate Fund.[20] However, payment shares for each nation were not determined.

Secretary General Ban Ki-moon declared the accord to be a "significant achievement."[21] President Obama called it an "unprecedented breakthrough."[22] But much of the media was highly critical, such as the *Süddeutsche Zeitung*, which stated "The fight against global warming has been set back by years."[23] The effective result of Copenhagen was a commitment to try again at the global conference in Mexico City in December, 2010.[24]

ATTACK ON FREEDOM AND OUR WAY OF LIFE

So, in order to stop global warming, extreme changes are in order. Rajendra Pachauri, the current Chairman of the IPCC, lectures us:

> ...We have embarked globally on a path of unsustainable development. Our lifestyles, the way we produce goods and services, are all part of a system that is completely unsustainable. I see solutions to climate change leading to a much larger philosophical shift in the way human society develops. We need a new matrix to define what human progress is.[25]

Climatism intends to change our "unsustainable lifestyles," including our travel, consumer goods, homes, diets, and families. Change is required, either voluntary, or coercively if necessary. Let's look at some of these "solutions" that will personally affect each of us.

They'll Control Air Travel—Fly Less, but Pay More

Air travel is a visible target for climate activists. Why should consumers be able to travel so fast across the world, emitting so much carbon dioxide in the process? Netherlands introduced a "departure tax" in 2008 for passengers at Dutch airports, which added 11 to 45 Euros to the cost of each flight. The Dutch government recently suspended the tax because it caused a steep decline in air traffic at the major Dutch airports.[26] Lord Turner, chairman of United Kingdom's Parliament committee on climate change, anticipates future measures:

> In absolute terms, we may have to look at restricting the number of flights people take.[27]

The largest travel-tax plan has been proposed by the developing nations. More than 50 countries have proposed a mandatory tax on international flights and shipping fuel to raise over $10 billion. The amount raised would be distributed to the developing countries to help them fight climate change.[28]

They'll Control Your Automobiles—Drive Less, but Pay More

Gordon Brown, U.K. Prime Minister, promised in 2008 that all cars sold in the U.K. by the year 2020 would be electric or hybrid.[29] This is quite a promise. Of more than 2 million cars sold in 2008 in the UK, only about 15,000 were hybrid and less than 1,000 were electric.[30,31] Electric and hybrid models suffer from high price, small size, limited features, and in the case of small electric cars, safety issues. Hybrid and electric car global sales were about 500,000 in 2008, or less than 1% of the world market of 60 million,[32] despite the crude oil price shock which boosted diesel and gasoline prices.

Of course, since most British consumers won't voluntarily sacrifice for the good of the planet, Mr. Brown's government is in the process of enacting a large array of subsidies and restrictions to "encourage compliance." For example, the U.K. government recently announced it would soon begin educating motorists on "eco-driving techniques." Those passing their driving test will also be tested on whether they can drive in an environmentally-friendly way.[33]

The European Commission is on a crusade to impose restrictions on European car makers. In 2007, Stavros Dimas, the European Environment Commissioner, proposed legal requirements that "the average new car in 2012 emit no more than 120 grams of carbon per kilometer." This is estimated to add more than £3,300 to the cost of each vehicle.[34] Specialty car manufacturers such as Land Rover and Porsche are expected to be especially hard hit. The EC has also recommended automobile advertisements contain warnings on CO_2 emissions similar to those for cigarette advertising.[35] Cigarettes have been linked to cancer by rigorous double-blind tests, but no such proof exists for man-made emissions and global warming. But then, when did real data matter?

California is the leading U.S. state for activist global warming policies. AB32, the California Global Warming Solutions Act was passed 2006, requiring greenhouse gas emissions cuts, and empowering the California Air Resources Board (CARB) to develop the implementation plan. CARB has recommended or is considering a number of measures, including setting vehicle fees based on emission levels, charging for passenger-vehicle insurance by the mile, fees for travel on congested roadways, and other measures to reduce vehicle miles traveled and increase costs.[36] CARB has even considered regulating car colors and reflectivity, because darker colors absorb more heat. They also recently mandated that oil-change technicians inflate tires to proper pressure to boost fuel mileage.[37]

Possibly the biggest impact was the announcement by CARB of the world's first "low-carbon mandate" for transportation fuels, which will go into effect in 2011. Ethanol does not appear to be able to meet this tough new standard and fuel refiners do not have a cost-effective method to comply.[38] Billions of dollars in increased fuel costs are projected for California drivers.

They'll Control Your Appliances Purchases

Fifteen states in the U.S. have now set appliance energy-efficiency standards that are more stringent than U.S. Government standards. These standards cover washing machines, refrigerators, televisions, lighting, and other appliances.[39] The California Energy Commission (CEC) is considering regulations that will make many of today's large-screen TVs illegal. The CEC wants a 49% reduction in consumed big-screen power by 2013.[40]

Are these measures all bad? No, there is obvious benefit in using less power and reducing electric bills for consumers. New York estimates its efficiency standards will save consumers almost $300 million per year, while reducing carbon emissions by 870,000 metric tons.[41] However, such claimed benefits are illusory.

The "white goods" appliance industry is very competitive. Efficiency mandates invariably raise product costs, which more than cancel the efficiency savings. If manufacturers can save energy and meet efficiency demands from the consumer, they are already doing it. Bureaucrats, who have never run a business, believe they can mandate a standard and therefore save consumers money, but they are sadly mistaken. They assume consumers and manufacturers are ignorant and that government officials know better.

Buying energy-efficient appliances, dialing down thermostats, and saving money through energy conservation are worthwhile pursuits. But imposing regulations to reduce carbon dioxide emissions by 870,000 tons per year is counter-productive. Saving 870,000 tons of CO_2 plus a dollar *might* get you a cup of coffee.

They'll Tell You What to Eat and How Much to Weigh

Shame on those of us who eat meat. According to Climatism, production of meat is a major source of global warming. Some have called hamburgers the "Hummers of food," aligning beef with the much-maligned low-mileage vehicle.[42] A new group called the Veg Climate Alliance exhorts us to change our diet for the climate.[43]

In a letter to *The Independent*, Dr. Pachauri of the IPCC and musician Sir Paul McCartney, both vegetarians, have teamed up to urge the world to end our carnivorous habits:

> Unfortunately, with higher incomes, societies, even in developing countries, are turning to greater consumption of animal protein, which reduces the availability of food grains for direct consumption by impoverished human beings. Already 60 per cent of food crop production in North American and Western Europe is being diverted for production of meat.[44]

McCartney has joined other celebrities in a call for all to observe "Meat-Free Monday" to tackle climate change. Mr. McCartney's daughter, Stella, also a life-long vegetarian, identifies another villain:

> Whether you eat meat or not, you can be part of this decision to limit the meat industry destroying our planet's resources.[45]

It seems the meat industry has joined the coal industry as a target of Climatism.

Sweden is testing a new system of labels for grocery and restaurant foods, displaying the amount of "CO_2 per kg of product." These labels state that a chicken sandwich (0.4 kg of CO_2 emissions) is better for the planet than a hamburger (1.7 kg). Sweden is also imposing regulations on farmers to reduce emissions.[46]

Climatist scientists have determined that fat people have larger carbon footprints than thin people. Phil Edwards, lecturer at the London School of Hygiene and Tropical Medicine, finds in a 2009 study that it takes more energy to transport heavier people, requiring more fuel, which creates more greenhouse gases. The study also found that obese people consume more than thin people.[47] Now you can diet, lose weight, and also save the planet. As Albert Einstein said, "Only two things are infinite, the universe and human stupidity, and I'm not sure about the former."

They'll Regulate your Family

Once people become only carbon footprints, rather than special beings with a heart and a soul, it's a short step to limiting people to achieve greenhouse gas reductions. Dr. John Guillebaud at University College London and Dr. Pip Hayes, from Exeter, writing in the British Medical Journal, state:

> We must not put pressure on people, but by providing information on the population and the environments, and appropriate contraception for everyone...doctors should help bring family size into the arena of environmental ethics, analogous to avoiding patio heaters and high carbon cars.[48]

Sir Jonathon Porritt has warned "couples who have more than two children are putting an irresponsible burden on the environment." This is not a fringe scientist, but the Chairman of the Sustainable Development Commission for the United Kingdom. Sir Porritt says:

> I am unapologetic about asking people to connect up their own responsibility for their total environmental footprint and how they decide to procreate and how many children they think are appropriate.[49]

The Center for Environment and Population, a U.S. group focusing on "the U.S. population's environmental and climate change impacts,"[50] has issued a report, warning:

> Because Americans are high resource consumers in a country with a large, rapidly growing population base, the U.S. has a much bigger "per-person" impact on global climate change than any other nation.[51]

China implemented a one-child policy in 1979. In 30 years, an estimated 300 million live births have been prevented by the policy. Although denied by the Chinese government, the program also resulted in forced sterilizations, forced abortions, the killing of disabled, orphans, and female babies, and other coercive policies.[52] Now China points to the program as a success in the fight against global warming. Su Wei, a senior Foreign Ministry official in China's 2007 climate delegation said that avoiding 300 million births "means we averted 1.3 billion tons of carbon dioxide in 2005."[53]

Dr. Paul Murtaugh and Dr. Michael Schlax of Oregon State

University point out that reproductive choices have a larger impact on carbon dioxide emissions than lifestyle choices:

> Under current conditions in the United States, for example, each child adds about 9,441 metric tons of carbon dioxide to the carbon legacy of an average female, which is 5.7 times her lifetime emissions.[54]

Andrew Revkin, Environmental Reporter for the *New York Times*, has proposed giving carbon credits to couples that limit themselves to having one child. Revkin raises the question:

> ...probably the single-most concrete and substantive thing an American, young American, could do to lower our carbon footprint is not turning off the lights or driving a Prius, it's having fewer kids...we'll soon see a market in baby-avoidance carbon credits similar to efforts to sell CO_2 credits for avoiding deforestation...[55]

With such a high potential reduction in carbon emissions, it's only a matter of time before our elected officials dive deeply into population control to stop global warming. There are many implications to this population-limitation attitude. First, any couple choosing to have several children is "environmentally irresponsible." Marriage itself, as the basic unit for child rearing, now has a carbon footprint stain. Being single is climate beautiful. Abortion gets high praise for preventing carbon footprints. If Climatism is allowed to continue to grow, look for coercive population control measures even in the Land of the Free.

A partner to human population control is soon to be house-pet control. A recently published example is *Time to Eat the Dog? The Real Guide to Sustainable Living* by Robert and Brenda Vale.[56] There are more than 500 million house pets worldwide, including a U.S. total of about 200 million, and growing. Can the world tolerate the carbon footprint of our pets? Those of you with large, obese dogs that eat meat should be especially concerned.

RATIONING AND THE CLIMATE POLICE

If Climatism wins the day, it's likely many of these policies will not be optional, but enforced. George Monbiot recommends a framework to get each citizen to comply:

> Every citizen is given a free annual quota of carbon dioxide. He spends it by buying gas and electricity, petrol and train and plane tickets. If he runs out, he must buy the rest from someone who has used less than his quota.[57]

Dr. Justin Kenrick, sociologist at the University of Glasgow, called for carbon rationing for Scotland in 2008.[58]

The U.K. Environment Agency is establishing a team of about 50 inspectors to audit businesses. These climate police will wear green uniforms and carry audit cards with the power to search company premises to enforce the Carbon Reduction Commitment regulation, which comes into effect in 2010.[59] Maybe this is where all those green jobs will come from?

Think this won't happen in the U.S.? In California, the Berkeley City Council ran into opposition over its 145-page Climate Action Plan in April 2009. Among other measures, the plan called for every home to receive an independent review of its energy efficiency and for *homeowners to pay for upgrades*. City officials backed down under vocal protests from citizens.[60] But this failed attempt to force consumers to cut carbon emissions may be only a small setback.

SWALLOW THIS SNAKE OIL

The "snake oil" remedies of Climatism are wide-ranging, extreme, and hugely expensive. They call for a massive reorganization of our energy usage that will impact every aspect of our lives. If the climate alarmists succeed, government bureaucrats will reorder, regulate, and ration our transportation, work place, homes, diet, and families. Our freedoms and way of life will be impaired and diminished.

So get rid of your travel vacations, your SUV, your pets, your hamburgers, your big-screen TV, and your big family. Enjoy your bicycle, your tofu, your book on recycled paper, and your fluorescent lights. Then you can feel good about yourself. But science tells us this is all futile. Global warming is due to natural cycles of the Earth, driven by the sun. Climatist emissions-reduction remedies are useless. We'll look at the economic impacts of these policies in Part III.

Part III

Renewable Energy "Solutions"

"The people never give up their liberties but under some delusion."
Edmund Burke (1784)

CHAPTER 12

ENERGY:
THE LIFEBLOOD OF PROSPERITY

"Coal is a portable climate." Ralph Waldo Emerson (1860)

nergy is the lifeblood of prosperity. Application of energy and the use of improved fuels have driven mankind's industrial revolution, including our trains, planes, and automobiles, the age of computers, and the internet revolution. Every part of today's civilization in developed nations, including transportation, homes, diet, consumer goods, communications, and health care, depends upon reliable and concentrated sources of energy.

Climatism demands a massive societal change in how we use energy. We must, we are told, forego the use of hydrocarbon-based energy. Instead, wind power, solar power, and biofuels should be implemented across the board. As we will show in this section, these policies are possible only with huge government subsidies; they are economically destructive and a vast misallocation of resources.

Prosperity is strongly related to energy usage. Figure 89 (next page) shows income per person and energy use per person for major nations of the world. The citizens of the U.S. use the most energy per person and also have one of the highest incomes of the major nations. The developed nations of Europe and Japan also have high incomes and energy use. In contrast, the people of China, India, and other developing nations have markedly lower incomes and energy use. In other words, economic prosperity is strongly connected to energy use.

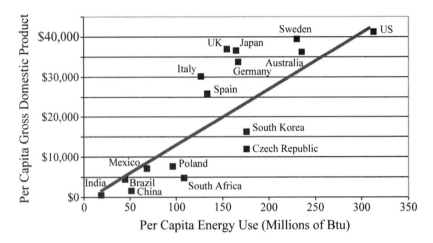

Figure 89. Prosperity and Energy Use. Plot of per person GDP and energy use for major nations in 2005. Prosperity is strongly correlated to energy use. (Adapted from Huber 2005, data from EarthTrends)[†]

As we discussed in Chapter 11, Climatism calls on developed nations to cut greenhouse gas emissions 50% to 85% by 2050. Developing nations are also called upon to constrain emissions. Unless cost-effective energy with low or zero emissions is available, adherence to these demands will stall the growth and reduce the prosperity of both developed and developing countries. It's astonishing that most nations of the world are intent upon sacrificing economic development for a false scientific belief.

THE HYDROCARBON REVOLUTION

Mankind has achieved an energy revolution during the last 300 years that could be called the "Hydrocarbon Revolution." The improving state of the world we discussed in Chapter 8 has been powered by use of hydrocarbon energy. Without the application and refining of hydrocarbon energy, the improvements in human prosperity, lifespan, and way of life would not have been possible. Three elements of energy use and refinement were essential to the Hydrocarbon Revolution: 1) the use of coal, 2) the refining of oil, and 3) the harnessing of electricity, a concentration of energy created primarily from hydrocarbon sources. A hydrocarbon, by the way, is an organic compound consisting entirely of the elements hydrogen and carbon. Natural gas and crude oil are our primary hydrocarbons. Coal is usually included as a hydrocarbon (and we will include it here), despite containing small amounts of oxygen and sulfur.

In the early 1600s, before the dawn of the Industrial Revolution and the age of modern energy, life was difficult in Europe. Society was plagued by smallpox and other infectious diseases, resulting in early childhood death. Life expectancy was about 30 years. Energy usage had remained mostly unchanged over the previous 1,000 years. Transportation was by horse-drawn vehicle. Grain was sown by hand and harvested by handheld sickle. Most food was grown locally, since it was difficult to transport and preserve. The difficult growing conditions of the Little Ice Age placed the population at the mercy of harvest failures and famines. Wood and coal was burned for cooking and heating. Vegetable- or whale-oil lamps and candles were used for light. Residents of London, Paris, and other major cities were required by law to hang a lamp out on the street at nightfall. Life was cold, dark, and short prior to mankind's use of hydrocarbons.[2]

Most historians place the birth of the Industrial Revolution in Great Britain during the second half of the 1700s and first half of the 1800s.[3] Two factors that enabled the Industrial Revolution were the use of mechanical machines and the use of hydrocarbon energy to power these machines. Development of steam-driven machines actually began at the end of the 17th century.

Thomas Savory, an English military engineer, is credited with inventing the world's first steam-powered engine. Patented in 1698, it delivered about one horsepower (746 watts) of power. Savory's engine, the first of its kind, was employed to pump water out of coal mines. However, the engine was not a success because its boiler was prone to explosions.[4]

Thomas Newcomen, an English blacksmith, improved on Savory's design and invented the first piston-based steam engine about 1710. The Newcomen engine was safe, delivered about five horsepower, and was able to pump water from coal mines with a depth of 150 feet. In one application in 1711, the engine replaced a team of 500 horses previously used to pump out a mine. Machines and energy were in and horses were on their way out.[5]

James Watt, a Scottish inventor and engineer, is regarded as the father of the modern steam engine. Watt improved on Newcomen's design by adding a second chamber, called a condenser, where the

---

Energy and Power Review

Energy is the amount of work that can be done by a force. Energy is measured in joules (J), megajoules (MJ: millions of joules), gigajoules (GJ: billions of joules) or British thermal units (Btu), a measure of heat energy. One Btu equals 1,056 joules. Total world energy usage can be measured in quads (quadrillion Btu). One quad equals 1,000,000,000,000,000 Btu.

Power is the amount of energy delivered per unit of time, measured in watts (W) or horsepower (hp). One watt is the amount of power in delivering one joule of energy per second. One horsepower equals 746 watts.

Electrical power is measured in megawatts (MW: millions of watts) and gigawatts (GW: billions of watts). Electrical energy is measured as power delivered over time, in kilowatt-hours (kWh), megawatt-hours (MWh), or gigawatt-hours (GWh).[6]

Some intuitive examples:
- A healthy person can briefly exert over one horsepower of power
- A weightlifter who lifts a 150-kilogram barbell over his head expends about 3,000 joules of energy
- A laborer can exert about 100 watts of power over an 8-hour day and daily expends 800 watt-hours or 2.88 megajoules of energy
- A 12-gallon (45-litre) tank of gasoline contains about 1.6 GJ of energy[7]

---

steam could cool. This allowed the piston chamber to remain hot, improving the cycle speed and efficiency of the engine and saving 75% on coal costs. Watt and his partner, Mathew Boulton, placed the new design into production in 1775 and sold it to coal mining companies. Watt also invented the horsepower unit of measure. A century later, the scientific community named the watt unit of power for him.[8] See the text box for a review of energy and power.

What made these inventions possible? Many factors, but a critical ingredient was the use of coal to fuel the new engines. Coal contains almost double the energy density of wood and provided the necessary power to run the new engines. In addition, while wood was traditionally burned to directly produce heat or light, coal was now used to create steam. The energy of coal, transformed into the gas pressure of steam, provided the power.

The history of hydrocarbon energy and technology is very much a "tag-team" between mechanization and energy, with advances in each field enabling advances in the other. This is described in *The Bottomless Well* by Peter Huber and Mark Mills.[9] The story of steam engines and coal is a perfect example. Early coal mining provided the power to enable the steam engines of Savory, Newcomen, and Watt. Their steam engines were first used to improve coal mining, which reduced the price of coal. Coal was then used to drive the Industrial Revolution in iron smelting and the development of iron and steel machine tools. Improved raw iron and metal tools enabled engineers to design better steam engines.

Near the end of the first phase of the Industrial Revolution, coal power was applied to transportation. The world's first steam-powered railway was the Stockton & Darlington Railway, which opened in England in 1825.[10] However, transportation required a fuel with better energy properties to enable lighter and smaller vehicles.

While the steam engine was moving into commercial use, the internal combustion engine that powers today's automobiles was moving along a slower development curve. The internal combustion engine operates by quite a different principal than the steam engine. Your car is actually powered by a series of controlled explosions of a "ready-to-explode" fuel-air mixture injected into the piston chamber. While gasoline engines use a spark plug ignition system, diesel engines rely only on the heat and pressure of the engine itself to ignite the fuel-air mixture. It is this fine mixing of air and fuel down to an atomic level that creates the explosive power. One ton of crude oil, when used as gasoline, releases about ten times as much energy as a ton of the explosive TNT.[11]

About 1780, the Dutch scientist Christian Huygens built an engine that used gunpowder as fuel. But the invention was too dangerous to be practical. Then in 1824, French physicist Nicholas Carnot published a book describing the principles of the internal combustion engine.[12] Another Frenchman, Jean Joseph Étienne Lenoir, built the world's first working two-cycle internal combustion engine in 1859. The engine used gas from coal as fuel, along with a battery to ignite the gas in the combustion chamber.[13]

During the same period, inventors around the world attempted to design vehicles powered by steam from coal, coal gas, and hydrogen, but designs were hampered by lack of a suitable fuel. The energy with properties needed to power the internal combustion engine was to come from refined crude oil, the second major element in the Hydrocarbon Revolution. Colonel Edwin Drake is recognized as the pioneer of oil exploration. In 1859, he sank a well 69 feet to strike oil at Oil Creek, Pennsylvania, and launched the U.S. petroleum industry. The Standard Oil Company was established in 1870 by John D. Rockefeller. The company built its first refinery in Pennsylvania to turn crude oil into kerosene, its first major product.[14]

Karl Benz, a German engineer, is credited with producing the first practical gasoline-fueled automobile in 1885, a three-wheeled design.[15] Two other German inventors, Gottlieb Daimler and Wilhelm Maybach followed with the first four-wheeled design in 1886.[16] These and future automobiles ran on gasoline, a high-powered liquid fuel refined from crude oil.

The third major element of the Hydrocarbon Revolution was the invention of electrical power. Electricity is not an energy source by itself (unless we learn to harness lightning), but is a concentrated form of energy produced from coal, natural gas, oil, or non-hydrocarbon sources such as hydropower, nuclear, solar, or wind. Electricity is our most flexible and valuable energy resource, able to power everything from the smallest microprocessor, to fiber optic communications, to a hospital's intensive care unit, to the lighting of our largest cities. Today, our information age continues to advance, courtesy of electricity produced from hydrocarbon fuels.

Thomas Edison switched on the world's first commercial electrical power service in 1882, delivering electricity to 59 customers in New York City from his Pearl Street Power Station. Electricity was developed by six huge coal-burning steam engines that each powered a "Jumbo Dynamo" generator with an output of 100 kilowatts.[17] In today's dollars, Edison's electricity cost about $5 per kilowatt-hour, compared to ten cents per kWh for today's U.S. customers.[18]

A good example of the Hydrocarbon Revolution and the use of coal, petroleum, and electricity is the energy history of the United States. Figure 90 shows U.S. energy consumption and the growth of hydrocarbon fuels. In 1850, total U.S. energy usage was about two quads (quadrillion Btu), consisting almost entirely of wood (and some animal muscle power). This usage grew 50 times to 101 quads by 2007. Coal surpassed wood in the 1880s, becoming the primary fuel for steam engines (including railroads), industry, residential and business heating, and early generation of electricity. Wood consumption declined to a low in 1950, concurrent with re-growth of U.S. forests. The upturn of the "wood and biofuels" curve in Figure 90 is due to the increased use of ethanol vehicle fuel during the late 1900s. Petroleum use began in the late 1800s and rose rapidly as the

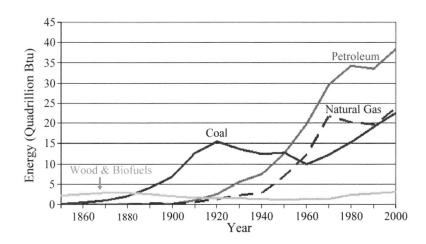

Figure 90. United States Hydrocarbon Energy Consumption 1850-2000. U.S. consumption of hydrocarbons and wood/biofuels, measured in quads (quadrillion Btu). Nuclear, hydroelectric, wind, and solar totaled 11 quads in 2000, but are not shown. (U.S. Census Bureau, 2009)[19]

primary fuel for the U.S. automobile boom, but also was used initially for residential heating and industry. Today, over 70% of U.S. energy from oil is for transportation.

Natural gas usage rose rapidly in the mid-1900s, becoming the primary fuel for residential heating and many industrial uses. Gas has high energy density and can be piped directly to residences and commercial buildings. This provides a major convenience advantage over oil and coal, which are delivered by truck. As a result, natural gas replaced oil and coal in the majority of heating applications.

Electricity was the fastest growing form of energy consumption in the United States in the second half of the 20th century, growing from about five quads to 40 quads from 1950 to 2008. As electricity demand increased, so did the use of coal and natural gas, today meeting 49% and 21% of the electricity generation need, respectively (Figure 91, next page). Nuclear power generation of electricity began in 1957. Nuclear has grown to supply 20% of U.S. electricity demand, despite concerns about safety and waste disposal. Renewable fuel use remains small, supplying only 9% of electricity demand in 2008, with two-thirds of renewables from hydroelectric power sources.[20]

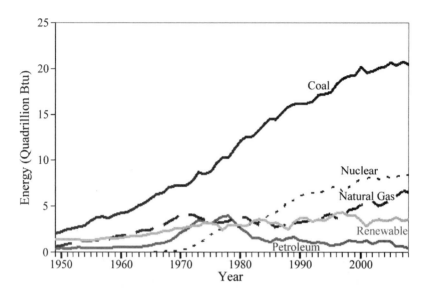

Figure 91. United States Electricity Production by Fuel 1949–2008.
Electricity production is measured in quads. Renewables include hydroelectric,
biomass, geothermal, wind, and solar. (DOE/EIA, 2008)[21]

WORLD ENERGY USAGE

A breakdown of world energy usage in 2006, according to the
International Energy Agency, is shown in Figure 92. Total world use
of energy is similar to that of the U.S., with oil the most used,
followed by coal and gas. The "combustible renewables and waste"
wedge captures the wood and dung fuel that is still used by much of
the developing world. Note that solar and wind power is in the other
category, much less than 1% of world usage. In Chapter 3, we
learned that carbon dioxide was a trace gas comprising very much less
than 1% of our atmosphere. *Climatism is asking us to solve the "very
much less than 1% CO_2 problem" with vast deployments of wind and
solar solutions that today are less than 1% of the world energy supply.*

Climatists demand that we eliminate the 26% of the world's
energy supplied by coal. James Hansen labels coal-burning power
plants "factories of death."[22] Figure 93 shows that many of the world's
largest nations rely on coal for electrical power. A change to other
fuels would entail major economic costs for these countries.

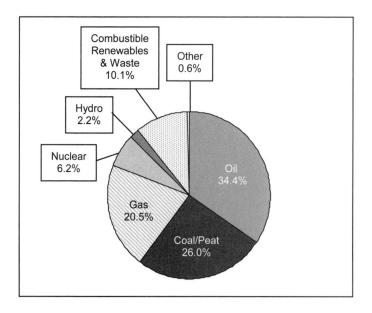

Figure 92. 2006 World Total Primary Energy Supply. "Combustible renewables and waste" includes wood and dung fuel used by much of the developing world. "Other" contains wind, solar, and geothermal. (IEA, 2008)[23]

Country	Percent of Electricity from Coal
South Africa	97%
India	91%
People's Republic of China	86%
Australia	85%
United States	53% (49% by 2008)
Germany	51%
United Kingdom	41%

Figure 93. Electricity from Coal in 2006 for Selected Nations. Major nations rely heavily on coal for electrical power. (Data from IEA, 2008)[24]

In addition to eliminating coal, Climatists demand that we curtail the 34% of the world's energy supplied by oil. Most environmental groups also oppose nuclear power and many oppose the use of hydroelectric power. That sums to approximately 70% of today's world energy usage. The 20% of world energy provided by natural gas also emits carbon dioxide, but at a lower rate than coal and oil.

Mankind didn't just choose hydrocarbons by chance, or make a wrong choice among alternatives. Coal was the lowest-cost, highest-energy alternative to produce steam and electricity. Along with nuclear, it *remains* the lowest-cost fuel for electricity today. Petroleum was refined into diesel fuel, gasoline, and aviation gas: liquid fuels of high-energy density, able to make an explosive mixture with air to power the internal combustion engine and later the aircraft turbine. Natural gas is a high-energy density hydrocarbon most easily piped to buildings, becoming the preferred fuel for home and industrial heating. Can the renewable energies, wind, solar, and biofuels, really replace these proven hydrocarbon fuels?

ENERGY USE GROWS DESPITE ATTACKS

The oil industry is a favorite whipping boy for U.S. political leaders and consumers. Gasoline prices are the most visible in the economy, with a sign on every street corner. Whenever crude oil prices rise, forcing price rises at the gas pump, Congress calls for a "windfall profits tax" and another price-fixing investigation of the oil industry

But, in fact, the energy industry has been seriously under attack for decades (text box). According to overpopulation and anti-affluence alarmists, the larger population and energy growth, the

---

The Ideological Attack on Energy

"If you ask me, it'd be little short of disastrous for us to discover a source of clean, cheap, abundant energy because of what we would do with it."
—Environmentalist Amory Lovins[25]

"The prospect of cheap fusion energy is the worst thing that could happen to the planet." —Jeremy Rifkin, Greenhouse Crisis Foundation[26]

"Our insatiable drive to rummage deep beneath the surface of the earth is a willful expansion of our dysfunctional civilization into Nature." —Al Gore, *Earth in Balance*[27]

"Giving society cheap, abundant energy would be the equivalent of giving an idiot child a machine gun." —Paul Ehrlich[28]

larger the environmental damage to Earth. These ideologues advocate control of greenhouse gases to suppress world energy usage.

Thirty years ago, some predicted that electricity growth had come to an end in the U.S. In 1980, the Union of Concerned Scientists (an organization later to adopt Climatism) announced:

> Because saturation levels for most major appliances are achieved, only minor increases in electricity consumption [will] occur.[29]

Environmentalist Amory Lovins stated in 1977:

> Many analysts now regard modest, zero, or negative growth in our rate of energy use as a realistic long-term goal.[30]

The Union of Concerned Scientists, Lovins, and others advocated "improving efficiency" to move to zero energy growth. But they failed to imagine personal computers and the internet revolution. U.S. electricity usage grew 80%[31] from 1970 to 2007 and total energy use was up 30% over the same period.[32]

Global energy usage grew at an even faster rate (Figure 94). According to the International Energy Agency, from 1973 to 2006, world total primary energy supply grew 92% and world electricity supply *more than tripled.*[33] Forecasts for the future show more growth

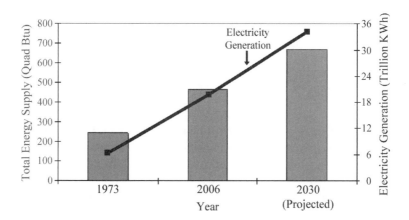

Figure 94. Growth in World Energy and Electricity Use. Actual growth from 1973 to 2006 and projected for 2030. Columns are total energy supply (left axis). Line is electricity generation (right axis). (EIA, 2008 and U.S. DOE, 2009)[34]

ahead. From 2006 to 2030 total energy usage is expected to increase another 44% with electricity generation increasing an additional 77%.[35] Indeed, the International Energy Agency warns in its 2009 study, *Gadgets and Gigawatts–Policies for Energy Efficient Electronics*, that the growth in information technology could "ruin attempts to halt global warming."[36] It doesn't appear we will reach energy "saturation levels" any time in the near future.

WORLD ENERGY RESERVES

A number of organizations in the energy field project that the world will run out of hydrocarbon energy in the very near future. The 2008 oil price shock added weight to their point of view. A 2007 report by Dr. Werner Zittel of the Energy Watch Group, a German organization, warns:

> ...world oil production has peaked in 2006...The world is at the beginning of a structural change of its economic system. This change will be triggered by declining fossil fuel supplies and will influence almost all aspects of our daily life.[37]

The organization goes on to criticize the International Energy Agency:

> The International Energy Agency, anyway until recently, denies that such a fundamental change of our energy supply is likely to happen in the near or medium term future.[38]

It's interesting to note that the Energy Watch Group also supports the Climatist view:

> Climate change will also force humankind to change energy consumption patterns by reducing significantly the burning of fossil fuels. Global warming is a very serious problem.[39]

"Peak oil" is a term originally used by Marion King Hubbert in 1956, to describe a bell-shaped curve for U.S. oil production. Colin Campbell, founder of the Association for the Study of Peak Oil and Gas (ASPO), defines peak oil as:

...the maximum rate of the production of oil in any area under consideration, recognizing that it is a finite natural resource, subject to depletion.

The ASPO believes a gap in demand between world oil supply and consumption will appear between 2010 and 2015. The ASPO also advocates the global warming views of Climatism.[40]

Is the world about to run out of oil and gas? Maybe, but skepticism is in order. Just as anti-energy forces tried to persuade us that demand had peaked in 1980, many organizations now forecast we will run out of hydrocarbon energy in the short term. These "peak oil alarmists" also forecast scenarios of economic collapse that sound strangely familiar. Dave Cohen, writing for ASPO states:

Although it is impossible to predict the future, extrapolating present trends...leads to "The Perfect Storm" sometime in the next decade. At the *tipping point*, oil prices exceed the pain tolerance of a sufficient number of global consumers, causing economies to roll over into severe recession [our emphasis].[41]

Mr. Cohen found another *tipping point* for catastrophe, this time due to oil depletion. He must be talking to James Hansen.

Huber and Mills point out that mankind's technology for producing oil continually improves to keep up with the difficulty of accessing oil from more remote locations. They point out:

Over the long term, the price of oil has held remarkably steady, even as the distance from the well-head to the oil has increased from hundreds of feet to miles. Today's production costs in the deep waters of the North Sea are not very different from costs in southeast Texas a century ago.[42]

Deep water wells are now drilled in a water depth of over 8,000 feet (2,440 meters). Well depth on land can reach down to 30,000 feet (9,150 meters) or almost six miles.[43] Horizontal wells can be drilled to a distance of more than three miles from the vertical shaft.[44] Ninety percent of the wells drilled in Prudhoe Bay, Alaska are horizontal.[45] Improvements in drilling technology have allowed energy companies to access hard-to-reach oil to meet market demand.

Whenever we start talking about projections and catastrophic scenarios, it's valuable to look at trend data. Surprisingly, it appears that *world proven reserves of both oil and natural gas are rising faster than consumption* (see figure 95). In fact, in 1980 the world had about 28 years of oil supply, based on proven reserves and the annual usage rate of 63 million barrels per day. By 2008, this had grown to 43 years of supply at the higher oil consumption rate of 85 million barrels per day. Similarly, despite higher annual consumption, world proven reserves of natural gas increased from about 49 years of supply in 1980 to 58 years of supply in 2007.

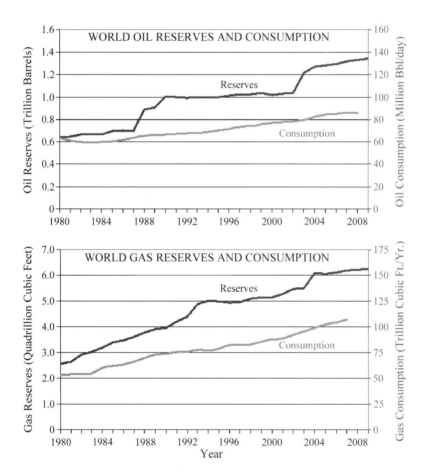

Figure 95. World Oil and Gas Reserves and Consumption 1980–2009. The top chart shows world oil proven reserves (left axis) and daily consumption (right axis). The bottom chart shows world natural gas proven reserves (left axis) and annual consumption (right axis). (DOE, 2009)[46]

How can world proven reserves be rising faster than consumption if we are running out of energy? Alarmists claim that world reserve totals are inflated because energy companies and nations have an incentive to over-estimate reserve totals. OK, but didn't the same incentive to over-estimate exist in 1980? Beware when energy alarmists who trust Climatism warn of energy disaster scenarios.

Indeed, it appears that yet more oil is available for world use. Early in 2009, Dr. Chuanmin Hu, an oceanographer at the National Oceanic and Atmospheric Administration (NOAA), was looking for algae blooms in the Gulf of Mexico using satellite images. He and his team discovered long black streaks on the photographs (Figure 96). They determined these to be petroleum slicks from natural seepage from the ocean floor. Dr. Hu found similar streaks on more than 50 photographs of the same area taken from 2000 to 2008. To show up on satellite photographs, these slicks must be fairly large, indicating a sizable amount of oil to be accessed.[47]

For many years, oil and gas have been bubbling up from the floor of the ocean near Santa Barbara, California, inaccessible because of U.S. Federal and State drilling bans. Scientists have now estimated

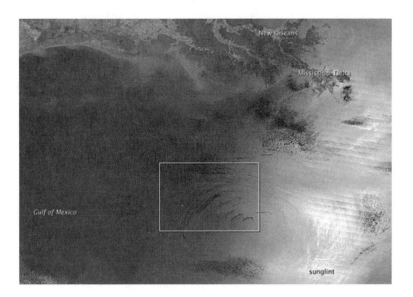

Figure 96. Natural Oil Slicks in the Gulf of Mexico. Photograph of the northern Gulf of Mexico, with the southern coast of Louisiana at the top. The dark streaks in the inset are petroleum slicks from natural seepage from the ocean floor. (NASA Earth Observatory, 2009)[48]

the flow at 70,000 barrels of petroleum and 3 billion cubic feet of natural gas per year. Every four years, the seepage from this one source of oil equals the 280,000 barrels spilled in 1969 when the Exxon Valdez ran aground in Prince William Sound, Alaska.[49]

Coal is the other of the three major hydrocarbon energy sources. Alarmists sometimes claim we need to build wind and solar farms because we are at "peak oil." But less than 5% of the electricity of the world is generated by oil. Coal is by far the largest fuel source for the world's electricity. Until electric car fleets grow to a significant size, the need for wind and solar power must be compared to coal.

The world has vast reserves of coal. The U.S. Department of Energy estimates 2007 world coal usage at 7.1 million short tons (907 kg per short ton) and 2005 world proven reserves at 929 million short tons.[50] This is 131 years of supply at 2007 world rates of usage, with more recoverable coal likely to be discovered during the next century.

SOME BASICS ABOUT ELECTRICAL POWER

There are many misconceptions about today's electrical power system. Some say that if you need more electrical power, you just put another outlet in the wall. Others believe we can easily store electrical power. It's important to discuss some basics of electrical power generation and delivery.

Three main components of a simple electrical power system are shown in Figure 97. Electricity is generated at the electrical power plant. In almost all cases, electricity is created by a spinning electrical

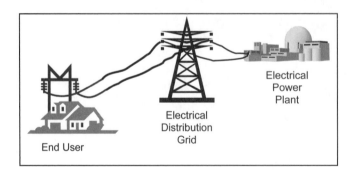

Figure 97. Basic Electrical Power System.

generator at the plant. Steam created by burning hydrocarbon fuels usually drives the generator. Heat from a nuclear reactor, falling water in a hydroelectric dam, or wind is also used. Photovoltaic solar systems directly generate voltage and convert this to electrical power.

Generated electricity is stepped up to very high voltage (hundreds of kilovolts) and delivered to the electrical distribution grid. Power is moved through the grid at high voltage to minimize transport losses. Distance to the end user is typically 100 miles, with a usual maximum distance of 300 miles (483 km) to reduce transmission losses. Transformers on the user end step down the voltage to 120 or 240 volts, depending upon local electrical network standards.

Note that there is no storage element shown in the network of Figure 97. There is no cost-effective large-scale storage system like your flashlight battery. *Electricity is generated as it is needed.* Therefore, power output from the electrical power system must be continuously adjusted to meet the demands of end users. When you switch on a light, you add "load," or demand, to the electrical system. The system must have enough instantaneous generating capacity to deliver power simultaneously to your house and to all other users. If the total load exceeds the generating capacity, a "brownout" occurs, resulting in a fall in voltage across the distribution network. Such a lowering of voltage can damage computers, televisions, or other sensitive equipment throughout the network.

Because electrical power is not stored, electrical networks are designed for peak power demand. Peak demand varies each day and by season and location. Peak demand seasons are winter in the United Kingdom and Australia, but summer for much of the United States. Electricity demand within a season often varies by 50% from minimum to maximum. The need to meet peak power demand requires utility system designers *to know exactly how much power can be absolutely relied on* from the generating plants at their disposal.

Computerized control centers are used to monitor system load and to bring generators on line to meet increased demand. In the U.S, 150 Control Area Operators coordinate efforts to meet the needs of the distribution grid.[51] Operators use power plants in three different roles to meet the demand: baseload power, intermediate power, and peaking power.

Baseload-power plants are run 24 hours per day to meet the minimum power need. These plants usually run at full power and deliver the lowest cost electricity to the network. Coal and nuclear plants are used as baseload plants because they are the lowest cost and most stable at full power. Intermediate-power plants are also called "load-following" plants, since they are placed on- or off-line to meet changing demand. Gas-turbine plants are often used for intermediate power. Peaking-power plants are older coal-, gas-, or oil-fired plants that are brought on-line only in peak-demand situations. Peak-power plants usually deliver the most expensive power in the system, so they are used as a last resort.[52] Hydroelectric-power plants are a flexible source of power that can be used in a baseband-, intermediate-, or peaking-power role, depending upon the amount of water stored in the reservoir of the hydropower dam.

There are actually many methods to store energy for generating electricity, but they are currently all very expensive for large-scale electric utility systems. Probably the most cost effective is pumped-storage hydroelectricity. In this case, an intermittent source such as wind is used to pump water from a low elevation to a higher elevation, such as into a dam reservoir. Water from the reservoir can be then released continuously to provide reliable electrical power. This method adds 25% to 50% to cost of a non-stored system and requires a dam and pumping system available to store energy.

ENERGY SUMMARY

Energy, and especially hydrocarbon energy, has been the lifeblood of mankind's economic prosperity for the last 300 years. Without the effective use of the coal, petroleum, and electricity of the Hydrocarbon Revolution, our modern information technology-based society would not have been possible.

World energy consumption continues to grow steadily, with electricity the fastest growing segment. Of world energy production today, 81% is hydrocarbon-based, while wind and solar provide less than 1%. Many warn of a near-term decline in oil production, but trends show that the years of supply of oil and gas are increasing. In addition, the world has more than a 100-year supply of coal.

Climatism calls for us to forego hydrocarbon use and switch to solar, wind, and biofuel alternatives. Let's see if these sources can carry the load.

CHAPTER 13

RENEWABLE ENERGY:
REALITY FAR SHORT OF PROMISES

"Facts are stubborn things; and whatever may be our wishes, our inclinations, or the dictates of our passions, they cannot alter the state of facts and evidence." John Adams (1770)

A l Gore proclaims that wind and solar power are "free." The Union of Concerned Scientists states:

All the energy stored in Earth's reserves of coal, oil, and natural gas is matched by the energy from just 20 days of sunshine.[1]

Even the U.S. Department of Energy says:

Every minute the sun bathes the Earth in as much energy as the world consumes in an entire year."[2]

It's great Earth is getting so much sunshine, but *such statements are meaningless.*

Here are two other assertions that are true, but about as useful:

- About six million tons of gold is dissolved in the ocean, enough to give every person on Earth a kilogram.[3]

- The Antarctic Ice Sheet contains 25 million cubic kilometers of fresh water, enough to supply the world with fresh water for 6,000 years at current usage.[4,5]

We've now solved the energy crisis, world poverty, and the water shortage. But, like the solar energy assertions, the benefits of these examples *cost more to deliver than their value.*

Wind and solar energy have two *major* deficiencies. First, they are *dilute.* Recall that gasoline, when burned as fuel, delivers ten times the energy of the equivalent weight of TNT. Petroleum, natural gas, and coal are *concentrated* forms of energy. For solar and wind sources to deliver significant levels of power, acres and acres of solar cells or thousands of wind towers are required. Without massive government subsidies, these energy sources can't compete with hydrocarbon fuels.

The second major deficiency is that wind and solar are *intermittent.* Two-thirds of the time, the wind doesn't blow hard enough to allow wind turbines to generate significant power. Solar power systems do not deliver power at night or on cloudy days. But electricity generation is not a part-time job. Recall from last chapter that electrical power grids do not store energy, but must deliver it immediately as users demand. Until low-cost energy storage systems are developed, the intermittent nature of wind and solar power is unsuitable for large-scale electricity generation. In fact, our best energy storage is in the chemical bonds of hydrocarbon fuels.

Therefore, wind and solar power are anything but free. They must be evaluated on their ability to deliver cost-effective power to the user. The economics show that wind and solar industries exist only because of enormous subsidies from governments of the world, driven by the misguided idea that global warming can be halted.

Governments of the world, infected by climate madness, compete with each other to see who can install the most green energy. Contrary to good economics, nations subsidize construction of thousands of 300-foot-high wind-turbine towers in the fruitless quest to cut carbon dioxide emissions. Developers pave acres of land with solar cells to deliver energy at a price *several times higher* than reliable hydrocarbon alternatives. Farmers divert vast areas of prime cropland to produce ethanol and biofuels that are less efficient than gasoline, while at the same time raising world grain prices. Let's look at these policies in detail.

SOLAR ENERGY: PERPETUAL 1% SOLUTION

A headline from the August 22, 1978 *Wall Street Journal* proclaims:

> Solar Power Seen Meeting 20% of Needs By 2000; Carter May Seek Outlay Boost[6]

Such articles are typical of many past predictions for a solar-powered world (see text box). Yet, despite the rosy predictions, in 2008 solar power produced *only 0.09%* of all energy consumed in the United States and generated *only 0.2%* of all electricity.[7,8] World use of solar power energy is also far less than 1% of total energy used.

Sunlight arrives in very dilute form. As we mentioned in Chapter 5, about 1,366 watts per square meter of solar energy enters the top of our atmosphere. On a clear day, when the sunlight is perpendicular to Earth's surface, about 1,000 watts per square meter reaches the surface after absorption and scattering by the atmosphere. For latitudes of Southern Europe and the U.S., this is reduced to about 800 watts at midday, since the angle of the sunlight is not quite perpendicular.[9] The efficiencies of solar systems are about 10% to 20%. After subtracting some power transmission losses, delivered power is about 100 watts per square meter. This means that only a single 100-watt light bulb can be powered for every card table-sized surface area of a solar-power system, and only at noon on a clear day.

For a lesson in solar reality, let's look at the Queanbeyan Solar

---

A Few of the Many Failed Solar Forecasts

"The private sector can be expected to develop improved solar and wind technologies which will begin to become competitive and self-supporting on a national level by the end of the decade if assisted by federally sponsored R&D." –Booz Allen & Hamilton, 1983[10]

"...the consensus as far as I can see is after the year 2000, somewhere between 10 and 20 percent of our energy could come from solar technologies, quite easily." –Scott Sklar, Solar Energy Industries Association, 1986[11]

"Within a few decades...the United States might get 30 percent of its electricity from sunshine..." –Christopher Flavin, Worldwatch Institute, 1990[12]

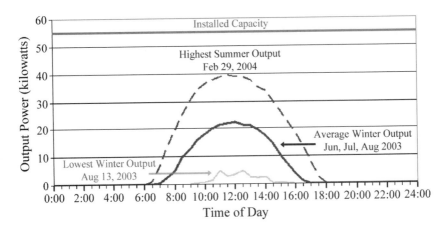

Figure 98. Output Power from Queanbeyan Solar Farm. Output power at each hour of the day for June–August, 2002 winter average and also a day of highest summer output and lowest winter output. (Adapted from Lang, 2009)[13]

Farm, located in Southeast Australia near Canberra. Queanbeyan is a small solar facility built in 1999. The farm uses 720 fixed solar panels with a total installed capacity of 55 kilowatts. But since sunlight varies greatly by time of day, season, and cloudiness, average delivered power was only 7.6 kilowatts during the typical year of 2003–2004, just 13.7% of installed capacity.[14] This average output, as a percentage of total potential output (13.7%), is also called the "capacity factor." Figure 98 shows a graph of average system output power by time of day. Due to low sunlight angles at morning and evening and no sunlight at night, the system delivers only 11 kW for *a maximum of six hours per day* during the winter. According to Country Energy, the owning utility, Queanbeyan produces 100,000 kilowatt-hours of power annually, enough to power "20 energy-efficient Australian homes."[15] However, the farm actually produces 66,400 kWh, enough to power about ten *normal* Australian homes and only between the hours of 0900 and 1500. This solar miracle costs $A50,000 ($US46,000) for each "6-hour" home.[16] However, it saves "100 tonnes of greenhouse gases annually," so our Climatist friends are pleased. But science shows that the CO_2 emission reduction isn't worth so much as "one shrimp on the barbie."

There are three points to note here. First, solar power doesn't work at night or at early or late hours in the day. Second, to find real delivered power for a solar system, divide installed capacity (also called "nameplate power") by a factor of five to seven. Third, the intermittent power of solar is very expensive.

Recall our discussion on baseload- and peak-power plants from last chapter. Solar plants cannot be used as baseload plants because they can't run around the clock, nor do they produce power at low cost. But using solar for peak load is also problematic. Peak needs are often in the evening, when solar power is nowhere to be found.

The solar industry developed several innovations to improve capacity factor and reduce the intermittent nature of solar power. Figure 99 shows both a parabolic trough design and a concentrating solar plant with a central tower. The parabolic trough uses reflective surfaces to concentrate sunlight on a central tube and heat a fluid flowing through the tube. The central tower system reflects sunlight from a large field of mirrors to heat a fluid in the top of the tower. Both systems use motors to tilt the reflecting surfaces to track the sun as it moves across the sky. Both systems use heat from the fluids to create steam to drive electrical generators. The fluid is also used as a limited storage system with three to six hours of heat storage capacity. These innovations add improvements, but also add cost. The parabolic trough system is generally considered to be commercial, while the central tower system is still viewed as experimental.

The Solar Energy Generating Systems (SEGS), consisting of nine parabolic trough systems, were built in California's Mojave Desert from 1985 to 1991. Eight of the systems are still operating. Together, these form the largest solar electricity generating plant in the world. Each of the eight remaining systems has an auxillary natural gas boiler

Figure 99. Innovations in Solar Power Plant Design. The left photo shows a parabolic trough solar field. The right shows the Barstow, California Solar Two concentrating solar plant with a central tower. (Desertec-UK, 2009)[17]

to provide power when solar is not available. For most of these facilities, natural gas generates 25% of the power. The nine systems in California have an installed capacity (nameplate power) of 354 megawatts, for a rated energy of just over 3,000 gigawatt-hours per year. Actual delivered electrical energy is about 600 megawatt-hours per year, or 20% of installed capacity.[18]

REALITIES OF SOLAR POWER ECONOMICS

Despite improving the concentration of sunlight and adding limited storage and gas-fired backup, solar power doesn't compete well with other electrical power sources. Figure 100 compares the SEGS solar power system with the Diablo Canyon Nuclear Power Plant, both located in California. The SEGS system has an installed power of 354 megawatts, compared to 2,240 MW for the two co-sited Diablo plants. But because Diablo runs at over 94% of rated power compared to 20% for SEGS, the Diablo facility delivers an average power of 2,121 MW compared to only 75 MW for SEGS.[19] Electricity cost per kilowatt-hour is difficult to accurately measure, given government subsidies and cost of capital over the life of the plant. But the Institute for Energy Research estimates actual costs at $0.11 per kWh for nuclear and $0.26 for solar thermal, for power plants to be placed into operation in 2016. According to the Institute, natural-gas and coal-fired plants are even less expensive than nuclear.[20] This makes electricity from a new solar-thermal plant of a design like SEGS more than twice as expensive as nuclear or hydrocarbon systems.

But, the most striking comparison is in the land area required. To deliver the same amount of electrical power, an SEGS-type system requires *sixty times* the land of the Diablo nuclear facility. This computes to 180 square kilometers or 70 square miles for the solar power option. This is the size of a square of land more than eight miles on a side. To match the Diablo plant, a solar plant would need to blanket an area the size of Madison, Wisconsin,[21] or the metropolitan area of Milan, Italy,[22] with concentrating solar troughs.

Luz International was the California company who built the nine SEGS solar systems. According to some, Luz "represented a successful merger of business interests and social responsibility." In 1991, Luz

	SEGS I-IX Solar (and Gas) Plants	Diablo Canyon Nuclear Plant
Location	Mojave Desert, CA	San Luis Obispo, CA
Entered Service	1986–1991	1985–1986
Installed Capacity	354 Megawatts (75% solar, 25% gas)	2240 Megawatts
Actual Average Delivered Power	75 Megawatts (21% of Capacity)	2121 Megawatts (94.7% of Capacity)
Site Footprint	1600 Acres (6.5 sq. km.)	750 Acres (3.0 sq. km.)
Approximate Electricity Cost	$0.26 per Kilowatt-hour	$0.11 per Kilowatt-hour

Figure 100. Solar and Nuclear Plant Comparison in California. Details for nine SEGS parabolic trough solar plants (left photo) and two co-sited Diablo Canyon Nuclear units (right photo) in California. Solar requires 60 times the land of nuclear and delivers electricity at more than twice the price.[23,24]

filed for bankruptcy. Cancellation of federal tax credits and lower competing energy prices caused the failure.[25] In short, Luz's solar business model could not compete. The SEGS systems were purchased at discounted prices and are run today by Sunray Energy and Florida Power & Light.[26]

Nevertheless, California is pushing forward with mandates to boost the use of wind and solar power. The state imposed a moratorium on nuclear plant construction 33 years ago, but still gets about 15% of its electricity from existing nuclear plants like Diablo.[27] About 32% of power is imported, primarily from low-cost coal plants like the Navajo Generating Station in Arizona. For example, the Los

Angeles Department of Power and Water now delivers 76% of its electricity from coal and natural gas, including imports of 40% from coal plants outside the state. However, Mayor Antonio Villaraigosa announced that Los Angeles will boost its power by renewables from 20% in 2010 to 40% in 2020.[28]

Why are Californians willing to pay double the price of hydrocarbon and nuclear for renewable energy? It's because they've adopted the projections of global warming catastrophe. In March 2009, the California Energy Commission paid for a study from the Pacific Institute for the California Climate Change Center. According to the study:

> ...general circulation model scenarios suggest very substantial increases in sea level as a significant impact of climate change over the coming century...Under medium to medium-high greenhouse-gas emissions scenarios, mean sea level along the California coast is projected to rise from 1.0 to 1.4 meters (m) by the year 2100...We estimate that nearly $100 billion...worth of property...is at risk of flooding.[29]

Actually, 1.4 meters is only five inches per decade, so even with all the alarming talk, it seems like normal seawall-shoring efforts can handle the model-projected sea level rise. Note that overpopulationist Anne Ehrlich is a director at the Pacific Institute.[30]

The Union of Concerned Scientists, famous for their misguided 1980 prediction that "only minor increases in energy consumption will occur" is also an advisor to the California Climate Change Center. With advice from the Union, the Change Center released the study "Our Changing Climate" in 2006, which warned of severe heat, reduced mountain snowpack, rising seas, increased air pollution, poorer agricultural conditions, and worsening forest fires.[31] Oceanfront property in San Francisco must be a bargain.

Despite the poor economics and because of climate alarmism, governments across the world are providing massive subsidies to stimulate solar generation projects. A recent example is the announced solar voltaic plant in the West Pullman neighborhood in Chicago. This photovoltaic system will directly convert sunlight to electricity, like the Queanbeyan Solar Farm in Australia. But while Queanbeyan is at 35° south latitude, Chicago is at 42° north latitude—not exactly the Sun Belt. The Chicago solar plant has a planned nameplate capacity of ten megawatts, but this will provide less than two megawatts of average delivered power.

The economics of the Chicago project are astonishingly poor. The costs of solar voltaic systems are even 50% higher than solar thermal systems, like the SEGS.[32] The plant will cover *41 acres* at a cost of $60 million, but system operator Exelon will only pay about $13 million. Taxpayers will subsidize $47 million, or 78% of the project. At 15% capacity factor, the plant will *intermittently* power only 1,200 average U.S. households, at the exhorbitant price of over $50,000 per home.[33]

HOME SOLAR SYSTEMS

Climatists promote the idea that if each homeowner just does their part, such as installing a home solar-electricity system, we can solve the climate crisis. National governments have responded with all sorts of incentives and subsidies to encourage home rooftops to be blanketed with solar cells. The U.S. Government provides a 30% tax credit toward not only the cost of a home solar-electric system, but also solar water heaters, fuel-cell systems, residential wind systems, and heat pumps.[34] Almost every state in the Union provides incentives from both state and local governments. But solar systems still don't make economic sense.

For example, the city of Austin, Texas has a "Solar America Cities Partnership" to encourage residences and businesses to install solar power. Austin will pay for more than 50% of the cost of a rooftop solar-cell installation for an apartment complex. For one project, they paid $100,000 of a total installed cost of $178,500. According to the Austin City Council, the project:

> ...is equivalent to the planting of 707 trees or 35 acres of forest...or the removal of five cars from Austin roadways...[and] will save 20.6 tons of carbon dioxide...[35]

Austin also provides a rebate to homeowners who install a solar-electric system. Let's look at the economics of a typical installation. For a 2.5 kilowatt residential photovoltaic system and assuming a generous capacity factor of 17%, the home saves about $410 over the course of a year for a typical household electricity usage (see text box). After the various governments pay 65% of the cost of an installation,

Example: Rooftop Solar System in Austin, Texas	
Installed 2.5-Kilowatt Residential Solar System (2,100 sq. ft. home)	$ 25,000
U.S. Residential Energy Tax Credit (30%)	- 7,500
Dallas Solar PV Rebate ($3.50 per watt)	- 8,750
Installed Price to the User	$ 8,750
Annual Electricity Savings at 17% Capacity Factor (electric rate of $0.11 per kilowatt-hour)	$ 410
Homeowner Payback	21 Years
Rate of Return on $25,000 investment	1.6% per year

the breakeven for the homeowner is still 21 years. The annual payback on the $25,000 invested by the government-homeowner partnership is only 1.6% per year.

Who pays for these miracles of climate control? Austin taxpayers pay for them. Austin is spending $4.5 million on its Solar PV Rebate Program in 2008–2009.[36] If you're an Austin taxpayer, why should you pay for this distortion of economics? Yet, this misuse of funds is happening not only in Austin, but all over the world.

Germany is the leading country in terms of installed solar electric power. In 1991, Germany enacted a "feed-in tariff" to encourage residences to install solar panels. A feed-in tariff is a marvelous regulation that forces electrical utilities to buy green electricity generated by users at above-market prices. German utilities were required to buy solar-generated power at a whopping 57 eurocents ($0.80) per kilowatt-hour for 20 years, almost triple standard German electricity rates and almost five times the cost of coal-fired electricity. Germany is located at 50° north latitude, a very poor location for sunlight intensity. Nevertheless, despite efficiencies much worse than our Queanbeyan Solar Farm example, the feed-in tariff caused an explosion of rooftop solar systems. More than 500,000 systems have been installed in the last 17 years, with a total installed capacity of 5,334 MW.[37] The German feed-in tariff was reduced 10% in 2009, but still remains far about the market rate for electricity.

Fifty-three hundred megawatts sounds like a lot of power, but the average delivered power of the 500,000 systems is less than the output of one large coal-fired power plant. German users pay higher electricity rates to support feed-in tariffs and cut carbon emissions in the belief they are helping to save the planet. Electricity in Berlin costs 21 eurocents ($0.29) per kWh, the second highest rate in Europe.[38] After all this subsidized solar mania, *solar power still produces only 0.5% of Germany's electricity.*[39]

Solar power has uses, but it's not large-scale electricity generation. According to the Solar Energy Industries Association, pool heating continues to be the largest application in the United States. In 2008, the U.S. was first in the world in total solar energy capacity at 9,183 megawatts, but only 17% of this was for generation of electricity.[40] In addition to pool heating, solar is effectively used in low-power remote location applications, especially when coupled with battery storage. Solar-powered telephone call boxes along highways are an example of this. Just to clarify: *We are advocates for solar power.* Let's remove the subsidies and mandates, however, and let solar power compete with other energy on the basis of economics, without misguided beliefs about clean energy, green power, and global-warming fantasies.

Howard Hayden, author of *The Solar Fraud*, sums up the problem with solar power:

> The point is that solar energy is dilute, and no amount of research or technology can change that fact. Nor, for that matter, can a well-funded crash program make the sun shine at night...Solar energy...is utterly inadequate to meet present needs, let alone to grow as demand increases.[41]

DESERTEC

Nevertheless, a consortium in Europe is planning the most ambitious project of all, a huge solar installation named Desertec. In July 2009, twelve companies, spearheaded by German firms, signed a Memorandum of Understanding to establish the Desertec Industrial Initiative (DII).[42] The Desertec project intends to build a massive solar collection system in the Sahara Desert of North Africa. The initial plan calls for construction of a string of concentrating thermal solar tower plants (shown in Figure 100), equipped with salt storage systems. High-voltage transmission lines are planned to carry power from North Africa to Europe via Gibraltar, Sicily, and Turkey.

Start-up for Desertec is aimed for 2019, with a goal of providing at least 15% of Europe's electricity needs. As part of the project, wind power systems are planned for Morocco and the Middle East. Desalinization of sea water will be used to provide fresh water for local communities. The project is expected to cover 6,500 square miles of desert and cost €400 billion ($560 billion).[43]

Greenpeace has already proclaimed the project "economically viable." Press releases point to the "success" of the SEGS solar projects in the Mojave Desert (that forced Luz into bankruptcy). In addition to the usual economic challenges of generating solar electricity, Desertec will need to send power more than 1,000 miles to Central Europe, boosting transmission losses from 7% to about 15%.[44] For Desertec to get off the ground, it will require the Mount Everest of all subsidies. The fact that a project like Desertec is even being considered is *a tribute to the depth of Europe's climate madness.* See quotes from Desertec leaders in the text box. It's not often you can build a power plant for "the lasting protection of human life."

If Desertec is actually built, it will be a vast misallocation of resources. Suppose the subsidy for Desertec totals €200 billion. That's an enormous amount to be spent on the misguided notion that mankind can turn a "carbon switch" and control the climate. Rather than spend €200 billion on Desertec to switch European electricity generation from hydrocarbons to renewables, suppose that enormous sum was put directly toward solutions to major world problems? Major strides could be made to reduce malnutrition and

---

Desertec to Combat Global Warming?[45]

"The establishment of DII is a giant leap by industry for the lasting protection of human life." –Max Schön, President of the German Association of the Club of Rome.

"It is time that everyone gets involved to preserve our planet and leave a better world for future generations." –Malik Rebrab, Chief Executive Officer Cevital

"The Initiative shows in what dimensions and on what scale we must think if we are to master the challenges from climate change..." –Caio Koch-Weser, Vice Chairman, Deutsche Bank

"We are pursuing a visionary plan. If it is successful, we will make a major contribution to combating climate change." –Dr. Torsten Jeworrek, Board of Management Munich Re

disease and to increase access to clean water and sanitation for many of the poorest in developing nations.

WIND POWER: THE CHOICE OF CLIMATISM

Wind power is the fastest growing renewable energy in the world. Berkeley National Laboratory reports that the U.S. invested $16.4 billion in 2008 and added 8,558 megawatts (nameplate power) of new capacity. Worldwide, 28,000 MW (nameplate) of wind power was added in 2008.[46] This means about 19,000 300-foot-tall steel-and-concrete wind towers were installed, monuments to Climatism.

Wind power has become the preferred energy source of Europe and the U.S. to fight climate change. Solar energy is too expensive and delivers too little energy. For locations above 40° north latitude, sunlight energy is very thin, making solar power completely impractical for Scandinavia, the U.K., and Northern Europe, as well as Japan, Canada, and the Northern U.S. Environmental groups continue to oppose nuclear power, despite its ability to generate electricity with zero greenhouse gas emissions. Biomass is renewable, but suffers from high collection costs, and is therefore only practical for small-scale solutions. Hydroelectric power is a good low-cost renewable, but most good river locations in the developed world already have a hydro plant. Carbon Capture and Storage (CCS) of emissions from coal and gas plants is proposed, but is very expensive and not ready for commercial deployment.

While wind power may be the best solution according to global warming alarmists, it's still a poor choice for electrical power. Figure 101 (next page) compares the proposed London Array offshore wind turbine field and the planned high-efficiency coal-fired plant at Kingsnorth in the U.K. A consortium of three firms plans to build the 1,000 MW London Array in the Thames Estuary of the North Sea, just east of Kent and Essex. The field will cover an enormous area of 245 sq. km. and deliver electricity only about 37% of the time, due to the intermittency of the wind.[47] In fairness, proponents of wind power say that wind towers can be added to locations without occupying the entire land area. But, wind power is still not a bargain. According to the 2008 House of Lords study, electricity from London Array will be double the price from a coal-fired plant.

	London Array Wind Turbine Field	New Kingsnorth Coal-Fired Plant
Location	North Sea, 12 Miles Offshore, UK	Kent, UK
Service Start-up	Planned 2012	Planned 2012
Installed Capacity	1000 Megawatts	1600 Megawatts
Actual Average Delivered Power	370 Megawatts (37% of Capacity)	1360 Megawatts (85% of Capacity)
Site Area	245 sq. km. (341 Turbines)	1.6 sq. km. (2 units)
Approximate Electricity Cost	16.2 pence per Kilowatt-hour	8.1 pence per Kilowatt-hour

Figure 101. Wind and Coal Generator Comparison in the U.K. Details for the planned London Array offshore wind turbine field (left photo) and the planned high-efficiency Kingsnorth coal-fired plant (right photo) in the U.K. The wind turbine array requires 563 times the area of coal and delivers electricity intermittently at twice the price. (Bowen, London Array, 2009)[48,49]

The costs of electricity from wind and solar power are clearly significantly more than from hydrocarbon sources. Figure 102 shows analysis from two authoritative sources, comparing renewables and traditional fuels. The Institute of Energy Research, using 2009 U.S. Department of Energy data, shows coal, natural gas, nuclear, and hydroelectric costs to be about 8–11 cents per kWh, versus 14 cents for on-shore wind, 23 cents for offshore wind, 26 cents for solar thermal, and 40 cents for solar photovoltaic.[50] A November, 2008 study by the U.K. House of Lords shows coal, natural gas, and nuclear at 8–9 pence per kWh, with renewables at 14 pence for

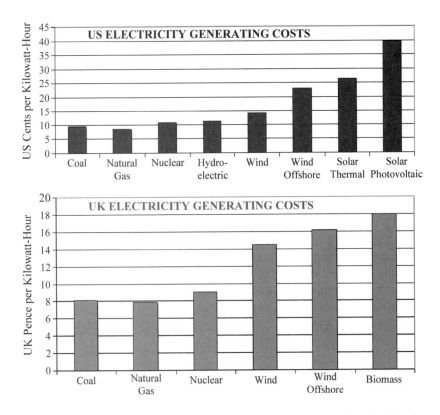

Figure 102. U.S. and U.K. Electricity Generating Costs by Technology. Comparative cost estimates to generate electricity by technology in the U.S. (top chart) and the U.K. (bottom chart). U.S. data is the levelized cost of new generating technologies placed in service in 2016 from the Institute for Energy Research, with core data from the U.S. Department of Energy, 2009.[51] UK data is from the House of Lords, 2008.[52] Cost data does not include carbon permits.

wind, 16 pence for offshore wind, and 18 pence for biomass.[53]

But, there are two major additional issues impacting a switch to electricity production from wind. First, since electricity generation is a system where electricity is generated immediately, rather than stored, coal- or gas-fired stations must remain turned on to backup intermittent wind systems. The House of Lords study states:

> At present, National Grid...keeps a number of power stations running at less than full capacity, providing about 1 GW of spinning reserve...if renewables provided 40% of electricity generation—the share the company believes would be needed to meet the EU's 2020

energy target—its total short term energy requirements would jump to between 7 and 10 GW.[54]

This means that, where a single one gigawatt coal or gas plant is kept running as a reserve today, 7 to 10 hydrocarbon plants would need to be running as standby for a system with 40% wind power.

The second and greater issue is that wind power, because of its inherent intermittency, can't be counted on when peak power demands occur. Meteorologists tell us that the coldest days of the year generally occur during winter high-pressure systems and the hottest days of the year during summer high-pressure systems. These are days when energy demand is the highest. But days of high pressure are also usually times when the wind doesn't blow. This means that, when peak demands occur, wind power is not available to meet peak needs.

Figure 103 shows the percentage of wind power output for the aggregate of more than 15,000 wind turbines in Germany for each day of December, 2007. Note that for 3–4 days of the month, the percentage of rated power output was only about 2%. This is data for the whole wind-power backbone of Germany. In addition, similar data for U.K. wind turbines show low output on the same dates, indicating that an "Arctic high-pressure area" had settled over most of Europe. This means most of the turbines in Europe were not producing on the same days, making power-sharing ineffective.[55]

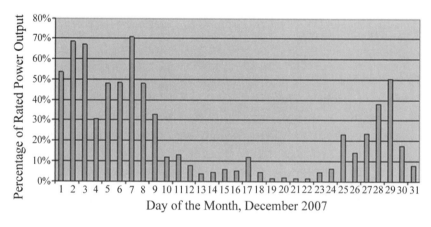

Figure 103. German Wind Power Output, December 2007. Percentage of rated power output for wind power in Germany for each day of December, 2007, a typical month. Note at least 6 days with less than 5% power output. (Adapted from Hyde, data from REISI, 2007)[56]

These periods of low output are not infrequent. Denmark is one of Europe's leaders in wind-power generation. Its "wind carpet" West Denmark system of more than 4,000 wind turbines–more than one-half the nation's wind power–had output less than 1% of rated capacity for 54 days during 2002.[57] It's clear wind power generation is too sporadic to be counted on.

Therefore, even when many monstrous wind towers are constructed, since electric power systems are designed to meet peak demand, additional coal and gas plants will also be needed. The House of Lords comments on peak needs:

> The intermittent nature of wind turbines...means they can replace only a little of the capacity of fossil fuel and nuclear power plants if security of supply is to be maintained. Investment in renewable generation capacity will therefore largely be in addition to, rather than replacement for, the massive investment in fossil fuel and nuclear plant required...[58]

Evidence of the inability to use wind in place of coal-fired plants can be seen in the examples of Germany and Denmark. By 2009, Germany had more than 19,000 wind turbine towers installed and Denmark had more than 5,200, remarkable for a nation of its size. But neither country has been able to replace a single coal-fired plant.

So, wind power will be 80–100% higher in cost, will require much greater standby power generation, and new hydrocarbon plants will need to be built anyway to cover peak load requirements. Nevertheless, the U.K. and governments across the world are forcing construction of wind power to meet the demands of Climatism.

WIND POWER: BUILT ON SUBSIDIES AND MANDATES

The U.K. government has mandated that 15% of the energy supply of the nation be from renewable sources by 2015. Since electric cars and wind-powered homes are even less economical, most of renewable energy must come from electricity generation. The U.K.'s Low Carbon Transition Plan calls for 6% of the current electricity generation from renewables to rise to 31% by 2020.[59]

Let's estimate how many wind turbine towers will be needed. The U.K. consumed an average electricity demand of 44 GW (gigawatts) in 2008. Say 25% of that power will need to be delivered by wind in 2020, the majority of the 31% from renewables, or 11 GW of average power. If each wind turbine operates at an average capacity factor of 35% of nameplate power, the U.K. will need about 31.4 GW of nameplate wind power. At 2.0 megawatts (MW) per turbine, *about 15,700 wind turbines will need to be in place!* There are about 2,600 wind turbines in the U.K. today, so the plan calls for growing the number by a factor of six.

Wind turbines today are much bigger than windmills of the past. A 2 MW wind turbine from Czech manufacturer KV Venti sits on top of a 105-meter pole that is taller than the Statue of Liberty and almost twice as high as the London Tower Bridge. The maximum height of the pole and rotor is 195 meters, which is taller than the Washington Monument and easily taller than the famous cathedral in Cologne. To get the best wind, designers prefer to put these towers on the top of hills. As a result, British citizens will be able to see these monsters from afar. The skyline of the English coast and countryside will be dominated by great views of wind turbines. Each turbine requires a base of almost 1,500 tons of concrete and a minimum land area of about five acres. The minimum distance between two turbine poles is 140 meters, so 15,700 turbines stretched out in a line would reach 2,200 kilometers, or the distance from the northern tip of Scotland to just short of Gibraltar at the south of Spain.[60]

Since wind turbine towers are more expensive, less dependable, and more visibly intrusive than traditional power plants, subsidies and mandates are required to get the utilities of the world to construct them. Feed-in tariffs are used as a direct production subsidy for wind power in Austria, Denmark, Germany, Portugal, South Korea, Spain, and other nations. Germany's feed-in tariff for wind power was at a level about 23% over the market price for electricity for the first five years of operation.[61] Spain's price for wind power was a whopping 36% to 109% over the market price, depending upon market conditions.[62]

The United Kingdom uses an indirect system of Renewable Obligation Certificates (ROC) rather than direct subsidies. One ROC is generated for each megawatt of renewable energy power generated. Utilities must either generate or acquire certificates to avoid a tax of £4.41 on each MWh of electricity supplied.[63] Electricity from renewables was mandated at 9.1% for 2008-2009, rising to

15.4% in 2015-2016.[64] The U.K. National Audit Office estimates the total financial assistance to the renewables industry in 2010 will total £1 Billion. British citizens will pay 5–6 pence per kWh of additional cost to support wind power.[65] Paul Golby, the chief executive for the U.K. division of international energy utility EON, said:

> Without the renewable obligation certificates nobody would be building wind farms. This is the balance we are trying to strike: protecting the environment and the cost of building wind farms.[66]

The United States is also building an impressive array of mandates and subsidies for renewable energy. At the state level, subsidies and mandates are spurring the growth of wind power. As of June 2009, 29 states and the district of Washington, D.C. had enacted Renewable Portfolio Standard (RPS) legislation. RPS policies mandate that the utilities of a state purchase electricity from renewable sources. For example, Connecticut's RPS requires 6% of electrical power from renewables in 2009, growing to 20% in 2020, from solar, wind, biomass, hydroelectric, and other renewable types. Electric utilities in Connecticut must purchase such renewable energy in state or import, or pay a fine of 5.5 cents per kWh. Fourteen states also have mandatory solar power provisions.[67]

States are also providing many financial incentives for wind and renewables, including property tax exemptions, sales tax exemptions, and "public benefits funds." Public benefits funds are collected from electricity consumers by a surcharge on their bill to subsidize and promote renewable energy. Sixteen states and Washington, D.C. have established public benefits funds that will total $7.3 billion over the next eight years. A complete summary can be found at the Database for State Incentives for Renewables and Efficiency.[68]

At the U.S. national level, the most significant of the subsidies is the Renewable Electricity Production Credit, commonly called the production tax credit (PTC). The PTC provided a subsidy of 2.1 cents per kWh in 2008 for the first 10 years of operation for each wind power system placed into service. Note that this is not for research and development, but a direct subsidy for electrical power produced. Other federal policies have added to the subsidy, including an investment tax credit alternative to the PTC, favorable accelerated depreciation, and the provision of interest-free loans to wind projects. The PTC was extended through December, 2012 early in 2009.[69]

Figure 104 shows U.S. Government energy subsidies by fuel for electricity production. Subsidies for renewable energy totaled almost $5 billion of the total outlay of $16.6 billion.[70] Renewable subsidies are both the largest and the fastest growing category, and most of this is for wind power.

As a result of all these incentives, wind-power electricity is subsidized at between 30% and 100% of the cost of installation and operation in developed nations. **The world is spending over $100 billion per year to install wind power, with electricity output as erratic and unpredictable as the wind, delivered at 80 to 100% higher cost than hydrocarbon or nuclear alternatives. All spent in vain, in the *futile* effort to reduce greenhouse gas emissions.**

Electricity that powers our television and light bulbs is delivered at standard voltage and current. No one can tell the difference between

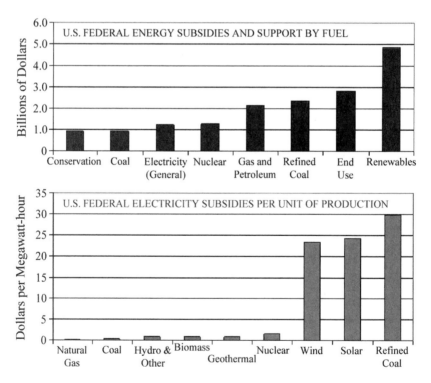

Figure 104. U.S. National Energy Subsidies in 2007. Total U.S. national energy subsidies by fuel (top chart) and for electricity per unit of production in dollars per MWh (bottom chart). U.S. energy subsidies exceeded $16 billion in 2007, with almost $5 billion for renewables, the largest category. (DOE, 2009)[71]

"green electricity" from a wind farm or "dirty electricity" generated in a coal-fired plant. Would you be happy to repurchase your identical car at twice the price?

As in the case of solar power, *we're advocates for wind power*. Wind has been cost-effectively used to pump water and drive grain mills. **But it's not a reliable source of electrical power.** Let's remove the subsidies and stop the vast misallocation of global resources. Allow wind power to compete on its own merit with other energy sources.

CARBON CAPTURE AND STORAGE: A FOOL'S ERRAND

Carbon Capture and Storage (CCS) is the concept that emissions can be reduced from coal- and gas-fired power plants by capturing the CO_2 gas as it escapes from combustion and then piping it to underground reservoirs for storage. Recall from Chapter 11, that three of the 15 "emissions reduction wedges" involve CCS. Some Climatists realize that since coal is the most-used world fuel for electricity generation, wind and solar are unlikely to be short-term replacement solutions. Even Denmark, with its 5,000 wind turbines, still gets 50% of its electricity from coal-fired generators.[72]

In May, 2009, U.S. Secretary of Energy Steven Chu announced $2.4 billion in funding for CCS:

> To prevent the worst effects of climate change, we must accelerate our efforts to capture and store carbon in a safe and cost-effective way. This funding will both create jobs now and help position the United States to lead the world in CCS technologies, which will be in increasing demand in the years ahead.[73]

U.K. Climate Change Secretary Ed Miliband reportedly favors clusters of CCS systems and a national grid for transporting and storing the "polluting" emissions.[74] However, CCS remains experimental and many years away from deployment.

Coal is as natural as the good black soil in a farmer's field, both created by natural processes of Earth. The *incomplete combustion* of coal can and has created gases harmful to humans, including carbon monoxide, sulfur dioxide, and carbon particulates. But mankind has been steadily reducing these pollutants for many years through use of improved coal-fired processes. From Chapter 8, these traditional air pollution levels follow an Environmental Kuznets Curve and are declining in most developed nations. As we also discussed, carbon dioxide is a harmless gas and not a pollutant. Nevertheless, let's examine the idea of capturing CO_2 from hydrocarbon power plants.

Carbon Capture and Storage adds three major costs to production of electricity from power plants. First, the CO_2 must be captured from the plant combustion. This involves absorption techniques that are still experimental. CCS plant costs are increased an estimated 40% to 50% when capture is added. In addition, energy is expended to capture the CO_2, so 10% to 40% more fuel must be burned to achieve the same power output.[75]

Second, the captured carbon dioxide must be transported by vehicle carrier or pipeline to a storage location. Since about two tons of CO_2 are created for every ton of coal burned, enormous quantities of CO_2 will need to be transported and stored. Transportation of CO_2 is a proven technology, but the pipeline infrastructure linking power plants to storage areas does not exist today and would need to be constructed. Third, storage areas would need to be identified and prepared under the sea or land in depleted oil wells or salt mines.

Cost estimates for CCS vary, but typical numbers for a coal plant indicate a 50% cost increase to capture 90% of the CO_2.[76] However, to meet these numbers, technological breakthroughs are usually assumed. Like the United States, other governments are already spending billions of dollars based on erroneous assumptions and misinformation. Dr. Michael Economides of Texas A&M University doubts that CCS will ever be implemented, because of leakage issues and the huge volume of storage required. In a 2009 study, he states: "Our very sobering conclusion is that underground carbon dioxide sequestration via bulk CO_2 injection is not feasible at any cost."[77]

Carbon Capture and Storage can never be cost effective because it brings zero benefits to society. In fact, sequestration of carbon dioxide may be harmful to the environment. Physicist Freeman Dyson of Princeton University talks about the importance of carbon dioxide:

> The fundamental reason why carbon dioxide in the atmosphere is critically important to biology is that there is so little of it. A field of corn growing in full sunlight in the middle of the day uses up all the carbon dioxide within a meter of the ground in about five minutes. If the air were not constantly stirred by convection currents and winds, the corn would stop growing.[78]

By sequestering CO_2, we are not only putting carbon underground, but also storing two oxygen atoms for each carbon atom. Does it make sense to store our carbon and oxygen away underground?

In any case, it's unlikely mankind will be able to sequester enough CO_2 to even be measureable in our atmosphere. Analysis of carbon isotopes shows that 96% of the CO_2 in the atmosphere is from natural carbon cycle processes. If climate madness isn't reversed, the pipelines created for Carbon Capture and Storage will join wind towers as foolish monuments to Climatism.

BIOFUELS: GOVERNMENT-FINANCED EXPLOSION

In his 2006 State of the Union message, President George Bush declared:

> America is addicted to oil...tonight I announce the Advanced Energy Initiative...We must also change how we power our automobiles. We will increase our research in better batteries for hybrid and electric cars and in pollution-free cars that run on hydrogen. We will also fund additional research in cutting-edge methods of producing ethanol, not just from corn but from wood chips and stalks or switch grass.[79]

Government promotion of biofuels and ethanol is not new. As Robert Bryce points out in *Gusher of Lies*, every U.S. president since Richard Nixon in 1974 has proposed initiatives to achieve energy independence.[80] The Energy Tax Act of 1978 established the U.S. "gasohol" industry, providing a subsidy of $0.40 per gallon of ethanol blended with gasoline.[81] Ethanol has been the proposed substitute for imported oil for several decades in the U.S.

The primary biofuels are ethanol and biodiesel. Ethanol is blended with gasoline and biodiesel is blended with diesel fuel as partial substitutes for fuel based on petroleum. Ethanol is made from crops or plant matter, while biodiesel is made from vegetable oil or animal fat. See the text box (next page) for more information on biofuels.

A quick look back at the breakdown of world energy supply (see Figure 92) shows that more than a third of the today's world energy usage is oil. Despite the best efforts of Prime Minister Gordon Brown to preach the virtues of electric cars, the world is tied to the petrol-

---

Ethanol and Biodiesel

Ethanol is a clear alcohol that can be used as a fuel in spark-ignition engines. It's typically blended with gasoline in amounts from 5% to 85%. Ethanol provides only about two-thirds of the energy content of gasoline and therefore reduces vehicle gas mileage.

Most ethanol is produced today from corn or sugar cane, but sugar beets, sorghum and other plant material is also used. Ethanol from corn and other starches is produced by various break-down processes, followed by fermentation to yield fuel. Ethanol from sugar cane involves decomposition processes and then heating to distill the ethanol. Research is underway on future methods, called "second generation" processes, to produce ethanol from cellulosic material, such as wood, grasses, and crop waste.

Biodiesel is blended with diesel fuel for use in diesel engines. The energy content of biodiesel is better than ethanol, reaching 88 to 99 percent of that of diesel fuel. Biodiesel is usually produced from vegetable oil, such as soybean oil, or animal fat. The fat or oil reacts with alcohol and catalysts to produce the fuel. Initial production began with cooking oil and other low-value oils, but vegetable oils are increasingly used as industry volumes have increased.[82]

---

driven automobile in the short term. Windmills, solar fields, conversions to gas-fired power plants, and carbon capture do nothing to address greenhouse gas emissions from world consumption of oil. So, man-made global warming alarmists have turned to biofuels as a solution to reduce world oil consumption.

Figure 105 shows the explosion in world biofuel production since 2000. In 2008 the United States both consumed and produced between 50% and 55% of the world's ethanol. The European Community is the largest producer and consumer of biodiesel, producing about two-thirds of the world's total in 2008.[83]

There is little question that the effort to try to control global warming has been a major driver in the explosion in demand for biofuels. In 2003, the European Union established the Biofuels Directive, with the goal of replacing 2% of vehicle fuel supply by 2005 and 5.75% by 2010, in order to contribute to objectives of:

> ...meeting climate change commitments, environmentally friendly security of supply and promoting renewable energy sources.[84]

European biodiesel production expanded by a factor of five from

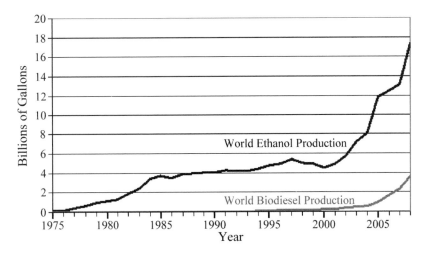

Figure 105. World Biofuel Production 1975–2007. Rapid ramp in world production of ethanol and biodiesel since the year 2000. One gallon equals about 3.8 litres. (F.O. Licht, Lines, 2009)[85]

2003 to 2008, after the directive was issued. In 2007, the EU boosted the directive to an ambitious 2010 target of 10%.[86]

Former U.K. Prime Minister Tony Blair made a point to visit new biodiesel plants and issue press releases. After the start of production at the Argent Energy Plant in Motherwell, Scotland in 2005, the Prime Minister issued a press release including the following:

> The prime minister has frequently stressed his personal commitment to combating global warming and has made climate change one of his two priorities for the UK's presidency of the G8. Biofuels and biocrops have a potentially important role to play, and the prime minister is very glad to see production and marketing of these environmentally-friendly fuels taking place.[87]

Chancellor Angela Merkel of Germany, former President Jacques Chirac of France, and most other European leaders have strongly promoted biofuels in the name of a cleaner environment.

Former U.S. President George Bush viewed ethanol as a means to reduce U.S. carbon emissions, without the pain of accepting the Kyoto climate change treaty. In an interview with *EV World* in 2006, Bush said:

...much of my position was defined early on in my presidency when
I told the world I thought that Kyoto was a lousy deal for America
...It meant that we had to cut emissions below 1990 levels, which
would have meant I would have presided over massive layoffs and
economic destruction...we will invest in new technologies that will
enable us to use fossil fuels in a much wiser way...it means that
we've got to figure out how to use ethanol more in our cars.[88]

Why not boost the production of biofuels? Biofuels are
proclaimed as a path to improved energy security and increased
agricultural development, along with being a sustainable,
environmentally friendly fuel. *But the world is paying a heavy price for
biofuels, both in terms of subsidies and the impact on world food prices.*

Biofuels are very heavily subsidized. The most visible financial
support is in direct payments for production. The U.S. Government
pays tax credits of $0.51 per gallon to ethanol producers (reduced to
$0.45 per gallon in 2009) and $1.00 per gallon to biodiesel
producers.[89,90] The U.K. government pays a rebate of 20 pence per
litre ($1.25 per gallon) to producers for both ethanol and biodiesel.[91]
Most other developed nations have some form of direct subsidy. But
these are only part of the iceberg. Other elements of financial support
include exemption from vehicle fuel excise taxes, subsidies for
production of agriculture feedstocks, grants to build biofuel
production facilities, and payments for research and development. As
in the case of wind and solar power, state and provincial governments
offer additional incentives that add to national government subsidies.
For an extreme example, the State of Kentucky provides a $1.00 per
gallon tax credit for biodiesel, on top of the $1.00 federal government
credit.[92]

Figure 106 shows an estimate of average total government support
in 2007 from the International Institute of Sustainable Development.
Total U.S. support was over $1.00 per gallon for ethanol and over
$2.00 per gallon for biodiesel, amounting to about 65% of the
market price of biofuel. Average financial support in the European
Union was over $3.50 per gallon for ethanol and over $2.50 per
gallon for biodiesel, totaling over 90% of the price of biofuel.[93]

In addition to subsidies, the United States and other nations
mandate certain levels of biofuel usage. The Energy Independence
and Security Acts of 2005 and 2007 established a Renewable Fuel
Standard (RFS) with percentage mandates for biofuel production on

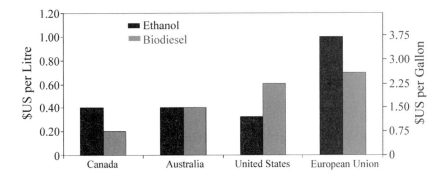

Figure 106. Biofuel Subsidies in Developed Nations. Estimated total government financial support for ethanol and biodiesel in US dollars per litre (left axis) and U.S. dollars per gallon (right axis). (IISD, 2007)[94]

U.S. vehicle fuel refineries. The RFS states that U.S. refiners must blend 11.1 billion gallons of biofuels into vehicle fuels in 2009, rising every year to 36 billion gallons by 2022.[95] Conventional biofuels are capped at a maximum of 15 billion gallons, with second generation biofuels, such as those produced from cellulose, to become the majority by 2018. To meet these demands, U.S. industry ramped ethanol capacity to 7.2 billion gallons by early 2008, with an additional 6.2 billion gallons of capacity under construction.[96]

With this explosive growth, the levels of subsidies are rising through the roof. In 2009, U.S. financial support from all areas for production for about 9 billion gallons of biofuels is about $12.7 billion. Total world subsidies for production of about 18 billion gallons will approach $20 billion. If the U.S. is to achieve its goal of 36 billion gallons by 2022 and other developed nations are to achieve goals of 20% of fuel by 2030, *global biofuel subsidies are on target to reach $100 billion per year.*

However, this ramp is providing surprisingly little in the way of energy security. The production of over 9 billion gallons of biofuels in 2008 was only 5% of the total gasoline and diesel fuel consumed by the 240 million vehicles on the road in the U.S.[97, 98]

Two factors are working against biofuels as a viable replacement for gasoline and a means to reduce dependence on oil. The first is that biofuels require energy to process corn or other feedstock into fuel. For ethanol from corn, this means energy used to produce corn seed, fertilizers, and insecticides. Fuel is required for farm machinery and transport to the ethanol processing plant. Electricity is required at the farm and electricity and natural gas are used at the processing plant. Net energy estimates for ethanol run all the way from an energy loser to an 80% energy gain. The U.S. Department of Agriculture (USDA) places the number at a 34% energy gain.[99]

The second factor against ethanol in particular, is that the energy density of ethanol is only about 66% that of gasoline. Gasoline has an energy content of about 115,000 Btu per gallon versus 76,000 Btu for ethanol.[100] This means that a vehicle that gets 30 miles per gallon using gasoline will get about 29 mpg using a gasoline blend containing 10% ethanol. Not much change here. However, the same vehicle will get only 21.4 mpg if it uses the E85 blend (85% ethanol).

These two factors severely reduce the gasoline-replacement value of ethanol. If we take the USDA estimate of a 34% energy gain, along with the energy density level of only 66% of gasoline, we get a net energy gain of 22% for each gallon of ethanol produced from corn. *This means the U.S. needs to produce almost five gallons of ethanol to replace the energy of one gallon of gasoline.* Not much of a gain for the average total U.S. subsidized support of about $1.20 per gallon.

Despite the poor net gain in energy, the U.S. Department of Agriculture believes there is benefit from producing ethanol, because we can use other forms of energy such as coal and natural gas, instead of oil to do so. Hosein Shapouri, of the USDA Office of the Chief Economist states:

> …producing ethanol from domestic corn stocks achieves a net gain in a more desirable form of energy, which helps the United States to reduce its dependence on imported oil.[101]

Ethanol may be made from many natural substances, but sugar cane appears to be the best. In Brazil, 70% of automobiles sold run on gasoline blends with 24% ethanol made from Brazilian sugar cane. Marcelo Oliveira of Washington State University determined that sugar cane produces 3.67 units of energy for each unit expended in the production of cane-ethanol.[102] Second generation ethanol processes using cellosic feedstock such as corn residue are expected to produce good energy outputs for each unit of energy input, but these processes are not cost competitive today. In any case, the issue of lower energy density and reduced mileage compared to gasoline remains when using either sugar cane or feedstock from cellulose.

Biodiesel delivers better energy ratios than ethanol. A gallon of diesel has an energy density of about 130,000 Btu, about 13% more than gasoline. Biodiesel has a density of about 118,000 Btu, almost 3% more than gasoline.[103] Therefore, mileage from biodiesel is as good as or better than pure gasoline. In addition, for each unit of energy used to produce biodiesel, about 3.5 units of energy are produced by biodiesel fuel.[104] But the big disadvantage of biodiesel is that fuel yields per acre are much less than yields for ethanol. U.S. corn crops provide about 375 gallons of ethanol per acre compared to only about 52 gallons of biodiesel per acre of soybeans.[105]

Both ethanol from corn and biodiesel continue to require huge subsidies to compete with petroleum fuels, except when the price of crude oil spikes to high levels. In addition to mounting subsidies, the explosion in production of biofuels has a major negative side effect.

DON'T BURN FOOD!

A February 27, 2008 article in *Time* reported:

> Rocketing food prices some of which have more than doubled in two years have sparked riots in numerous countries recently. Millions are reeling from sticker shock and governments are scrambling to staunch a fast-moving crisis before it spins out of control. From Mexico to Pakistan, protests have turned violent. Rioters tore through three cities in the West African nation of Burkina Faso last month, burning government buildings…Days later in Cameroon, a taxi drivers' strike over fuel prices mutated into a massive protest about food prices, leaving around 20 people dead…Indian protesters burned hundreds of food-ration stores in West Bengal last October…[106]

The unrest around the world was in response to a doubling in the price of wheat, corn, and other grains during 2007. Price increases were blamed on a number of factors, including $100 per barrel oil, but also the diversion of farmland to produce biofuels.

Figure 107 (next page) shows the ramp of U.S. corn use for Ethanol fuel compared to corn exports. The U.S. has traditionally provided 50% to 70% of the world's corn exports. It's a tribute to U.S. farmers that exports have continued to remain high over the last

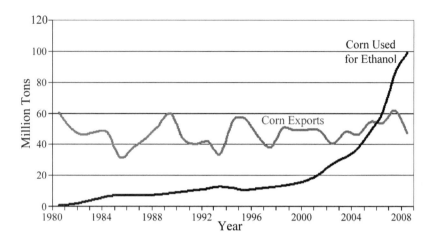

Figure 107. U.S. Corn for Exports and Ethanol 1980-2009. U.S. corn used for ethanol and for export in millions of metric tons. Data is for fiscal year beginning September 1 of the previous year. (USDA, 2009)[107]

eight years while corn use for ethanol fuel has increased by a factor of five. But, most experts agree that the demand for biofuel grains has boosted world food prices.

Figure 108 shows that ethanol vehicle fuel now uses almost one-third of U.S. corn production. Corn used for ethanol has far surpassed both corn used for food and exports. In view of U.S. mandates to again double corn ethanol use, it's questionable whether U.S. farmers will be able to continue to maintain export levels.

Should we be using farmland to replace petroleum on such a massive scale? Recall from Chapter 11 that one of the emissions reduction wedges proposes to use biofuels to eliminate one gigaton of carbon dioxide emissions per year. Climatists call for using *one-sixth of the world's crop land* for biofuels to achieve this reduction. Dennis Avery of the Hudson Institute warns:

> ...the world's food and feed demand is set to more than double by 2050. That means that good cropland will become very scarce around the world. Human society is already farming about 37 percent of the global land area, and already using almost all of the good-quality land. Additional farmland will have to come at the expense of forest and wild species, and is likely to incur heavy penalties in terms of soil erosion, drought risks, and endangered wild species.[108]

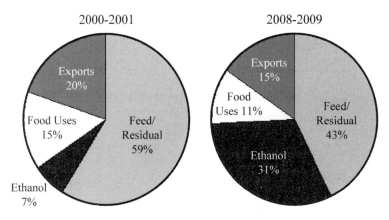

Figure 108. Uses of U.S. Corn 2000–2001 and 2008–2009. Ethanol for fuel has become almost one-third of U.S. corn production. (USDA, 2009)[109]

U.S. plans to produce 36 billion gallons of ethanol by 2022 are more than goals; *they are mandates written in law.* European biodiesel targets have similar legal standing. These mandates on fuel refiners will cause the price of biofuels to rise. Whenever farmers can get a better price from biofuels than grain exports, they will do so.

Each bushel of corn produces about 2.7 gallons of ethanol. A sport utility vehicle using E85 fuel consumes about 25 gallons of ethanol in a single tank of fuel. Therefore a single tank of E85 blend uses over nine bushels of corn, which some have claimed is enough calories to feed a person in the developing world for a full year.[110]

The bottom line is that biofuel subsidies and mandates are distorting agricultural markets and reducing world food supplies. **We are headed down a road where people will starve in the name of energy security and futile efforts to control Earth's climate.** We should take Dennis Avery's advice: "Don't burn food!"[111]

FUEL ECONOMY MANDATES AND ELECTRIC CARS

As we discussed in Chapter 11, two wedges proposed by Climatism to reduce greenhouse gas emissions are: 1) doubling fuel economy to 60 mpg and 2) using hydrogen fuel-cell cars. These goals are unlikely to be achieved, but they will result in the imposition of onerous mandates from Climatist political leaders. Both Europe and the U.S. are enacting fuel economy legislation to reduce CO_2 emissions.

The United States announced a new National Fuel Efficiency Policy in May, 2009. New fuel economy standards were established for all new cars and trucks sold starting in 2012. These standards culminate in a required average fuel economy of 35.5 mpg in 2016, an increase from the current national standard of 27.5 mpg. A White House press release says the program is:

> ...projected to save 1.8 billion barrels of oil over the life of the program...with...a reduction of approximately 900 million metric tons in greenhouse gas emissions.[112]

Administration officials say the increase in fuel efficiency will add about $1300 to the cost of a new car.[113]

The U.S. Government projects that the 28.2% improvement in mileage standards will result in reduced gasoline use, thereby reducing crude oil consumption and emissions. But it's not clear these projected gains will actually occur. Figure 109 shows the history of U.S. road vehicle miles per gallon and fuel consumption. Despite a 39% increase in fuel economy from 12.4 mpg in 1960 to 17.0 mpg in 2005, U.S. fuel consumption increased 202% during the same period.[114] An important reason for the increase is that users drive more often and faster, in order to get other tasks done quicker. Therefore, *consumption rises with increasing vehicle efficiency*, offsetting

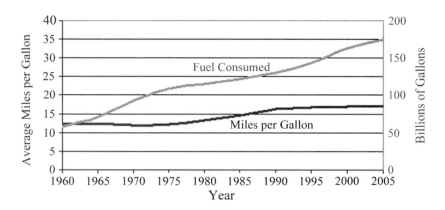

Figure 109. U.S. Use of Vehicle Fuel and Miles per Gallon 1960–2005. Motor vehicle fuel consumed and average mpg for personal passenger vehicles, buses, trucks, and motorcycles. Despite increases in fuel economy, fuel consumption continues to rise steadily. (U.S Bureau of Transportation Statistics, 2009)[115]

most of the gains from the mileage improvement. Huber and Mills discuss efficiency in detail (of which mpg is a measure), concluding:

> Efficiency may curtail [energy] demand in the short term, for the specific task at hand. But its long-term impact is just the opposite... efficiency fails to curb demand because it lets more people do more, and do it faster—and more/more/faster invariably swamps all the efficiency gains.[116]

In order to meet mandates of the Kyoto Protocol, the European Union has established standards based directly on carbon dioxide emissions. The passenger automobile fleet average to be achieved by 2015 is 130 grams of CO_2 emitted per kilometer driven (g/km). The target for 2020 is 95 g/km. Current European car emissions are about 160 g/km. Car manufacturers unable to comply must pay fines that rise from €5 to €95 *per car* by 2019.[117] Emissions standards are also proposed for light commercial vehicles (vans and minibuses), which have so far been exempt from the regulations.

A report by the European Climate Change Program estimates the cost of compliance for cars to get to 130 g/km is €3,000 ($4,200) per car.[118] The cost of compliance for vans and minibuses to meet to a proposed 175 g/km is a whopping €6,000 ($8,400) per vehicle.[119] European vehicle manufacturers are concerned this will make them uncompetitive in world markets. **Europeans will pay €3,000 to €6,000 more per vehicle for CO_2 emissions reductions that won't have a pence worth of effect on the climate.**

Electric and hybrid gasoline–electric cars will not be sold in significant volumes in the near term. These vehicles were only about 1% of the 2008 world sales of 60 million automobiles. Despite tax credits and other incentives, hybrid and electric vehicles remain unattractive to most buyers due to deficiencies in size, acceleration, driving range, safety, and other features. The Toyota Prius, the most successful of the hybrids, is an engineering marvel. But even with its revolutionary design, it's still priced at over $30,000, a very expensive alternative to other cars on a feature-to-feature basis. According to a survey in 2007 by CNW Marketing Research, the main reason 57% of Prius buyers bought the car was "it makes a statement about me."[120] Beyond looking green to your neighbors, it's difficult to match the power density of gasoline or diesel with electric alternatives.

In 2003, President George Bush proposed the Hydrogen Fuel Cell Initiative:

> With a new national commitment, our scientists and engineers will overcome obstacles to taking these [hydrogen fuel cell] cars from laboratory to showroom, so that the first car driven by a child born today could be powered by hydrogen, and pollution-free.[121]

Hydrogen fuel cell cars hold promise for the future, but that future is still far over the horizon. Hydrogen fuel cells burn pure hydrogen and emit nothing but water, the ultimate in low-emission technology. But pure hydrogen needs to be created, stored, and distributed. Hydrogen does not exist in nature except in compounds, so we don't have hydrogen readily available to burn in hydrogen fuel cells. Energy must be used to create hydrogen by breaking down water or another hydrogen-containing compound. More energy is required to create hydrogen than it contains as fuel. Hydrogen is also much more difficult to store and transport than natural gas.[122]

Hydrogen fuel cells remain experimental, still a long way from commercial application in automobiles. In May of 2009, the United States discontinued federal funding for fuel cell vehicle development.[123] Hydrogen fuel cell cars have become just another emissions wedge that will not be realized in the near term.

RENEWABLES ARE ANYTHING BUT FREE

The governments of the world have adopted Climatism and discarded reality. Acres of solar cells sit idle at night and on cloudy days. Thousands of 300-foot-high wind towers interrupt the vistas of coastline, field, and hill, standing motionless for two-thirds of their existence. Thirty percent of America's corn crop is burned up on U.S. highways, while people in developing nations struggle to get enough to eat. Billions of dollars in government subsidies are spent every year to fund solar, wind, and biofuel industries, *which could not compete and would not exist without these subsidies.* Yet, these renewable sources supply only a pitifully small amount of the world's energy needs. All this in the absurd attempt to control a trace gas and stop global warming.

CHAPTER 14

CHAINS FOR
THE DEVELOPING NATIONS

"Of all tyrannies, a tyranny sincerely exercised for the good of its victims may be the most oppressive... those who torment us for our own good will torment us without end for they do so with the approval of their own conscience." C. S. Lewis

W

e in the developed world enjoy the benefits of hydrocarbon-based energy. Coal, natural gas, and oil are the basis of our prosperity and our way of life. But much of the developing world does not yet have these benefits.

The proponents of Climatism claim their policies are meant to prevent death and suffering in developing nations. Rajendra Pachauri, chairman of the UN's IPCC, states:

> ...99% of the casualties linked to climate change occur in developing countries. Worst hit are the world's poorest groups. While climate change will increasingly affect wealthy countries, the brunt of the impact is being borne by the poor, whose plight simply receives less attention.[1]

However, Climatist remedies call for developing nations to deploy ineffective solar and wind renewable energies and expensive carbon capture and storage schemes, denying these countries the benefits of low-cost hydrocarbon energy. In doing so, Climatism is actually

293

shackling the developing world, reducing economic growth, and dooming millions of people to lives of disease and suffering.

CHINA AND INDIA: THE DILEMMA OF CLIMATISM

As we discussed in Chapter 11, climate model "oracles" warn that atmospheric levels of carbon dioxide must be stabilized at 450 ppm to hold global warming to a 2°C temperature rise over 1990 levels. The 30 nations of the Organization for Economic Cooperation and Development (OECD) have a combined population of about 1.2 billion people.[2] From Climatism's point of view, greenhouse gas emissions from these 1.2 billion are responsible for most of the current atmospheric CO_2 concentration of 385 ppm.

China, with 1.3 billion people, and India, with 1.2 billion people, are rapidly industrializing.[3] If the 2.5 billion of China and India develop along the same hydrocarbon-energy path taken by OECD nations, global emissions of CO_2 will continue to rise, even if emissions from today's industrialized world are cut to zero. The remaining three billion people on Earth are also building industrial economies and growing their greenhouse gas emissions.

The growth of China is amazing. Shanghai is an astounding city, with dozens of skyscrapers and a population that now exceeds 20 million.[4] Shanghai's magnetic levitation train speeds visitors from its Pudong International airport to the city at 500 kilometers per hour (300 mph).[5] Beijing adds almost one million new vehicles each year, creating major traffic problems. From 2000 to 2008, China's Gross Domestic Product increased 150%! Annual GDP growth has been about 10% for the last 30 years. China now has more than 70 million vehicles, over three times the number in 2000. Annual vehicle growth is 20%, with 12 million vehicles added every year and growing.[6] This annual increase approaches the total number of vehicles in use in Netherlands and Belgium combined.

Airports are a particularly good example of the magnitude of the problem faced by Climatism in its quest to reduce emissions. Greenpeace has opposed airport expansions in Europe for years. In 2008, the organization purchased a plot of land to block a proposed third runway for Heathrow Airport in England.[7] True believers have climbed on top of a British Airways aircraft to unfurl banners denouncing climate change.[8] Climate activists chained themselves together on a runway at Stansted in the U.K.[9] But all of this is dwarfed by the building of airports in Asia.

China opened 47 new *airports* between 1990 and 2002 and plans to build another 43 airports by 2012. China's target is 244 airports in operation by 2020. India has announced plans for 500 operational airports by 2020.[10] Efforts to block a runway in the U.K. are comparatively insignificant.

Figure 110 shows the growth in carbon dioxide emissions for the U.S., Europe, China, and India. According to the U.S. Department of Energy, China became the world's biggest CO_2 emitter in 2006. From 2000 to 2006, China's emissions doubled and India's emissions increased 30%.[11] In comparison, installation of 5,000 wind turbines in Denmark has only a trivial effect on world emissions.

From the Kyoto Treaty signing in 1997 to 2006, world CO_2 emissions increased 26%. Over the same period, Europe's emissions *increased* 4%, even though all European nations signed the treaty and agreed to an average *reduction* of 5%. Despite the dislocations of massive renewable energy deployment, the Kyoto Treaty has done little to stop increases in global greenhouse gas emissions.

The dilemma of Climatism is that, regardless of how many windmills

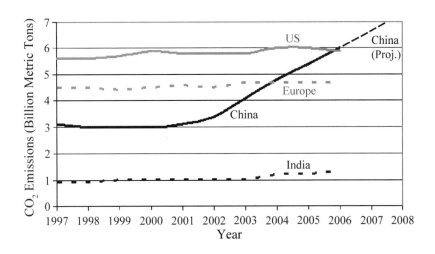

Figure 110. Carbon Dioxide Emissions 1997–2006. Emissions for U.S., Europe, China, and India. China passed the U.S. in 2006 as world leader. Multiply CO_2 emissions by 12/44 to get gigatons of carbon. (EIA, 2008)[12]

Germany or Denmark build, unless China and India adopt severely limiting polices, global greenhouse gas emissions will continue to rise rapidly. Therefore, the alarmists will negotiate and plead, offer transfer payment bribes of billions of dollars, and eventually use coercive trade barriers to get China, India, and the other developing nations to play ball. Climatism demands that its ideology be imposed on the developing world.

DEVELOPING NATIONS IN DIRE NEED OF ENERGY

Two energy elements essential for economic growth are *electrical power and hydrocarbon-fueled vehicles.* Population, Gross Domestic Product (GDP), and per-person electricity and vehicle usage for developing regions are shown in Figure 111. India and non-OECD Asia and Africa have the lowest GDP, the lowest electricity usage per person, and also a low level of motor vehicle use. For comparison, the OECD (developed) nations have a per capita GDP of $29,950, an annual electricity usage of 8,381 kilowatt-hours per person, and use over 500 motor vehicles per 1,000 persons.[13,14,15] Greater use of electricity and fuel-powered vehicles is needed in developing nations to boost commerce and economic development.

Compared to Africa and Southeast Asia, South America has been an example of effective use of electrical power. Over 90% of South America's 420 million people have access to electricity, compared to

Nation/Region	Population (millions)	GDP per Capita ($US)	Electricity Usage per Person (kWh)	Vehicles per 1,000 Persons
Former USSR, Other Eurasia	342	9,237	4,481	164
Middle East	197	10,421	3,163	146
China	1,321	4,553	2,060	24
India, Other non-OECD Asia	2,151	3,438	667	26
Central and South America	460	8,167	1,777	115
Africa	944	2,480	557	31

Figure 111. Developing World Electricity and Vehicle Use in 2006. The two regions with the lowest per person electricity usage also have the lowest per person GDP. Per capita Gross Domestic Product numbers use purchasing power parity estimates. (DOE/IEA, International Energy Agency, UN, 2009)[16,17,18]

less than 60% access for India and Southeast Asia, and only 25% access for sub-Saharan Africa (Africa south of the Sahara desert).[19] Hydroelectric power provides more than 60% of electricity in South America, the highest percentage for any region in the world. Natural gas and petroleum are also primary fuels, but wood and charcoal are still used widely in industry. The largest dam in South America in terms of electrical output, and second largest in the world behind China's Three Gorges Dam, is the Itaipu Hydroelectric Dam shown in Figure 112. The Itaipu Dam provides a remarkable 20% of the electricity used by Brazil and 80% of that used by Paraguay. This *single* installation has an annual output of more than 90 billion kilowatt-hours, *which is equal to the average output of 40,000 wind turbines, one-third of the installed turbines in the world.*[20]

South American nations are also growing the use of gasoline-powered vehicles, with over 100 motor vehicles in use per 1,000 persons.[21] Motor vehicle usage dominates transportation in most parts of the continent. National rail and highway networks are being expanded in Brazil and Argentina and have been deployed to a lesser extent in Chile, Columbia, Ecuador, and Uruguay. Brazil has even spanned parts of the Amazon basin with roads.[22] In addition to the use of gasoline, Brazil has a growing industry based on ethanol made

Figure 112. Itaipu Hydroelectric Dam. Itaipu dam operates on the Paraná River on the border between Brazil and Paraguay, supplying 20% of the electricity of Brazil and 80% of the electricity of Paraguay. (Oates, 2007)[23]

from high-energy sugar cane.

Almost one billion of the 2.1 billion people in India and Southeast Asia do not have access to electricity. The countries with the lowest electrification rates are Bangladesh (32%), Cambodia (20%), and Myanmar (11%).[24] Access is only part of the story. India and the nations of Southeast Asia are looking to strongly boost electrical power production during the next 30 years to support industrial growth. Hydroelectric projects are planned for provinces near the Himalaya Mountains and the Mekong River on the Southeast Asian peninsula. Nuclear power plants are also in consideration for Indonesia, Vietnam, and Thailand.[25]

India, Pakistan, and Indonesia are the three most-populated nations of South Asia. Ninety percent of India's electricity is generated by coal-fired plants. Indonesia and Pakistan use gas and oil as primary fuel sources. In all three nations, wood is burned as fuel by citizens for heating and cooking.

Each of the three nations is underdeveloped in terms of motorized transport, with vehicle usage per thousand persons in India (18), Pakistan (13), and Indonesia (42).[26] The majority of the small number of vehicles in each country are two-wheeled motor scooters or motorcycles. Despite the low penetration of vehicles, Indian cities are heavily congested with pedestrian and vehicle traffic. The congestion is caused by poor road systems and the lack of urban buses or an inter-city rail network.

Middle and Western Africa is the least economically developed region in the world. Only 10–20% of the people have access to electricity in nations such as Angola, Democratic Republic of the Congo, Ethiopia, Kenya, Madagascar, Mozambique, Rwanda, Tanzania, and Uganda.[27] Electricity access in rural areas is often as low as 2% (see Figure 113). Motor vehicle use is about 10 vehicles per 1,000 persons across the region.[28]

Reliable, low-cost electrical power is the foundation of modern industry. Electricity is essential to allow African companies to compete in world markets. Not only access, but reliable power, free from outages, is critical. Itai Madamombe, writing for *Africa Renewal* of the UN, tells the story of an outage at a food factory in Ghana:

> ...food tins at the Prime Pak canning factory were positioned on the assembly line, ready to be sealed before export. Without warning, the machines came to a screeching halt...Thirty per cent of the

Figure 113. Africa at Night. Composite of satellite photographs showing Africa, Middle East, and Europe at night. Note the lack of lighting for the billion people of Africa. (NSN, 2009)[29]

consignment spoiled.[30]

Vijaya Ramachandran of the Center for Global Development, published results of a survey of private sector businesses in Africa in 2008. She summarizes:

> Central to the issue of growth is the development of the private sector. Without the creation of jobs and businesses, there is no real chance for many Africans to raise their standard of living. Extensive surveys of private sector businesses carried out over the past decade show that the poor performance of the private sector can mostly be attributed to the high costs of the business environment…two key constraints identified by these surveys: the lack of power and roads.[31]

Since 2003, GDP growth has increased in sub-Saharan Africa. A number of resource-poor nations have been growing at the healthy rate of over 5% per year. This has boosted the demand for electrical power, increased the frequency of outages, and overwhelmed power-generating systems. Electrical outages average five hours long, and 12-hour outages are not uncommon in middle Africa. Outages occur between 50 and 100 days per year for many nations.[32]

The shortage of power has caused the use of alternative solutions. In the summer of 2007, the government of Kenya called on manufacturers to move production to nighttime hours to reduce the peak daytime load. Manufacturers were faced with the problem of getting workers to and from work in the dark, increasing business costs. Backup gasoline generators are commonly used throughout the region, raising the costs of manufacture for African firms. Energy costs for African companies average about 6% of total costs—double the cost for firms in China—a disadvantage in global markets.[33]

Citizens of developed nations take their road systems for granted. The National Highway System (NHS) of the United States is said to be the largest public works project in the world. Started in 1956 by President Dwight Eisenhower, it now consists of almost 47,000 miles of roadway. It's just 4% of the U.S. road system, but carries 40% of all highway traffic and 75% of heavy truck traffic. Counties that contain NHS highways host 99% of all jobs in the U.S.[34] The NHS has been an essential part of U.S. economic development during the last 50 years.

In comparison, roads in many parts of central and southern Africa are almost non-existent. My daughter recently traveled the entire 900-mile length of Mozambique and back again by SUV, a journey taking ten days of long, hard travel. Fifty percent of the trip was a bone-jarring ordeal over rutted, unpaved dirt roads. At one point, a ferry was used to get their vehicle across a river. The 50% of the road that was paved was filled with large potholes. And this was the main highway of Mozambique.[35]

Businesses in Africa incur high costs because paved highways and motor vehicles are not available. These limitations raise the price tag for obtaining production materials and shipping finished goods. Improved roads and available vehicles are key to economic growth.

We have provided this background to stress that low-cost, reliable electrical power and improved motor transport are vital to growth of the developing nations. Unfortunately, Climatism is exerting every effort to prevent free-market use of hydrocarbon energy in developing nations. The chains of forced use of unreliable green energy are being assembled link by link.

PAST CHAIN: TRAGEDY OF THE DDT BAN

Malaria is one of the world's great killers. According to the United Nations, between 190 million and 330 million cases of malaria occurred in 2006, and almost one million people died, the majority in sub-Saharan Africa.[36] The history of the pesticide DDT shows that a significant share of the ongoing plague of malaria can be placed at the feet of the environmental movement.

In 1962, environmentalist Rachel Carson authored the book *Silent Spring*, which claimed that man-made pesticides such as DDT were harming and killing not only animals and birds, but also humans.[37] Al Gore wrote the introduction for the 1994 edition, declaring "Without this book, the environmental movement might have been long delayed or never have developed at all."[38] Spurred by Carson's book, environmental organizations rallied to oppose the use of DDT, including the Environmental Defense Fund, which was established in 1967 to specifically pursue a ban on the pesticide.[39]

Dichloro-Diphenyl-Trichloroethane (DDT) is a remarkably effective pesticide against not only malarial mosquitoes, but typhus-carrying lice, farm pests, and other insects. Its effectiveness as a pesticide was realized by Swiss chemist Paul Muller in the late 1930s. Muller was awarded a Nobel Prize for his work with DDT in 1948.[40] DDT was used heavily to de-louse soldiers and civilians in World War II. The pesticide was released for commercial sale in 1945 and used widely for agriculture. DDT use peaked in the U.S. in 1959, but was exported widely, with world use increasing until the mid-1960s.[41]

Rachel Carson and environmental groups made many claims about dangers to wildlife and humans from DDT. It's claimed to be toxic to fish and some shellfish. It's claimed to cause eggshell thinning in birds of prey and to threaten the American Bald Eagle with extinction. Studies suggest DDT causes cancer, reproductive disorders, and other abnormalities in humans. But, with the exception of the thinning of raptor eggshells, and despite being one of the most heavily analyzed substances in human history, little scientific basis exists for most of these claims.

Certainly widespread spraying of DDT for agricultural use is not healthy for the environment. Environmental groups are correct in this regard. But environmental efforts caused a complete ban of DDT, removing limited indoor spraying as an effective tool against malaria.

Under intense lobbying from environmental groups and against the advice of his scientific advisors, William Ruckelshaus, head of the U.S. Environmental Protection Agency, banned DDT in 1972:

> The general use of the pesticide DDT will no longer be legal in the Unites States after today…the continued massive use of DDT posed unacceptable risks to the environment and potential harm to human health.[42]

The U.S. ban was accompanied by bans by most developed countries in the 1970s and 1980s. The UN's Stockholm Convention on Persistent Organic Pollutants in May 2001 targeted 12 pollutants, including DDT, for "reduction and eventual elimination," effectively extending the ban of DDT worldwide. One hundred and fifty-two parties signed the Convention by May 2004.[43]

International organizations such as the World Health Organization, USAID, the European Union, and environmental groups such as Greenpeace, Environmental Defense, Sierra Club and the World Wildlife Fund applied pressure to developing nations for the last 30 years to prevent DDT usage. For example, DDT spraying was discontinued several decades ago in Mozambique because 80% of the nation's health budget came from donors who refused to allow the use of DDT.[44]

Limited use of DDT can be highly effective in reducing malarial infections. When used in limited indoor spraying, the pesticide kills and repels malaria-carrying mosquitoes for months. Spraying twice per year is sufficient to eliminate in-home risk for people living in malarial-afflicted regions. DDT is also less expensive and effective for longer periods than any other alternative pesticide.

There are numerous global examples of the effectiveness of DDT in reducing malarial infections. Venezuela reported 8.1 million cases of malaria in 1943. With use of DDT, this number dropped to 800 cases by 1958. India had over 10 million cases of malaria in 1935, but fewer than 300,000 in 1969. In Italy, the number of cases dropped from 412,000 in 1945 to only 37 in 1968.[45]

Tragically, malaria cases rebounded in many countries after DDT use was discontinued. Cases in Sri Lanka dropped from 2.8 million in 1948, before DDT use, to 17 cases in 1964. But, after discontinuation of DDT use, cases rose again in just five years to 2.5

million in 1969.[46] South Africa used DDT through the 1980s with an average of only 20 deaths per year, discontinued use in 1996, and saw the death rate climb to over 400 in just 4 years. DDT use was restarted in 2000, dropping malarial deaths in South Africa below 20 again by 2002.[47] In total, the ban on DDT *likely resulted in more than ten million additional deaths from malaria over the last 30 years.*

After 30 years of opposition, the World Health Organization (WHO) reversed its stance in 2006 and advocated support for limited indoor spraying to combat malaria (text box). USAID and other organizations also ceased opposition to DDT or softened their positions. Although many organizations still oppose DDT use, we're optimistic that this tool against malaria can once again be put to use.

We've discussed the history and tragedy of DDT as a stark example of imposition of developed-nation ideology on the developing world. In the same way, Climatism seeks to shackle

---

World Health Organization Lifts DDT Ban in 2006[48]

In September, 2006, the World Health Organization (WHO) lifted its ban on the use of DDT to combat malaria. The statement by Dr. Arata Kochi, Director of the Malaria Department for WHO, is informative:

> "...Some people told me that there was a good reason why its [DDT's] wide scale use had been phased out. I was told the practice was unsafe for humans, birds, fish and wildlife; that the use of DDT in the United States in the 1950s had led to the near extinction of the bald eagle. I was told that indoor spraying with DDT was 'politically unpopular.' But I believe that public health policies must be based on science and the data, not on conventional wisdom of politics. As we examined the issue, we found that the scientific and programmatic evidence told a different story. We found that:

> - One of the best tools we have against malaria is indoor residual house spraying as it has proven to be just as cost effective as other malaria prevention measures.
> - Of the dozen insecticides WHO has approved as safe for house spraying, the most effective is DDT.
> - DDT presents no health risk when used properly indoors. Well-managed indoor spraying programmes using DDT pose no harm to wildlife or to humans.

> ...WHO is now recommending the use of indoor spraying not only in epidemic areas but also in areas with constant and high malaria transmission, including throughout Africa..."

developing nations in the chains of "sustainable development" and renewable energy. According to Climatism, the developing nations must comply to help stop climate change.

RENEWABLE FANTASY FOR DEVELOPING NATIONS

The United Nations, closely followed by environmental groups, is leading efforts to push developing nations to deploy costly and ineffective renewable energy technologies. It doesn't matter that renewables are two to three times more expensive, less reliable, and in some cases still not commercially viable. Ideology and dogma trump science and economics. After 20 years of building the ideologies of sustainable development and Climatism, *the UN has lost all touch with reality regarding energy and economics.*

An example is the 2006 publication from the UN's Economic Commissions for Africa titled "Sustainable Energy: A Framework for New and Renewable Energy in Southern Africa."[49] This document promotes renewable energy strategies for 238 million people in the thirteen nations in southern Africa, located between the Democratic Republic of the Congo and Tanzania in Central Africa, and South Africa. Excepting South Africa, these are some of the poorest nations in the world, with life expectancies of 30 to 40 years, clean water and sanitation available to less than one-half of the population, and electricity available to less than 20%.

The document calls for goals of 10% of energy from renewables by 2010 and recommends higher targets for 2020. *The document is very unbalanced.* It calls for the production of energy from wind, solar, biomass, and "micro-hydro," with no mention of large-scale sources that can really make a difference, such as electricity generation from coal, oil, nuclear, and major hydroelectric projects. It lauds the costly and foolish subsidy policies in use in Europe and the U.S. today, including feed-in tariffs and renewable portfolio standards, and encourages Africa to set up similar non-economic programs to boost the use of renewable energy:

> The framework recommends the adoption of measures to stimulate RETs [renewable energy technologies] such as obligations on the electricity industry to purchase renewable energy, supply a certain

proportion of their energy from renewable sources and the development of a guaranteed market...[50]

The UN's renewable fantasy is supported by many environmental organizations, including the Institute for Environmental Security, headquartered in Netherlands. The Institute published "Renewable Energy for Africa" in 2009, lamenting the nature of the energy used by the people of Africa, which is: "...imported, expensive, and environmentally degrading, such as coal, oil, firewood, and natural gases."[51] The report praises the wonders of "micro-hydro" with an example project in Kenya:

> The Tungu-Kabri micro-hydro power project in Kenya is one such project...Villagers worked once a week or more for 2 years to build it, and now it generates about 18 KW of electrical energy, enough to benefit 200 homes.[52]

Folks, they've got to be kidding. Eighteen kilowatts is only enough power for 180 100-watt light bulbs! This tiny amount of electricity won't provide 200 homes with a decent standard of living, let alone power any factories. The report goes on to recommend against large-scale hydroelectric power projects:

> Knowing as we do, that large scale hydropower plants that utilize massive dams and reservoirs can be quite detrimental to the surrounding ecosystem, small scale hydropower is a more environmentally friendly way of utilizing the great resource of water that is available in Africa, with the potential to provide the same amount of energy with less environmental degradation.[53]

Small-scale energy solutions do have a role to play in Africa. Photovoltaic solar power can provide electricity in locations not yet served by the power grid. A typical system consists of a frame of solar panels that generates 300–500 watts of power during sunlight hours and a battery to store power. This is enough energy to meet simple needs such as powering a personal computer and three or four light bulbs for a few hours, or charge a cell phone.[54] Like the UN and the Institute for Environmental Security, many environmentalists favor this limited source of energy for the developing world (text box next page). These energy sources are fine for remote locations, but should

---

Environmental Quotes on Energy for Africa

"To build a power plant and run lines to houses, to huts, to anything is a tremendous amount of work... how about... just giving them the service where they need it–on the roof of their hut."

–Actor and environmentalist Ed Begley, Jr.[55]

"I don't think a lot of electricity is a good thing... If there is going to be electricity, I would like it to be decentralized, small, and solar-powered."

–Gar Smith, editor of the environmentalist publication *The Edge*[56]

---

not be subsidized to meet some foolish carbon-free goal. In fact, "micro-sources" are wholly inadequate to meet the needs of people of Africa as they develop and industrialize.

To give you an idea of the scale of electricity needed for Africa's billion people, let's look at the proposed Mphanda N'kuwa dam on the Zambezi River in Mozambique. This medium-sized project will provide 1,200 megawatts of electrical power with an estimated investment cost of $3.5 billion, financed by the Export-Import Bank of China. If the project is approved, construction will start in 2011 and be completed by 2015.[57] Note that *more than 66,000 micro-hydro projects* like the one recommended by the Institute for Environmental Security would be needed just to equal the output of this one hydroelectric project. Yet, if completed, the Mphanda N'kuwa dam will add *less than one percent to the electricity consumed today by the 20 million people of Mozambique,* even with this nation's low level of electricity usage. **Many hundreds of large-scale hydroelectric, coal, gas, and nuclear projects are needed to boost the standard of living of the people of Africa.**

Yet, the environmentalists-turned-Climatists seek to delay or block large-scale energy projects. The organization International Rivers (IR) opposes the Mphanda N'kuwa dam, arguing the project will endanger the ecology of an "already precarious" Zambezi River basin. The IR website proudly displays a graph showing construction of large dams declining since 1985, and praises their own part in stopping such "destructive" projects.[58] Clearly, ecology is more important to IR than the lives and well-being of people.

Climatist proposals are similar for Southeast Asia and other developing nations. In 1999, Greenpeace advocated a restructuring of energy production in Southeast Asia, proposing an energy mix of 50% biomass, 34% small hydro, 16% wind, and 1% solar.[59] However, India, Indonesia, and Thailand still generate 80% to 90% of their electricity from coal, gas, and oil today.

Cambodia, one of the poorest nations in the region, relies on imports of diesel fuel and oil to generate electricity, at prices approaching five times the cost of electricity in the U.S. Only about 20% of the people in Cambodia and neighboring Laos (Lao People's Democratic Republic) have access to electricity. Both nations have proposed construction of a hydroelectric dam on the Mekong River in the Hou Sahong channel bordering both nations. The project is opposed by International Rivers and other organizations.[60]

Climatism and environmentalism—no matter how well-meaning or how many famous people rally around these viewpoints—actually result in the deaths of people living in developing nations. Without electricity and gas, more rudimentary fuels continue to be used by millions of people in Africa and Southern and Southeast Asia. Figure 114 shows two women carrying wood gathered for fuel, a typical scene in rural areas of developing

Figure 114. Women Carrying Wood for Fuel in Côte d'Ivoire. (Zenman, 2008)[61]

nations. Women and children walk miles each day to gather wood or fetch water. According to the World Health Organization, more than two billion people rely on dung, wood, crop waste, or coal for their basic cooking and heating needs, resulting in harmful indoor air pollution:

> Every year, indoor air pollution is responsible for the death of 1.6 million people—that's one death every 20 seconds…In sub-Saharan Africa, the reliance on biomass fuels appears to be growing as a result of population growth and the unavailability of, or increases in the price of, alternatives such as kerosene and liquid petroleum gas.[62]

To the extent that Climatism forces the use of expensive renewable energy sources rather than lower-cost large hydropower, coal, gas, and nuclear energy, economic growth is reduced, and access to electricity, improved water and sanitation, and improved health care is delayed. These impacts on developing nations are discussed by Paul Driessen in his book *Eco-Imperialism*. Driessen summarizes:

> To block the construction of centralized power projects, as not being "appropriate" or "sustainable" is to condemn billions of people to continued poverty and disease—and millions to premature death.[63]

STRANGLEHOLD ON GLOBAL CAPITAL

How can Climatists force the developing nations to use renewable energy? Why don't the developing nations just reject projects that don't make economic sense? The answer is that financial funding for major development projects is often required, and this capital comes from banks in the industrialized world. And Climatism has put a stranglehold on global capital flows.

The drive by activist groups for investment bank "responsibility" began in earnest in the early 1990s. The Friends of the Earth and other environmental non-governmental organizations (NGOs) campaigned against Merrill Lynch and Morgan Stanley because of their financial support for China's Three Gorges Dam Project. In 1998, the Friends of the Earth and the National Wildlife Federation pressured ABN Amro, a bank of Netherlands. They opposed the bank's financial support for Freeport, a U.S. mining company, with mines in Indonesia and New Guinea. As a result of this NGO campaign, ABN Amro published an environmental policy that same year.[64] As climate change became a global issue adopted by environmental–become Climatist groups, international bank lending for projects emitting greenhouse gases increasingly came under attack.

Under pressure from Friends of the Earth, BankTrack, and Rainforest Action Network, ten major banks, including ABN Amro, Barclays, and Citibank, developed the Equator Principles in 2003 and revised the Principles in 2006 (see text box). According to the Equator Principles web site, the principles are:

A financial industry benchmark for determining, assessing and managing social & environmental risk in project financing.[65]

The principles were originally established to provide environmental standards for global lending and to be a common ground between banks and environmental NGOs. But, increasingly they have become a hammer to pulverize banks into what Climatism deems responsible behavior.

The Equator Principles are meant to be voluntary guidelines for

---

The Equator Principles

The Equator Principles are 10 principles for lending by international banks, based on environmental and social standards used by the International Finance Corporation (IFC), the private sector lending arm of the World Bank. The Principles cover projects with capital costs of $10 million or more. Banks adopting the Principles will only lend on projects conforming to the Principles.[66]

Some of the key elements are:

- each lending project will be assessed "based on the magnitude of its potential impacts and risks in accordance with the environmental and social screening criteria of the IFC"
- the project assessment will be reviewed by an independent environmental expert

The IFC's performance standard on Pollution Prevention and Abatement lists a top objective "To promote the reduction of emissions that contribute to climate change."[67] So the Equator Principles are a Climatist tool to channel global capital flows into renewable energy and out of hydrocarbon energy. Sixty-eight banks or bank groups, comprising over 80% of the world's commercial lending capacity, have now adopted the Equator Principles.[65]

global banks, but adoption of the principles has been anything but voluntary. Tremendous pressure has been applied to major banks to accept the Principles and adopt "responsible" lending practices in the name of Corporate Social Responsibility. The Climatist groups call their tactics "rank 'em and spank 'em." The concept is to attack and defeat the largest banks first and then get the others to follow suit.

Rainforest Action Network (RAN), headquartered in San Francisco, led the charge to "greenwash" the world's investment banks. In 2000, RAN launched a campaign against Citigroup's lending practices using consumer boycotts, media blitzes, and confrontational attacks. After three years, Citigroup surrendered and worked with RAN to develop a corporate environmental policy that was a precursor to the Equator Principles. As a result, Citigroup announced a greenhouse gas reporting policy in 2004 for its financial projects. Thermal power projects now report estimated greenhouse gas emissions for the life of the project.[69]

RAN is not only creative in its campaigns but also ruthless. In 2005, RAN activists brought the fight directly to the home of J.P Morgan Chase CEO, William Harrison. They put up "Old West wanted posters" of Harrison in his neighborhood and harassed his neighbors.[70] They transported a group of second-graders from a Fairfield County elementary school to JP Morgan Chase offices in Manhattan to pressure the bank to "stop lending money to projects that destroy endangered forests and cause global warming."[71] J.P Morgan capitulated to RAN and adopted an aggressive policy on greenhouse gases:

> We will therefore work with our industry, clients and policy makers
> to establish a policy framework for direct and indirect greenhouse gas
> emissions reductions.[72]

Niger Innis, spokesman for the Congress of Racial Equality, a group in New York that advocates increased investment in developing nations, criticized J.P. Morgan as "guilty of political correctness and cowardice."[73]

The Export–Import Bank of China (China Exim) is not yet an adopter of the Equator Principles and has been an active investor in energy projects in Africa. China Exim invested in the controversial Merowe Dam on the Nile River in Northern Sudan and is also a planned lender to the Mphanda N'kuwa hydroelectric project on the

Zambezi River in Mozambique. Friends of the Earth, BankTrack, and International Rivers are all highly critical of China Exim and its lack of "social and environmental safeguards."[74] Rest assured, pressure is being exerted to bring China Exim and other Chinese banks into the Climatist fold.

The venture capital industry has their own version of the Golden Rule, which is: "He who has the gold, makes the rules." This "golden rule" also applies to global investment banking. By forcing the surrender of lending practices of Citibank, J.P Morgan Chase, Bank of America, and other major banks of the world, Climatism has established a stranglehold on global investment capital. Increasingly, this pressure will dictate energy usage by denying funds for hydrocarbon energy programs in developing nations (see text box).

BankTrack spells out their vision for international lending in their paper "A Challenging Climate: What International Banks Should Do to Combat Climate Change." They demand:

> ...banks should take steps to **disentangle themselves from activities and projects that substantially contribute to climate change.** Toward this end, they should
> - End support for all new coal, oil and gas extraction and delivery;
> - End support for all new coal fired power plants;
> - End support for the most harmful practices in other GHG-intensive sectors[75]

Climatists are demanding an end to international lending on

Financial Lending Under Pressure[76]

Oil Tar Sands, Alberta Canada: Rainforest Action Network and other groups are currently opposing Royal Bank of Canada support for the project.

Ilisu Dam, Turkey: Approved by Turkish government, but European governments and banks withdrew in 2009 due to activist pressure.

Phulbari Coal Mine, Bangladesh: Asia Development Bank withdrew funding in 2008 due to activist pressure and local opposition.

Nuclear Power Plant, Belene Bulgaria: Approved by government of Bulgaria, but European banks withdrew support in 2006 due to activist pressure.

coal, gas, oil, nuclear, and hydroelectric power projects, which together provide 90% of the world's energy supply. As a poor substitute, they are forcing Africa and other developing regions to use expensive and unreliable wind and solar energy that provides less than 1% of the world's energy supply. As Paul Driessen says: "That is a virtual guarantor of perpetual poverty."[77] The world-view of **Climatism grows, now backed by the capital assets of the world's banks.**

BRIBES FOR DEVELOPING NATIONS

Climatism has framed global warming as a moral issue, caused by the industrialized nations, but mostly impacting the developing nations. Barbara Stocking, chief executive of Oxfam GB, an environmental group, lectures:

> Climate change is…a gross injustice—poor people in developing countries bear over 90% of the burden—through death, disease, destitution and financial loss—yet are least responsible for creating the problem. Despite this, funding from rich countries to help the poor and vulnerable adapt to climate change is not even 1 percent of what is needed.[78]

The 1992 United Nations Conference on Environment and Development (the Rio de Janiero Earth Summit) developed the principle of "polluter pays." The OECD provides the definition:

> The polluter–pays principle is the principle according to which the polluter should bear the cost of measures to reduce pollution according to the extent of either the damage done to society or the exceeding of an acceptable level (standard) of pollution.[79]

Therefore, the UN and Climatist backers seek the industrialized world to make massive transfer payments to the developing nations. This might make sense if carbon dioxide was a pollutant and causing global warming, but it's not.

In addition to the moral argument calling for correcting of a gross injustice, there's another big reason for the transfer payments. Recall the Climatist Dilemma from earlier in the chapter. Climatism must get China and India to curb greenhouse gas emissions. If China and India don't join the crusade, it doesn't really matter what other countries do—global greenhouse gas emissions will continue to rise. So, massive bribes in the form of transfer payments are required. Of course, the UN is planning to manage the funds transfer.

Despite Climatist rhetoric that the transition to clean energy will create jobs and be of minimal cost, developing nations don't agree. Limitation of greenhouse gas emissions will entail major economic costs. Economist Satyanarayan Murthy and his team at the Indira Gandhi Institute of Mumbai have estimated the cost impact for India. Murthy's analysis shows that emission reduction achieved by switching to renewable energy imposes costs in terms of lower GNP and higher poverty. He finds that for a 30% reduction in carbon emissions over the next 35 years, India needs compensation of between $87 billion and $278 billion over the period.[80]

The size of the transfer payments being discussed is very big. For Africa, the payments would not only be for use of renewables, but also to compensate for global warming damage due to "droughts, floods, heat waves, and rising sea levels." African leaders are asking for $67 billion in annual global warming payments.[81] The island nations of the South Pacific Ocean have seen IPCC projections of sea level rise and want compensation. The Inuit people of the Arctic fear the results of Arctic Icecap melt and want compensation. All told, China, India and the developing nations have asked for an annual payment of one percent of GDP of the OECD nations, which would total *$300 billion per year.*[82]

This idea of huge payments to China for climate change is poison to U.S. taxpayers. At hearings in the U.S. House of Representatives on April 21, 2009, Al Gore was asked about the demands by China for transfer payments by Joseph Pitts, Representative from Pennsylvania. Mr. Gore knew full well of the demands from China and other nations, but he artfully dodged the question.[83]

Misdirection of resources is the real tragedy of Climatism. The Institute of Renewable Energy Industry of Germany, estimates the world spent $170 billion on renewable energy in 2008.[84] *This is $170 billion wasted to purchase intermittent and costly wind and solar power, and low-efficiency biofuels.* Application of this sum toward solving real world problems could have major positive impact in reduction of disease, poverty, and starvation; and it could improve access to clean water and sanitation, as well as boosting *reliable* sources of energy.

A great cry will arise when citizens of Europe, the U.S., and other developed nations realize their political leaders intend to transfer billions to China, India, and other developing nations. It's likely leaders of industrialized countries will back down from this plan. Without a huge bribe to curb emissions, developing nations will continue to industrialize, setting the scene for coercive measures.

SPECTRE OF CLIMATE PROTECTIONISM

It's a short step to label carbon-emitting nations "bad global citizens," "polluters," and worse. It's another small step to impose punitive tariffs on imported goods from those nations. Remember the Climatist belief that climate is the top priority. It's clear that Climatists value the goal of stopping global warming as more important than human lives, freedom, Western Civilization, and prosperity in the developing nations. We can add world trade to this list.

In April of 2009, John Dingell questioned Al Gore in the U.S. House of Representatives about adopting a cap-and-trade system and getting compliance by the developing nations:

> JOHN DINGELL: How do we see to it that those other countries do things...and that we are not the only country who is going to suffer the economic penalties of going forward on this [cap and trade] while other countries ride on our back?

> AL GORE: ...I believe that the provisions in this bill put in place a mechanism for dealing with any recalcitrant nation that does not go along, and I believe we have the legal authority under the WTO [World Trade Organization] to do that...[85]

Indeed, the legislation proposed in the U.S. includes a provision for a tariff on carbon-emitting nations. Senators from the Midwest avoid calling it a tariff and use the phrase "border adjustment":

> We believe that a border adjustment mechanism is critical to ensuring that climate change legislation will be trade neutral and environmentally effective.[86]

France introduced the concept of a "carbon tariff" into European Union climate negotiations in 2008. France argues that such a tariff will create a level playing field for European companies competing with firms from nations without a tax on carbon emissions.[87] The European Union proposed trade restrictions in its Climate Action and Renewable Energy package in 2008. The EU's position is that, if an international agreement is not reached to reduce emissions, then "compensatory measures" will be imposed on imports from emitting nations after 2013.[88] Since agreement was not reached at the December 2009 Copenhagen summit, expect intense pressure for carbon tariffs.

U.S. Energy Secretary Steven Chu, a true believer in Climatism, publically advocated the use of a carbon tariff in March, 2009 as a weapon to protect U.S. manufacturing. According to Mr. Chu:

> If other countries don't impose a cost on carbon, then we will be at a disadvantage…we would look at considering perhaps duties that would offset that cost.[89]

Some companies favor a carbon tariff as a lever against imported products of competitors. The U.S. steel industry is pushing for tariffs against China steel makers, claiming they emit two to three times as much carbon dioxide as U.S. industry. Scott Paul, executive director for the Alliance for American Manufacturing states:

> Chinese steelmakers enjoy unfair advantage in global trade due to the lack of enforcement of exceptionally weak pollution standards.[90]

Restriction of world trade imposes very real costs on all nations. Exporting nations are hurt because they are unable to produce and ship goods for which they have a competitive advantage to other nations. Importing nations are hurt because their citizens lose the benefits of consumption of lower cost imported goods. Economists tell us the Great Depression of 1929–1933 resulted in large part because of the U.S. Smoot-Hawley Tariff Act of 1930 and passage of protectionist measures by other nations.[91]

If the persuasion and international transfer payment bribes fail, Climatism will resort to punitive tariffs on China, India, and carbon-emitting nations of the developing world. All must pledge allegiance to Climatism or be punished.

SUMMARY

Developing nations are in urgent need of low-cost hydrocarbon energy to boost economic growth, raise standards of living, and reduce poverty and suffering. Unfortunately, Climatism is bent on imposing coercive policies to force deployment of intermittent and expensive wind, solar, and biofuel solutions that are wholly inadequate to meet the needs of these countries. Propaganda, control of international lending, and transfer payment bribes are already in play, with punitive carbon tariffs soon to be added. Link by link, the chains of "sustainable poverty" are locking into place.

CHAPTER 15

CLIMATISM IN ACTION: DEBACLES AROUND THE WORLD

"However beautiful the strategy, you should occasionally look at the results." Winston Churchill

The world accepted the theory that mankind is ruining the climate with greenhouse gas emissions at the 1992 World Summit in Rio de Janiero. For the last 18 years, Climatist-recommended renewable energy solutions have been implemented across the globe. On Sir Winston's advice, let's see how well these remedies are doing.

"ENLIGHTENED" CITY OF AUSTIN, TEXAS

Austin's Climate Protection Plan calls for using 30% renewable energy by 2020, one of the most aggressive plans in the U.S.[1] Part of this plan is a sub-goal of powering all Austin municipal facilities by 2012 with 100% renewable energy.[2]

GreenChoice, the renewable program operated by electrical power company Austin Energy, was initially hailed as a smashing success. GreenChoice is the largest green-energy program by volume of sales in the United States. It provides almost two million megawatt-hours of actual delivered power per year to residents and businesses, or about 12% of Austin's electricity.[3] Austin Energy also provides the only major program in the U.S. with separate purchasing options for electricity from renewables, GreenChoice, and electricity from hydrocarbon and nuclear plants. The program won numerous awards, including the 2005 Wind Power Pioneer Award from the U.S. Department of Energy.[4]

GreenChoice offered electricity in batches and signed users to five-year and ten-year contracts as renewable power became available. Electricity rates were frozen for the term of the contract, appealing to users concerned with expected price increases in hydrocarbon energy, even though GreenChoice was more expensive than regular rates. The perceived advantage of "clean energy" also appealed to some customers. Figure 115 compares the electricity surcharge of regular fuels and GreenChoice for residential customers from 2000 to 2009. Add a base charge of 3.55 cents per kilowatt-hour to the surcharge to get the total electric rate.

In 2009 on Batch 6, GreenChoice ran into serious problems. The GreenChoice surcharge rate increased to 9.5 cents per kilowatt-hour,

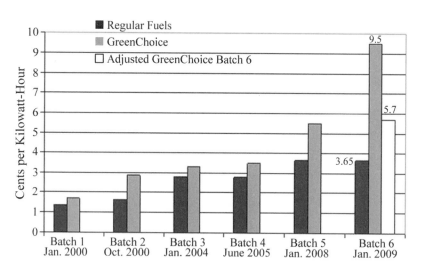

Figure 115. Electricity Cost of Renewables and Regular Fuels in Austin. Comparison of residential GreenChoice renewable electricity charge to regular fuel charge in Austin, Texas from 2000 to 2009. Add 3.55 cents base charge per kWh to get a typical total cost in 2009. Renewable rates are no longer competitive with hydrocarbon and nuclear, so Austin is subsidizing GreenChoice from other payers with the Adjusted Batch 6 Rate. (Austin Energy, 2009)[5]

significantly higher than the regular fuel electricity rate of 3.65 cents. After six months of promotion, only 1% of GreenChoice Batch 6 was sold.[6] With the base rate added, the total residential GreenChoice rate was 13 cents per kWh, compared to 7.2 cents for regular service. Customers would not pay the premium, particularly for a five- or ten-year locked-in rate.

In August 2009, the Austin City Council voted to reduce the GreenChoice surcharge to 5.7 cents per kWh and pass the difference in charges on to non-GreenChoice customers. This makes the Austin rate for "green" electricity only 30% more than for regular fuels. Austin may have greener electricity, but rates for customers are significantly higher than other cities in Texas, such as San Antonio.[7]

Recall also that wind and solar power are heavily subsidized prior to delivering electricity. Wind and solar receive a Production Tax Credit from the U.S. government of between 1.5 to 2.1 cents per kWh, depending upon the year they were placed into service. The state of Texas also provides a 10% tax deduction from state franchise (corporate) tax and an exemption from state property tax. *So the real cost of green electricity in Austin is more then 15 cents per kWh, or more than double regular electricity from hydrocarbons and nuclear.*

More price hikes are on the way. In 2009, Austin Power signed purchase agreements to buy 100 MW of biomass-generated power, as well as 30 MW from a new solar photovoltaic facility just east of Austin.[8] Biomass is estimated at double the cost of coal-fired, and solar is 2.5-4 times as expensive as coal-fired electricity (Chapter 13).

Some customers are also concerned about the reliability of wind and solar power. Austin is the site of semiconductor wafer-fabrication facilities for Advanced Micro Devices, Freescale Semiconductor and other high-technology firms. These facilities make high-value integrated circuit chips and depend upon highly reliable electrical power. A single power outage of several seconds can cost millions of dollars in production losses. Roger Wood of Freescale Semiconductor is concerned about Austin Energy's electricity generation plan:

> ...there should be reliability goals in the plan, just as there are carbon and renewable goals.[9]

Austin Energy is the showcase for a renewable energy system in the United States, but it's not price competitive. Even with large federal and state subsidies, Austin's electricity from renewables is twice as expensive as that from traditional fuels and the gap is widening. It's a shame the people of Austin have been sold on Climatism. Science shows their efforts are futile, both in terms of reducing atmospheric carbon dioxide concentration and global temperatures. In the words of movie character Dirty Harry: "That's a heck of a price to pay for being stylish!"[10]

RENEWABLE DEBACLE IN SPAIN

Spain is recognized by many as a world leader in renewable energy. The nation installed the most solar photovoltaic power and the fifth most wind energy in 2008, moving into second place for solar and third place for wind in terms of installed capacity.[11] In March 2009, President Barack Obama praised Spain as an example for the United States to follow: "Spain generates almost 30 percent of its power by harnessing the wind."[12]

While Spain's renewable energy growth is remarkable, let's look deeper to learn the real story. We'll follow analysis by Dr. Gabriel Calzada Álverez, an economics professor at the University of Rey Juan Carlos.[13] Spain's renewable history began with Royal Decree 2366/1994 in December 1994. The decree established a feed-in tariff structure for cogeneration and other facilities generating renewable electricity. Recall that a feed-in tariff pays the renewable electricity generator a rate above market price and forces retail electricity companies to buy the generated electricity. That is, the tariff provides a high price and a guaranteed market for renewable electricity.

The Royal Decree 436/2004 of March 2004 boosted the feed-in tariff price. The tariff for both wind-generated and solar photovoltaic electricity was set above the average pool price for electricity from other fuels. As shown in Figure 116, the tariff for wind varied from 50% higher to 97% higher than the pool price, while the solar rate was five to ten *times* the pool price.[14]

These generous tariffs produced an explosion of wind and solar system building. Wind power nameplate capacity grew from under 2,300 MW in 2000 to 15,000 MW by December 2008. Solar grew at an even faster rate, from 11 MW in 2004 to over 3,000 MW (nameplate power). High prices and guaranteed markets attracted a

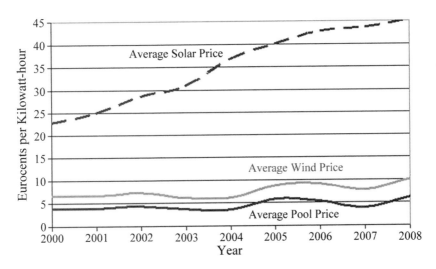

Figure 116. Electricity Prices in Spain 2000–2008. "Average Pool Price" is an average of coal, gas, nuclear, and hydropower. All prices were set by the government of Spain, including artificially high prices for wind and solar electricity. (Calzada, 2009)[15]

huge number of investors, all wanting a piece of the renewable energy action. Investors could achieve a rate of return of 17% per year, guaranteed by the government of Spain. This meant that an invested sum of €100,000 would grow to over €5 million in 25 years. According to Dr. Calzada, new renewable energy providers included "builders, real estate companies, hotel groups, and even truck manufacturers."[16]

To assure that rates to electricity customers were stable, the government of Spain paid the difference between the average pool price and the wind and solar price, accumulating what is known as the "rate deficit." The rate deficit subsidy quickly grew to €5.6 billion in 2008 and totaled €15.2 billion ($19.1 billion) from 2000 to 2008. However, since these subsidies are committed for 25 years, the total commitment to renewables is €28.7 billion ($36 billion).[17] The Spanish Prime Minister Jose Luis Rodriguez commented on the subsidies:

...it is true that renewable energy is expensive and can't be done without public sector support, it is an important investment in the future.[18]

Creation of "green jobs" has always been an objective of Spain's renewable energy program as well as an objective of the European Community. In 2008, European Commission President José Manuel Barroso described a goal of new EC proposals:

> Responding to the challenge of climate change is the ultimate political test for our generation...Our package not only responds to this challenge, but...is an opportunity that should create thousands of new businesses and millions of jobs in Europe.[19]

Spain's renewable industry build-out created a significant number of jobs, estimated by Dr. Calzada at 50,000, but at a horrendous subsidy cost of €571,000 ($716,000) per job.[20] *However, there is an unseen loss of private sector jobs that is larger than the renewable job gains.* Three factors cause this unseen job loss. First, because Spain's government invested in renewables, there is an opportunity cost of not investing in other areas of the economy. Second, Spain must either raise taxes or pursue debt financing to pay for the €28.7 billion in subsidies. Third, electricity rate increases driven by less efficient wind and solar sources reduce economic productivity. The combination of investment opportunity cost, paying for renewable subsidies, and more expensive electricity results in job losses in other areas of the economy. Calzada estimates:

> The study calculates that the programs creating those jobs also resulted in the destruction of nearly 113,000 jobs elsewhere in the economy, or 2.2 jobs destroyed for every "green job" created.[21]

Indeed, electricity rates have been climbing. Although part of this is due to price increases in natural gas, a large portion is due to the growing portion of less efficient wind and solar energy. The basic pool price of electricity climbed from 3.9 eurocents in 2000 to 6.3 eurocents in 2008, an increase of 61%.[22] Companies in energy-intensive industries, such as beverage, food processing, and metallurgy are finding it hard to compete from locations in Spain. Acerinox, the world's second-largest manufacturer of stainless steel, has been boosting capacity in the U.S. and Columbia rather than Spain, due to rising electricity prices.[23]

The boom in Spain's renewable industry was unsustainable and ended in 2008. Unable to cope with mounting deficits from subsidies, Spain issued Royal Decree 1578/2008 in September 2008. The decree established a quota system for new installations and reduced subsidy payments, particularly for solar systems. The solar industry lost 15,000 jobs within a few months of the decree.[24]

Spain's demand for photovoltaic solar cells made it the largest market in the world in 2008 at over €16 billion (see Figure 117). After the decree of September 2008, demand for photovoltaic cells dropped more than 80%. As a result, solar cell manufacturers in Europe and Asia suffered a precipitous drop in sales and have been shedding workers.

The program in Spain created enormous on-going budget deficits, raised electricity rates by 60%, caused a solar cell bubble and meltdown, and was estimated to cause the loss of two jobs for each green job created. While all this was happening, Spain's greenhouse gas emissions *increased* 13% from 2000 to 2006 and 53% from 1990 to 2006, the highest rate of increase in Europe.[25] Despite this debacle, Climatists view Spain's renewable energy program as a success, because the country now generates 10% of its electricity from

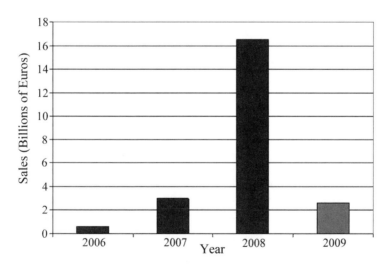

Figure 117. Photovoltaic Solar Market in Spain 2006–2009. Spain was the largest market in the world in 2008, but decreased by over 80% in 2009, due to subsidy reductions. (Asociación de la Industria Fotovoltaica, 2009)[26]

wind and solar. Does nothing else matter but reaching some arbitrary goal for renewable energy?

It's easy for a government to start an industry and create jobs by using enormous subsidies. Suppose tomorrow the U.S. decided to be the number one producer of electric cars. To do so, suppose they *fix the price* of electric cars at between two and three times the price of gasoline-powered cars (like the Spain feed-in tariff), and then *force U.S. consumers* to buy all of the electric cars that automakers can produce (like Spain's guaranteed renewable electricity mandate). This would probably produce world leadership within a few years (but also trigger a trade action from the World Trade Organization). Most people would regard such action as a major distortion of the economy. But this is very similar to what Spain has done to create a renewable energy industry.

RENEWABLES FALL SHORT IN PHILIPPINES

The Philippines is a developing nation with a 2008 population of 90 million. The nation has a per person Gross Domestic Product (GDP) of $3,539, automobile usage of about 34 vehicles per thousand persons, and about 85% of the people have access to electricity.[27,28,29] The state of economic development and energy usage in the Philippines is roughly equal to that of India, except only about 60% of Indians have access to electricity.

The Philippines is located on the northwest rim of the Pacific "Ring of Fire." The Pacific Ring of Fire is a circle running through Japan, Alaska, and the Western United States that contains the majority of the world's volcanoes.[30] As a result, there are many good locations for geothermal power plants in the Philippines. The country is the world's second largest producer of geothermal power behind the U.S., with a total installed capacity of about 1,500 megawatts.[31] Geothermal sources generate about 18% of the electricity of the nation, with 18% from hydroelectric, and the other 64% from gas, coal, and oil fuels. The country's total electrical capacity is about 6,000 MW.[32]

The government of the Philippines enacted a number of measures to encourage the use of wind, solar, and other renewables. The Renewable Energy Bill of 2008 was the most recent legislation to provide financial incentives. The incentives include seven-year tax holidays for renewable developers, exemption from value-added tax for sale of power, and a wide range of other tax benefits.[33] Berthold Breid, CEO of the Berlin-based Renewables Academy (RENAC), approves of the measures:

> With the coming into force of its Renewable Energy Act and its participation as a founding member of the International Agency for Renewable Energies this year, the Philippine government has demonstrated a clear commitment to renewable energy.[34]

Von Hernandez, executive director for Greenpeace Southeast Asia, also applauded the legislation:

> …this landmark legislation is expected to not only end our dependence on climate changing fossil fuels, but also help propel the Philippines towards a low carbon path of economic prosperity and genuine sustainable development. Through this law, we hope to see less and less development of dirty coal power plants and more investments in clean, renewable energy systems.[35]

Large-scale geothermal and hydroelectric power plants in the Philippines have been an economic success. These projects produce large amounts of power at competitive prices. *But the wind, solar, and mini-hydroelectric renewable energy projects recommended by Climatist groups have been either inadequate or failures.*

Off-grid mini-solar efforts have been underway for a number of years. An example is the USAID-funded program to provide solar cell and battery systems to 170 remote communities and 1,200 households on the island of Mindanao.[36] These programs provide valuable stop-gap electrical power to communities not yet connected to the electrical grid, extending electricity to more of the nation's people. But the total power output of all mini-solar systems deployed in the Philippines is less than one megawatt, much less than 1% of electricity demand, showing that these systems are inadequate for large-scale electricity generation.

The first major wind, solar, and mini-hydro projects deployed in the Philippines provide power to the Cebu Energy Development Corporation (CEDC). These are the Northwind Bangui Bay Wind Power Project, the Pangan-an Island Solar Electrification Project, and the Agua Grande Mini-Hydro Project. These systems were installed prior to many of the new renewable financial supports from the Philippine Government. Each of these projects is generating power below their designed outputs and operating at a loss.[37]

The Northwind Wind Power project invested $55 million in 2003 to erect 20 wind turbines, each with a nameplate power rating of 1.65 MW. Northwind was one of the first wind turbine products in Southeast Asia. According to Segundino Tiatco, Jr., plant manager for the Northwind facility, wind power is "undependable, unreliable, and intermittent." Mr. Tiatco says the facility was designed to produce 33 megawatts of electricity, but only generates an average 3–4 MW due to the unpredictability of the wind strength. He says that wind is not the answer to the power supply problem but is best used to "supplement" other electricity sources.[38]

The Pangan-an Island Solar Electrification facility is a small 25-kilowatt photovoltaic solar system with battery backup. It initially received a positive response from the community. However, the electricity price was set at an astronomical 50 Philippine Pesos ($1.02) per kilowatt-hour, and most citizens could not afford the electricity. The rate was subsequently cut to P23.50 ($0.48) per kWh (which is still five times the cost of electricity in the U.S.), making the project a money loser.[39]

The Agua Grande mini-hydroelectric project was built in 1983 with a nameplate generating capacity of 4.5 MW. But the maximum power output is a disappointing 1.6 MW, even in the rainy season. Due to the high investment cost of P3.45 million per kilowatt-hour, Agua Grande has also operated at a loss.[40]

The Philippine economy has been growing. GDP Growth was more than 4% every year from 2002 to 2008.[41] The growth of the economy, along with the goal of the Philippine government to provide electricity to every citizen, has boosted the demand for electrical power. For an example of this growing demand, we can again discuss Cebu Island. According to Jesus Alcordo, President of the CEDC, the electrical grid is now at a 30% deficit relative to demand. The deficit during mid-day peak-use hours is 90 MW and growing, just on the island of Cebu.[42]

To close the gap between supply and demand, CEDC is building two small coal-fired plants. The first is a 246 MW plant in Toledo City and the second is a 200 MW plant in Naga. Of course,

Climatist groups oppose these projects. Amalie Obusan, Greenpeace Southeast Asia Climate and Energy campaigner, declares:

> Pushing for coal-fired power plants in Cebu is not only irresponsible, but completely ignores that vast renewable energy potential of the island.[43]

Greenpeace international issued a statement stating that the Naga and Toledo City plants must not be used if the Philippine Government is really committed to the United Nations plan to cut down carbon dioxide emissions in the country.[44]

The lessons from the Philippine experience are 1) that wind, solar, and mini-hydro deliver less power than claimed, and 2) that these renewables are too expensive when not heavily subsidized. It's also clear that the quality of life of the people in the Philippines is unimportant to groups promoting Climatism. The only thing that matters is the quest to reduce those evil greenhouse gas emissions.

THE SORROW OF DENMARK

Denmark claims to produce 20% of its electricity from wind power, the largest percentage of any country in the world. (Spain is a distant second with just under 12% of electricity from wind.) With 5,200 wind-turbine towers, the nation also has the highest density of wind towers, with one tower for every 1,000 people. The wind system has a nameplate generating power of 3,100 MW.[45] Denmark has also fostered a wind turbine manufacturing industry and today claims one-third of the global wind turbine market. A promotional brochure from Energinet.Dk, the country's electricity transmission system operator, calls Denmark the "wind power champion."[46] Connie Hedegaard, Denmark's Minister for Climate and Energy, is proud of what Denmark has accomplished:

> The Danish way of introducing and adapting wind energy is what we call "the Danish Wind Case." It is my sincere hope that he rest of the world will profit by the Danish experience.[47]

But, let's take a closer look at this Danish wind "miracle." In 1990, the Danish government established goals for renewable energy sources, including a national goal for 1,300 MW of installed wind power capacity by 2000. At the same time, the government promoted research and development into wind power with subsidies and also established a favorable feed-in tariff, the first of a series of such tariffs. The tariff structure encouraged thousands of private individuals to invest in wind turbines and form cooperatives to deliver power to the electrical grid.[48]

As we saw in Chapter 13, wind power output varies unpredictably. Winds for the entire nation of Denmark can go from zero to 13 meters per second, full-rated power, in a few hours or less. Denmark's ability to use wind power effectively is in large measure due to exchange of electrical power with nearby Norway and Sweden. Both nations are blessed with large amounts of hydroelectric power. For 2006, hydroelectric power produced 97% of Norway's and 41% of Sweden's electricity.[49] Hydroelectric power can be quickly turned on or off, providing a good match for Denmark's highly variable wind power output. When the wind is strong, Denmark exports power to Norway and Sweden, which turn down their hydroelectric generators. When the wind fails in Denmark, Norway and Sweden turn up their hydroelectric generators and export power to Denmark. So in effect, the hydroelectric reservoir lakes of Norway and Sweden "store" energy from Denmark's wind power system.

The proximity of partner nations with large hydroelectric systems makes Denmark unique. Hugh Sharman, principal of Danish energy consulting firm Incoteco, has concluded that, on average, over 50% of the wind power energy of Denmark is exported to Norway and Sweden. As a result, only about 10% of Denmark's electricity comes from wind power, rather than the 20% claimed.[50] Without these northern partner nations to receive electricity during high periods of wind, the Denmark wind system would not be able to operate effectively. Other nations without nearby hydroelectric storage, such as the United Kingdom, will not have the same advantage.

So is Denmark's wind power program a success? In terms of CO_2 emissions, the results are less than planned. Even with all the wind farms, Denmark's emissions are down only 0.3% from 1990 to 2007–essentially flat.[51] It's still a long way to reach the country's 20% emissions reduction target by 2020.

In terms of cost, the 2008 electricity rate for households in Denmark was 28 eurocents per kilowatt-hour, *by far the highest in*

Europe, and about four times the U.S. price.[52] About 50% of this rate is taxes and a large portion of the taxes go to subsidies for the wind power system. Sharman estimates the annual subsidy from 1996 to 2004 at €257 million per year. Total Public Service Obligation subsidies for wind electricity have been averaging about 7–8 eurocents per kWh, which is roughly 50% over the spot price for hydrocarbon electricity.[53] In short, citizens of Denmark are paying *more* for electricity because wind power is a higher cost generation method.

But, there is yet another story to this subsidy. Since about 50% of the nation's wind power is exported, half of the subsidies paid by Danish consumers go to benefit the citizens of Norway and Sweden. Mr. Sharman estimates this "foreign aid subsidy" to total €970 million from 2000–2008. This export of subsidized electricity is not publically discussed in Denmark.[54]

Denmark is also running into another weakness of wind power systems. Some of the turbines are wearing out. Wind turbines are constructed in areas of maximum winds, always a rugged weather environment. Most of the country's wind turbines were built during two peak periods, 1988–1992 and 1996–2002.[55] Wind power manufacturers promise today's turbines will achieve 20–25 years of useful life. This compares to the useful life of 40–60 years for a hydrocarbon or nuclear facility. Wind turbines from the 1988–1992 build out are reaching end-of-life and will soon need to be replaced.

In summary, Danish wind power achieved only limited gains in emission reductions, but produced electricity rates that are the highest in Europe, and continues to penalize consumers who subsidize the 50% of wind power that is exported. Yet, Denmark plans to spend €12 to €15 billion to double wind capacity by 2025. Sharman is concerned about the nation's current path:

> The very fact that the wind power system, that has been imposed so expensively upon the consumers, can not and does not achieve the simple objectives for which it was built, should be warning the energy establishment, at all levels, of the considerable gap between aspiration and reality. Denmark needs a proper debate and a thorough re-appraisal...before forcing the country into a venture that shows high risk of turning into an economic black hole.[56]

Yet, it seems that Denmark has paid an even higher price than just higher electricity bills and taxes. Over the years, I've had the pleasure of visiting Denmark several times, both on business and holiday. In college, I traveled through Denmark by train with a good friend, lodging at youth hostels. We stayed at Ribe on the southwest side of the Jutland Peninsula. Ribe is said to be the oldest town in Denmark, founded in the 8th century by the Vikings. My companion and I spent one beautiful sunlit day walking five miles from Ribe to the North Sea and back. There was nothing to interrupt our pastoral view of the flat farmlands, low hills, and blue sky. But I'm not sure the same view is there today.

Figure 118 shows a map of Denmark with black dots marking the locations of the thousands of steel and concrete wind turbine towers. It's sad that field and farm, shoreline and hill, and almost all of

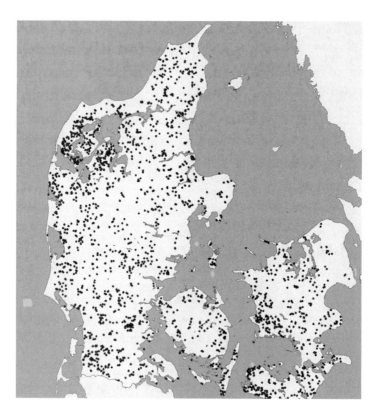

Figure 118. Wind Turbines in Denmark. Black dots mark the locations of almost 5,000 wind turbine towers. Light gray areas in the sea are offshore wind farms. (Adapted from Energinet.Dk, 2009)[57]

Denmark's vistas are interrupted today by these monuments to Climatism. But this is all so unnecessary. The 5,200 wind turbines of Denmark produce a total average output of *only* 817 megawatts.[58] **They could ALL be replaced by a SINGLE 1,000 MW coal-fired or nuclear plant.**

Why were all these towers built? Government leaders are tasked with making choices between alternatives. Because Denmark's leaders believe in Climatism, they chose to build 5,200 steel and concrete towers all over their beautiful nation, rather than one additional conventional power plant. Ms. Hedegaard tells us:

> ...Denmark should be a green and sustainable society with a visionary climate and energy policy...The answer to these challenges lies in the way we produce and consume energy and in our ability to adapt our society to climate change.[59]

CALIFORNIA: A BLIZZARD OF CLIMATE REGULATIONS

California is the epicenter for Climatism in the United States. Many of the world's leading environmental groups are headquartered in California, including the Sierra Club and International Rivers. Governor Arnold Schwarzenegger professes his true belief:

> We simply must do everything we can in our power to slow down global warming before it is too late...We can save our planet and also boost our economy at the same time.[60]

The Governor has aggressively promoted the use of renewable energy. In 2002, California first established a Renewable Portfolio Standard calling for the use of 20% renewable energy by 2017. After his election in 2003, Governor Schwarzenegger accelerated the 20% RPS requirement forward to 2010. In November 2008, he signed executive order S-14-08, requiring that all retail sellers of electricity use 33% renewable energy sources by 2020.[61]

California is one of the largest users of electricity in the world. In 2008 the state consumed 307,000 gigawatt-hours of electrical energy, about eight times the electricity consumed by Denmark and about 80% of that consumed by the United Kingdom.[62] Thirty-two percent of 2008 consumption was imported from other states, including almost all of the electricity from coal. Renewables provided 10.6% of the power, with most from geothermal and the least from solar. Figure 119 shows California's electricity consumption by fuel type.[63]

California made early efforts to invest in renewable power. Like the Philippines, the state is located on the Pacific Ring of Fire and generates about two-thirds of the geothermal power produced in the United States. Geothermal facilities in Napa and Sonoma counties north of San Francisco have been generating electricity since the early 1960s. Forty-three operating geothermal plants produced almost 4.5% of California's electricity in 2008.[64]

California was also an early pioneer in large-scale electricity generation by both solar and wind power. Federal and state income tax credits in the 1980s provided a subsidy of about 50% of the capital investment, stimulating the construction of the first commercial power plants for both wind and solar.[65] The nine Solar Energy Generating Systems (SEGS) in the Mojave Desert built by Luz International were part of this early investment.

Large-scale wind farms at Altamont Pass near San Francisco, and near Los Angeles at Tehachapi Pass and San Gorgonio Pass, made California the world's leader in renewable energy from wind power. By 1990, 17,000 small wind turbines (50 to 300 KW) with a nameplate capacity of over 1,600 MW were installed, providing 90%

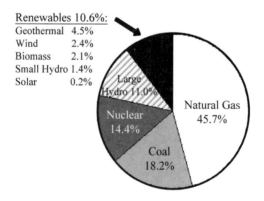

Figure 119. California 2008 Electricity Consumption by Fuel Type.
Consumption includes electricity produced in state and imported. In 2008, 32% of electricity was imported, including most produced by coal. (CEC, 2009)[66]

of the world's installed wind capacity at that time.[67]

California is counting on renewable energy to provide greenhouse gas reductions. The state's goals are first to reduce emissions to the 1990 level by 2020 and next to 80% below the 1990 level by 2050. But even with California's early pioneering efforts, emissions continue to rise. Data from the California Air Resources Board show emissions rose 12.3% from 1990 to 2000 and an additional 4.7% from 2000 to 2008.[68]

Despite an impressive early start, California is finding the switch to renewables difficult. Figure 120 shows that California renewable energy consumption peaked in 1992, then declined, and only recently surpassed the previous peak in 2008. It's clear the state will not be able to meet the 2010 goal of 20% from renewables, requiring renewable output of more than 50,000 GWh of electricity.

The state's renewables drive has run into several problems. First, electricity from geothermal sources has declined since 2005 and is up only 8% since 1997. Geothermal facilities are renewable in the short-term, but tend to lose energy output over the longer term. Second,

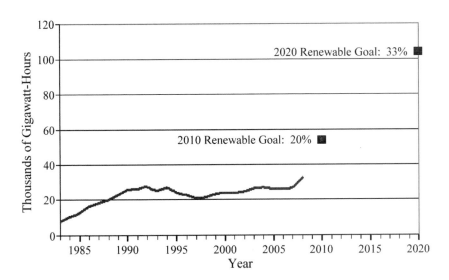

Figure 120. Renewable Energy Consumption in California 1983-2020. Renewables include electrical production from geothermal, biomass, wind, and solar, but exclude large hydropower. Small hydropower is included in graph data only for years 2001–2008. The state is far short of the 2010 goal of over 50,000 GWh. (CEC, 2009)[69,70]

electricity from biomass has been flat since 1997. Third, wind power ran into a significant "speed bump" in the 1990s.

The problem is that *wind turbines kill birds.* Not just sparrows, but large birds, bats, and often protected species will fatally collide with windmill blades. Beginning in 1988, a series of reports were published reporting a high level of bird fatalities at the Altamont Pass Wind Resource Area (APWRA) in California. BioResource Consultants conducted a four-year study of bird mortality at the pass for the California Energy Commission (CEC) and concluded:

> ...between 881 and 1,300 raptors are killed annually in the APWRA. For all birds combined, that number is estimated at between 1,766 and 4,721... Among these, researchers estimate that the APWRA turbines are annually killing 75 to 116 golden eagles, 209 to 300 red-tailed hawks, 73 to 333 American kestrels, and 99 to 380 burrowing owls.[71]

It seems that the Altamont Pass wind farm was built in a "major raptor migration corridor." Studies also show that rodents and other prey sometimes multiply in the area under the wind towers, attracting raptors and increasing the number of collisions.

The problem was deemed serious enough for Alameda County to stop permits for capacity expansion at APWRA. The California Energy Commission issued voluntary guidelines in 2007 for wind power developers to follow. The guidelines call for the use of "bird use count" studies and "raptor nest searches." Such studies must be reviewed as part of permitting requirements under the California Environmental Quality Act (CEQA).[72] Permitting requirements and legal opposition have slowed the growth of California wind power. California has dropped to third behind Texas and Iowa in terms of installed wind power capacity and was 19th in 2008 in terms of capacity additions.[73]

Bird kills and bat kills at wind farms are not just a problem in California, but also at every installed wind power system worldwide. Collisions are an issue in Spain, Norway, Minnesota, and many other locations. System operators are proposing novel approaches to solve the problem. One system in Texas proposes to detect migrating bird flocks with radar and then switch off the turbines.[74] Add more cost to the already high price of wind-generated electricity.

The Bioresource Consultant study for the CEC conservatively

estimated about one bird death for every two turbines per year, although other estimates are higher.[75] The U.S. Department of Energy estimated 2008 worldwide wind capacity at about 122,000 MW, which means about 120,000 wind turbines installed.[76] If one bird is killed per year for every two turbines, this means 60,000 birds killed in 2008. In 2009, the number will likely grow to 70,000 kills. At the rate wind turbines are being built, we'll soon be at 100,000 birds killed every year, many of them raptors. This is a large number by any standard.

Most environmental organizations have been remarkably silent about this. Daniel Beard of the National Audubon Society labeled turbines "condor cuisinarts" in 1999, but coverage in the news media has been rare.[77] The same environmental organizations that were driven by concern for birds in Rachel Carson's *Silent Spring* have now sold their souls to Climatism.

With little growth from geothermal and biofuel sources, and with wind power slowed by the bird kill issue, California is placing heavy weight on growth of solar-generated electricity. This includes both large-scale solar plants and roof-top solar systems. Over the last 30 years, California residents and business have installed more than 50,000 roof top solar systems, delivering 150 MW of nameplate and about 15 MW of average power.[78] The total of large-scale and roof-top systems generate about 100 MW of average power or about 838 MWh of electricity, *only 0.27 percent of California's electricity usage.*

California is boosting incentives for solar power generation. On top of the federal 30% tax credit, the state offers residents and businesses rebates that amount to 10% to 15% of the cost of a solar system. For years, the state has had a feed-in tariff requiring utilities pay the market price of electricity for small solar systems. But since this tariff has "not generated enough growth" in solar systems, California is proposing new methods to boost the feed-in tariff to above market prices, as used in Europe.[79]

Given California's Renewable Portfolio Standards mandates of 20% by 2010 and 33% by 2020 and the high level of federal and state subsidies, a growing number of large-scale solar projects are being proposed. In September 2009, 40 different projects were listed on the California Energy Commission web site as "under review" or "announced." Most of these projects use the parabolic trough solar thermal or solar tower systems we discussed in Chapter 13. If all 40 projects are approved, the actual average generated electricity would be about 2,400 MW of power, assuming an average output of 20% of rated power. Many of these proposed systems are hybrid solar and natural gas systems, so about 25% of this power is from natural gas backup.[80]

Twenty-four hundred megawatts is a large amount of electrical power. This would provide over 20,000 gigawatt-hours of electrical energy, compared to California's 32,532 of total renewable electricity today. But if these projects are implemented, other impacts will be also be huge. Roughly 210 square miles of land would need to be covered with solar troughs or mirrors, equivalent to a square 14.5 miles on a side. *This is an area equal to paving the entire San Francisco peninsula from the Golden Gate Bridge to San Mateo with solar cells.* Since solar is 2.5 to 4 times as expensive as conventional fuels, the more installed, the more electricity prices will rise.

California residents and businesses are struggling under a blizzard of global warming regulations. In September 2006, Governor Schwarzenegger signed AB32, the California Global Warming Solutions Act, the toughest greenhouse gas legislation in the U.S. The Act set greenhouse emissions targets for 2020 and 2050 and empowered the California Air Resources Board (CARB) to develop a "Scoping Plan" of regulations and a mandatory reporting system to track and monitor emissions.[81] CARB issued the Scoping Plan in December 2008, with implementation to begin in 2010.

The Plan contains many measures for reducing vehicle emissions. Most important is a Low Carbon Fuel Standard (LCFS), calling for a 10% reduction in carbon emissions from vehicle fuels. Vehicle regulations will be imposed on tires, engine oils, paints, window glazes, and cars by level of emissions. Fees on citizens include per-mile charges for vehicle insurance, fees for traveling on congested roadways, and fees on housing developments and businesses to pay for emissions reductions. Fees are added on "high global warming potential" chemicals. Heavy-duty trucks will see new requirements for aerodynamic trailers, lower rolling resistance wheels, and refrigerated vehicles. Additional regulations are in the works for cargo vessels, rail operations, and other goods transportation industries.[82]

The Plan calls for California participation in the Western Climate Initiative, a cap-and-trade system, with other western state partners. Power plants and other large industrial manufacturers will be required to purchase credits for carbon emissions by 2012. On-road transportation and residential and commercial natural gas will be added to the system in 2015.[83]

The CARB web site lists several objectives of the Scoping Plan, including to "achieve significant greenhouse gas reductions" and to "demonstrate national and international leadership." CARB actually believes they can raise taxes, smother citizens with regulations, and at the same time improve the California economy. The Scoping Plan projects the following benefits in 2020 relative to the "business-as-usual" scenario:

• Increased economic production of $33 billion
• Increased per capita income of $200
• Increased jobs by more than 100,000[84]

CARB is living in a fantasy world. A study by Varshney & Associates for the California Business Roundtable finds that AB 32 will cost the state $182 billion in lost output, 1.1 million in lost jobs, and a cost of $3,857 for each California citizen.[85] Even CARB's own review panel unanimously criticized CARB's assessment of the economic benefits, concluding the state "handpicked" data to improve projected results.[86]

California already has among the highest electricity rates in the U.S. The residential rate is 14 cents per kilowatt-hour, a 23% premium over the U.S. national average, and the industrial rate is 38% over the average.[87] State citizens will pay more and more for the "privilege" of green power.

The irony is that California citizens believe they are making a difference by cutting their greenhouse gases. If they are able to reduce their emissions to the state target of 427 million tons of CO_2 equivalent in 2020, *they'll have made less than a 0.4% change in world emissions.* Even this is meaningless, since man-made carbon emissions do not drive global temperatures.

THE U.K.: MANY SACRIFICES FOR ILLUSORY BENEFITS

The United Kingdom is bent on the path of radical carbon emission reductions in the name of Climatism. The government, scientific organizations, and major institutions are forcing citizens to march lockstep toward reorganized lifestyles, lower standards of living, and higher energy prices, all to reduce greenhouse gas emissions. Yet Britain's efforts will have an insignificant effect on overall world emissions.

Fifty years ago, coal provided 74% of the U.K.'s primary energy needs. The discovery of large reserves of natural gas in the North Sea during the 1960s, along with the introduction of nuclear power, started a transition away from coal. Natural gas replaced "coal gas," (gas manufactured from coal), for home cooking and heating. At the same time, nuclear power was introduced and served the growing demand for electricity. But by 1990, the Kyoto Protocol baseline year, coal still generated two-thirds of electricity in the U.K. However, during the 1990s, new environmental controls on sulphur dioxide and nitrogen oxides required electricity suppliers to install abatement equipment on coal-fired plants. At the same time, low prices for natural gas from the North Sea encouraged many suppliers to build new gas-fired plants in place of coal. By 2000, coal production of electricity had dropped to about one-third of the nation's needs, with natural gas generation increasing to provide another one-third.[88]

By switching from coal to natural gas, the U.K. has reduced greenhouse gas emissions by as much as any other nation. The Department of Energy and Climate Change (DECC) claims a 21% reduction in 2007 from 1990.[89] Since natural gas emits only about 45% of the carbon dioxide emissions of coal, the U.K. was able to make early large emission reductions primarily by switching one hydrocarbon fuel for another.

In July 2009, the DECC for Her Majesty's Government issued the Low Carbon Transition Plan, the national strategy for energy and combating climate change. Ed Miliband, Secretary of State of Energy and Climate Change, states in the foreword:

> The transition to a low-carbon economy will be one of the defining issues of the 21st century. This plan sets out a route-map for the UK's transition from here to 2020...every business, every community will need to be involved. Together we can create a more secure, more prosperous low carbon Britain and a world which is sustainable for future generations.[90]

The Low Carbon Transition Plan (LCTP) is quite ambitious, as

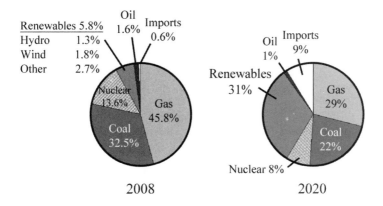

Renewables 5.8%
Hydro   1.3%
Wind   1.8%
Other   2.7%

2008

2020

Figure 121. U.K. Electricity Production by Source in 2008 and 2020 Plan.
U.K. plan to shift from 80% fossil fuels in 2008 to 31% renewables and reduced fossil fuels by 2020. (DECC, 2009)[91]

shown in Figure 121. It calls for renewables to grow from 6% in 2009 to 31% in 2020. It assumes flat electricity demand, but given the oppressive level of green regulations being imposed, zero electricity growth and zero economic growth are likely.

A closer look at the plan shows it to be unrealistic. U.K. average electrical power consumption in 2008 was about 44 gigawatts. With limited gains possible from hydroelectric, biomass, and solar, more than 90% of the growth in renewables must come from wind power. This means the plan needs a delivered wind power increase of about 10 GW by 2020. The U.K. has about 2,600 wind turbines in operation today, operating at 27.6% capacity factor and delivering about 1.1 GW of average power.[92] If we generously assume capacity factor rises to 35%, the nation will need to add 28.6 GW of wind turbine capacity to generate the 10 GW needed. At two megawatts per wind turbine, this means over 14,000 wind towers will need to be installed. *More than three wind turbines will need to be placed into service per day over the next eleven years to meet this goal.*

Opposition to onshore wind towers has been growing from British citizens, so its likely most of these towers will be constructed offshore. Because construction conditions in the North Sea and the seas around Britain are challenging, there will be months at a time when weather conditions do not permit tower building. Installation of electricity transmission lines will also be required, an expensive and time-consuming process.

As more and more wind turbines are built, Britain will need to deal with wind power at its worst. Britain does not have hydroelectric power to act as storage for wind systems, nor do nearby neighbor countries have much hydropower. Conventional power stations will need to be maintained in hot standby status for when the wind fails. Offshore wind power is also very expensive, sure to boost the electricity bills of the nation's citizens.

The LCTP is highly optimistic in terms of the impact on Britain's energy prices. The Plan states that the impact on U.K. electricity bills from climate change measures will be only 8% by 2020. But the new 25% of energy needs in 2020 from renewables will be mostly high-cost offshore wind power, which produces electricity at more than double the cost of hydrocarbon fuels. The LCTP also calls for the retrofit of all coal-fired facilities with Carbon Capture and Storage (CCS) equipment by 2020.[93] If proven feasible, this requirement will boost the price of the 23% of the nation's electricity from coal by about 50%. The combination of these policies will boost the cost of electricity by a minimum of 36% by 2020, not 8%.

U.K. electricity bills are currently low compared to Denmark and Germany, but they are rising at a faster rate. According to U.K.'s energy price regulator, Office of the Gas and Electricity Markets, the average customer electricity bill rose 75% from August 2003 to August 2009.[94] Tack a minimum 36% increase on these already rising rates for a maximum negative impact on British citizens and businesses.

In addition to renewable energy mandates, the Low Carbon Transition Plan and other U.K. government policies have many other mandates that will be costly to British citizens. Each citizen must pay to have a smart meter installed in the home to measure electricity usage by 2020. Regulations on housing construction to make the "whole house greener" will also boost the price of housing.[95] The British National Health System is removing "carbon intensive" meat from menus at hospitals, claiming this will "save millions of people round the planet from hunger, water shortages and coastal flooding."[96]

Transportation is facing a brace of new taxes and mandates. The Plan calls for new car emissions of only 95 grams of CO_2 per kilometer in 2020, in line with the mandate from the European Union.[97] Auto makers estimate additional per vehicle cost of £3,300 just to reduce emissions to 120 g/km, let alone 95 g/km.[98] Prime Minister Gordon Brown favors new taxes on air travel, stating "the security of our planet and our humanity is at stake" unless global warming is tackled.[99] Europe's Air Passenger Duty is £10 for short flights and £40 for longer flights, with plans in place to more than double the tax in 2010.[100]

All these measures will result in real costs to British citizens. Secretary Miliband now projects climate legislation will cost the U.K. £404 billion between 2009 and 2050. This equals about £20,000 for each British family over the next 40 years.[101]

Of course, few people want to incur these costs, so the U.K. government has fired up the propaganda machine to persuade citizens to adopt Climatism. Government pamphlets are circulating, urging people to "walk to work to stop climate change."[102] Ministers of Parliament have announced their personal participation in the "10:10 Campaign," a commitment to cut personal emissions 10% in 2010.[103] Lord May, President of the British Science Association, has called for faith groups to police social behavior.[104]

The British are an amazing people. Almost 70 years ago, the fall of France in June 1940 left most of Western Europe conquered and Britain standing alone against the might of Nazi Germany and Italy. It would be more than a year until the Soviet Union entered the war in July 1941, and then the United States in December 1941, as allies to the British cause. The citizens of the nation endured the Blitz of London, a nightly bombing by German aircraft. They also narrowly survived Germany's attempt at economic strangulation with the U-boat blockade. British citizens have shown they can endure any hardship with the right cause and leadership.

But, it's a shame they're now pursuing the pointless remedies espoused by Climatism. China's *annual increase* in CO_2 emissions equals the *total* emissions of the nations U.K. and Australia combined.[105] The climate sacrifices of the British people are relatively insignificant and would not be measurable in global temperature change, even if greenhouse gases were the primary cause of global warming.

THE EUROPEAN EMISSIONS TRADING CIRCUS

In December 2008, The European Union's Climate Action and Renewable Energy (CARE) package was approved by 27 European nations. By approving CARE, the nations accepted the following legally binding targets for year 2020:

1. Reduce European greenhouse gas emissions to 20% below 1990 levels, or 30% below 1990 if an international agreement can be reached.

2. Increase European use of renewable energy to 20% of total energy usage (not just electricity).

3. Boost biofuels to 10% of transport fuel usage.

Each nation agreed to individual specific targets for renewable energy usage. The United Kingdom agreed to grow renewable energy to 15% of its total, the largest renewable increase of any country.[106]

The nations also agreed to adopt a new Emissions Trading System (ETS) with a European Union (EU)-wide emissions target beginning 2013. The new system beginning in 2013 will be the "third phase" of the ETS. The planned EU-wide emissions target is viewed as an improvement from the individual country targets of Phase I (2005–2007) and Phase II (2008-2012).

The ETS is a carbon emissions trading system with the sole purpose to raise the cost of processes that produce greenhouse gas emissions. Also called a "Cap-and-Trade" system, it intends to establish a cost of carbon emissions to encourage (or force) companies and industries to adopt processes with lower emissions. The ETS is considered the primary policy tool for achieving a market-driven reduction in greenhouse gas emissions (see text box).

The EU Emissions Trading System can accurately be characterized as a disaster to date. Beginning with Phase I in 2005, most member nations set emissions targets higher than actual emissions (with the U.K.'s tough target a notable exception). Allowances, or permits, were then provided freely to participating companies. The price of allowances dropped sharply from €33 per ton of carbon to just €0.20

The European Union Emissions Trading System

In January 2005, the EU established the Emissions Trading System (ETS). The ETS is a carbon trading system designed to reduce greenhouse gas emissions. The system places a cap on the total emissions of participating industries, divides the cap amount into rights to emit, or "allowances," and allows participants to trade those allowances.

Participating companies must deliver allowances to regulating authorities proportional to the amount of their carbon dioxide emissions. To do so, they must acquire allowances through an auction or by purchase from other companies. This creates a market and results in a common price for carbon allowances. The theory behind the carbon trading system is to create an incentive for all participants to reduce emissions and to encourage the lowest cost emissions reduction methods across the system. The system calls for the cap to be tightened every year, to raise the price of allowances and the cost of emissions.[107] About 10,500 energy-intensive facilities in Europe are currently covered by the ETS.[108]

The ETS utilizes the Clean Development Mechanism (CDM) established by the United Nations. The CDM allows organizations in developed nations to gain "certified emissions reductions credits" through support of low-carbon development projects in the developing world. These credits allow greenhouse gas emitters to offset their own carbon emissions rather than implement costly emission-reduction programs.

The global carbon trading market grew to €92 billion ($131 billion) in 2008, with almost 5 billion metric tons of carbon traded. About 30% of the allowances traded were CDM certificates.[109]

---

per ton, eliminating any incentive to reduce emissions. Wealth transfer between nations occurred as U.K. firms were forced to buy permits from firms in France and Germany, which had a surplus.[110] Open Europe, a U.K. think tank, criticizes the ETS:

> The Emissions Trading Scheme (ETS) is supposed to be the EU's main policy tool for reducing emissions. But so far, it has been an embarrassing failure. In its first phase of operation, more permits to pollute have been printed than there is pollution. The price of carbon has collapsed to almost zero, creating no incentive to reduce pollution. Across the EU, emissions from installations covered by the ETS actually rose by 0.8%.[111]

Phase II has hardly been better. ETS plans call for eventual auctioning of permits, but permits are still being allocated free of charge. Electrical power producers in the U.K. are believed to have made windfall profits through a combination of selling allowances and passing costs on to consumers. The permit allocation system may actually encourage emissions since more permits have been allocated to coal-fired plants than gas-fired plants. But the biggest problem in Phase II is that member states can now import Kyoto Protocol credits in order to meet emissions reduction targets.

Under the Kyoto Protocol, participation in Clean Development Mechanism (CDM) projects in developing nations provides credits for use in place of allowances for emissions. CDM credits were not used in Phase I, but have swelled to 30% of the carbon trading market in Phase II. By a combination of allocated allowances and CDM credits, the ETS circus again provides little incentive for European firms to reduce emissions.

Intended to encourage developing nations to invest in low-emissions projects, the CDM system is rife with improper application and fraud. Kyoto credits are not awarded for actual emissions reductions, but for theoretical reductions that would have otherwise occurred. Smart project managers claim reductions for technologies that would be installed anyway. Xiaogushan dam in China was awarded $30 million worth of CDM credits, even though construction was well underway and had received loans from the Asian Development Bank.[112] Inspection companies have been accredited by the UN to police the market, but these firms are unreliable. DNV of Norway, the single largest auditor, was suspended in November 2008 for failure to properly audit CDM projects. SGS UK was similarly suspended in September 2009.[113]

Reductions of what are known as "exotic" greenhouse gases can be especially rewarding. The European aluminum industry successfully lobbied to have PFCs (Perflourinated Compounds) included in Phase III. Why would an industry lobby for an emissions restriction on itself? Aluminum processing emits PFC gas. One ton of PFCs is equivalent to 6,500 tons of CO_2.[114] Because aluminum firms are under global competitive pressure, the EU will likely provide large numbers of free allowances. By scrubbing out their PFC emissions, they'll be able to sell allowances to profit from ETS. According to Open Europe, about half of the money spent to date on Kyoto CDM credits has been awarded to clean up such "exotic" greenhouse gases.[115]

Carbon emissions trading is not a real market. Buyers and sellers establish the price of a bushel of corn, but buyers and sellers don't exist for a ton of carbon emissions. An artificial market for carbon emissions can only be established through the arbitrary and imperfect decisions of government officials.

As a result, carbon trading systems require *government officials to make an endless number of arbitrary decisions with the end result being a high impact on every part of the economy.* Regulators make decisions about which industries are to be covered by carbon trading and which are not. They determine the size of the cap and the number of carbon allowances. They decide how many free allowances are to be awarded, to which industries, and ultimately the size of the award to each company. Even measurement of carbon emissions is arbitrary. One can measure a bushel of corn, but emissions estimates must be made for every industrial activity, and regulators make these decisions. **In effect, bureaucrats are arbitrarily handing out allowances worth billions of Euros.** What a gold mine for national and EU officials! Such a system is fertile ground for intense industry lobbying, manipulation of emissions estimates, political payoffs, protection of favored companies, economic distortions of all sorts, and endless opportunities for corruption and abuse. **But, even if the system actually works, energy costs rise, jobs are lost, and economic growth stagnates.**

And talk about the rise of a global constituency! We now have tens of thousands, maybe hundreds of thousands of bureaucrats in the United Nations, the IPCC, the European Union, the environmental policy departments of all the provincial, state and national governments of the world, and on and on, who own their livelihood to Climatism. Do you think this constituency will accept the growing evidence that global warming is due to natural causes?

CLIMATISM IN ACTION

The evidence is clear. The results from across the world are gruesome and getting uglier every day. The demonstrated outcomes of Climatism are:

- rising electricity prices and loss of jobs;
- more expensive air travel, vehicles, houses, and consumer goods;

- multiplying regulations, higher taxes, economic distortions, and loss of freedoms;
- a huge expansion of national and international governments; and
- pastoral landscapes fouled by steel and concrete towers.

From California, to Denmark, to the Philippines and most everywhere in between, citizens are suffering from the punishing solutions of Climatism.

CHAPTER 16

ENERGY NONSENSE FOR THE GOOD OLD U.S.A.

"I am willing to love all mankind, except an American."
English author Samuel Johnson (1778)

T he leadership of the United States has finally accepted Climatism. President Obama, his executive branch staff, and the majority of the United States Congress believe man-made greenhouse gases are destroying Earth's climate. Secretary of State Hillary Clinton summarizes:

> ...American leadership is essential to meeting the challenges of the 21st century, and chief among those is the complex, urgent, and global threat of climate change. From rapidly rising temperatures to melting arctic icecaps, from lower crop yields to dying forests, from unforgiving hurricanes to unrelenting droughts, we have no shortage of evidence that our world is facing a climate crisis...Under President Barack Obama, the U.S. will take the lead in addressing this challenge, both by making commitments of our own and engaging other nations to do the same.[1]

Upon taking office in January 2009, the new administration and the 111th Congress moved quickly on new climate measures. In February, the White House proposed the budget for fiscal year 2010, with major measures to fight climate change. President Obama asked

Congress for a new energy policy:

> But to truly transform our economy, protect our security, and save our planet from the ravages of climate change, we need to ultimately make clean, renewable energy the profitable kind of energy. So I ask this Congress to send me legislation that places a market-based cap on carbon pollution and drives the production of more renewable energy in America.[2]

New measures for fiscal years 2012–2020 included $15 billion per year in subsidies for renewable energy, primarily wind and solar power. Another $63 billion in tax breaks and assistance are provided to individuals and businesses converting to clean energy technology.[3]

The Congress approved the President's budget on April 2, 2009 and began work on cap-and-trade legislation. The American Clean Energy and Security Act (ACES) was passed by the U.S. House of Representatives on June 26 and sent to the U.S. Senate for discussion.[4] Also called the Waxman-Markey Bill for sponsors Henry Waxman and Edward Markey, the Act proposes a carbon trading system and a broad host of new legislative mandates. The bill seeks to control every aspect of energy use by U.S. citizens. See the text box for a summary of some of the major provisions and impacts of ACES.

Proponents of Waxman-Markey (W-M) make many claims in order to sell this "snake-oil" to the American people. The primary ones are: 1) W-M will break U.S. dependence upon foreign oil, 2) W-M will generate green jobs and grow the economy, 3) W-M will cost only a postage stamp a day, 4) 20% of U.S. energy from renewables is possible by 2020, and 5) W-M will stop global warming. We've already discussed the fallacy of claim number five. But, let's take a look at each of the other four claims.

FALSE CLAIM #1: INDEPENDENCE FROM FOREIGN OIL

After the House of Representatives committee approval of ACES, committee chairman Henry Waxman stated:

> This bill, when enacted into law this year, will break our dependence on foreign oil…[5]

Waxman-Markey Clean Energy and Security Act:
Major Provisions and Estimates of Impacts[6]

- Goals for Greenhouse Gas Emission reductions below 2005 Year: 2012–3%, 2020–17%, 2030–42%, 2050–83%

- Nationwide Renewable Electricity Standard (RES) mandating power from renewable sources for utilities: 2012–6%, 2016–13%, 2020–20%. Current hydropower and nuclear power are not counted. "Quantified efficiency savings" can also be used to meet mandate.
  Impact: *Higher electricity bills and higher taxes and subsidies*

- Cap-and-trade system covering 85% of U.S. economy, beginning 2012. In 2012 only about 20% of allowances to be auctioned, rising to about 70% by 2030. The national cap-and-trade system will eliminate state and regional systems.
  Impact: *Higher electricity and gas bills, higher taxes, and higher costs for products and services*

- Carbon Storage Research Corporation with $1 billion in annual funding from electricity bills. New coal plants to achieve 50% or 65% lower emissions by 2025 with Carbon Capture and Storage.
  Impact: *Both higher short-term and long-term electricity bills*

- Program for plug-in hybrid electric vehicles (PHEVs). Utilities directed to develop plans to build charging-station infrastructure. Subsidies to consumers, state and local governments, and up to $25 billion in loans to U.S. auto manufacturers.
  Impact: *Higher taxes and subsidies for electric cars*

- National energy efficiency building codes to achieve a 30% reduction in energy use by 2012 and 50% reduction by 2015, with enforcement capability, and a building retrofit program. Labeling and subsidy payments for building efficiency.
  Impact: *Higher building costs, taxes, and subsidies, somewhat reduced by energy savings*

- New lighting standards and incentive programs to be established for lighting fixtures and appliances.
  Impact: *Higher lighting and appliance costs, somewhat reduced by energy savings*

- Direction to establish new greenhouse gas emissions standards for vehicles (similar to the European Union).
  Impact: *Higher vehicle costs*

- A wide array of government subsidies to reduce deforestation in developing nations, subsidize renewable energy, and promote energy research and green behavior.
  Impact: *Higher taxes and subsidies*

Secretary of Energy Steven Chu agrees:

> Such legislation will provide the framework for transforming our
> energy system to make our economy less carbon-intensive, and less
> dependent on foreign oil.[7]

It would be great for the United States to eliminate its dependence
on foreign oil. But, contrary to the claims of Waxman and Chu, the
ACES Act is not going to do it. As discussed in Chapter 12, the U.S.
uses about 100 quadrillion Btu (quads) per year of energy. This total
can be divided into three main areas: electricity generation (about 40
quads), transportation (about 30 quads), and heating (about 30
quads). The ACES Act is aimed primarily at two of these three areas,
electricity and heating. The major measures, the Renewable Portfolio
Standard and the cap-and-trade system, are intended to force utilities
to generate electricity from renewable sources. Other measures are
aimed at improving building efficiency, but these are likely to have
less overall impact on energy usage.

U.S. 2008 petroleum usage is shown in Figure 122. Seventy
percent of petroleum use was for transportation. *Renewable energy
from windmills and solar farms does nothing to reduce U.S.
transportation use of petroleum.* Another 11% of oil use was for
industrial, residential, and commercial petrochemicals and specialty
oil products, also not effected by ACES. The ACES Act, if passed,
will only affect the remaining 19% used for liquid propane gas, fuel
oils, and electricity. But, except for the 1% of petroleum used for
electrical power, ACES doesn't provide any alternatives. It's just
going to make the propane for your gas grill more expensive. Is this
how the U.S. will solve its imported oil dependency?

Figure 123 shows United States petroleum and petroleum product
production and imports for the last 28 years. In 1982, domestic
production was 72% and imports 28% of consumption. However,
despite the fact that energy independence has been the goal of every
president since Richard Nixon in 1974, imports have grown steadily
to become 57% of usage by 2008. In 2008, U.S. net annual imports
of oil totaled a massive four billion barrels, after falling from
an even higher number in 2007 due to the economic slowdown.[8]

Suppose the ACES is a major success in reducing use of liquid
petroleum gas and fuel oil (the 19% of petroleum usage impacted by
ACES) by 20% through higher prices. This would only reduce total

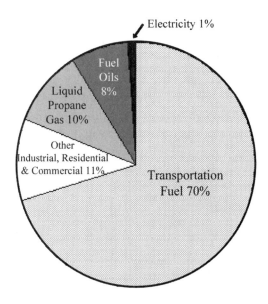

Figure 122. U.S. Petroleum Consumption in 2008. 81% of petroleum use was transportation fuel, lubricants, and petrochemicals. Only about 19% of consumption is affected by the Waxman-Markey energy bill. (EIA, 2009)[9]

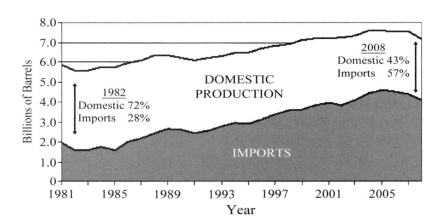

Figure 123. United States Petroleum Production and Imports 1981–2008. The top curve shows total annual U.S. consumption of petroleum products in billions of barrels. The white area is domestic production and the gray area is net imports. Imports now provide the majority of U.S. consumption. (EIA, 2009)[10]

petroleum usage by 4% and imports by about 2%, down to 55% of total consumption. Is this what Mr. Waxman means by "break our dependence on foreign oil?" Imports have been increasing their share of supply by about 1% per year, so within two years, the U.S. will be back at the same level of imports.

Waxman-Markey may even increase U.S. dependence upon foreign oil. U.S. refiners will be required to purchase carbon allowances, increasing their costs of producing domestic petroleum products. Refined foreign oil products will then look competitively cheaper, resulting in more purchases of imports, a tendency that further refutes the predictions of Waxman and Chu. Bill Durbin, head of carbon research and global energy markets at consultant Wood MacKenzie, comments:

> If you can import fuels without the same carbon costs as domestic refiners, you will have an advantage. Does that open the door for offshore refiners? I think it does.[11]

But wait, we're forgetting the planned huge subsidies for electric cars and plug-in infrastructure. Senator John Kerry bubbles:

> In a few years, you should be able to plug your American-made plug-in hybrid into the outlet in your garage, so that you never use a drop of gas on your daily commute. This won't happen overnight, but I promise you, it is closer than you think.[12]

This is wishful thinking without substance. About 500,000 hybrid and electric cars were sold worldwide in 2008, less than 1% of the world's total. The pure electric cars Mr. Kerry dreams about were a much smaller fraction of this 1%.[13] This is on par with the 0.6% of U.S. energy produced by wind and solar in 2008.[14] U.S. national and state government tax breaks, direct payments, and RPS mandates provide roughly a 50% subsidy to the wind and solar industries. Is the U.S. Government going to boost taxes and buy half of an electric car for every citizen? Electric cars are the eventual future, but not the near future. The Waxman-Markey bill will have a *negligible* positive effect, if any, toward breaking U.S. dependence on foreign oil.

However, there is a path to reducing U.S. oil imports from OPEC (Organization of the Petroleum Exporting Countries) with the proper energy policy. Figure 124 shows that imports from non-OPEC

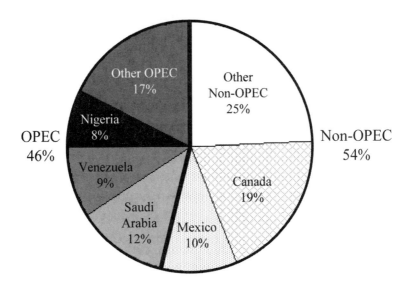

Figure 124. 2008 U.S. Petroleum Imports by Country. Non-OPEC nations now provide the majority of U.S. petroleum product imports. Canada is the fastest growing foreign supplier. (EIA, 2009)[15]

nations are now larger than imports from OPEC. Over the last 15 years, imports from non-OPEC nations have been growing faster than those from OPEC nations. OPEC provided 51% of U.S. oil in 1993, declining to 46% by 2008. Oil from Persian Gulf nations has declined in percentage from 21% in 1993 to 18% in 2008. Canada surpassed Saudi Arabia as the leading foreign supplier, providing 19% of imports, up from 14% in 1993. The best opportunity for near-term petroleum security is greater imports from friendly nations such as Canada, Mexico, and Brazil, coupled with increased domestic production, rather than windmills, solar fields and electric cars.

FALSE CLAIM #2: GROWTH AND GREEN JOBS

When introducing the ACES bill, Representative Waxman stated:

> This legislation will create millions of clean energy jobs…Our goal is to strengthen our economy by making America the world leader in new, clean energy and energy efficiency technologies.[16]

Senator Kerry agrees:

> It has the potential to provide jobs at every level of the economy, from the research scientists who discover new ways to harness the sun's energy to the construction workers who install solar panels.[17]

Economic growth, defined as an increase in goods and services, occurs for only two reasons. Either more people are delivering goods and services due to increased population, or the workers of a nation are delivering more goods and services per person. The amount of goods and services produced per person is called productivity. If we discount population increases, *a nation achieves economic growth by increasing productivity.* There is no economic growth if productivity is flat or declines.

The Climatist idea that a program to replace hydrocarbon energy with renewable energy can grow the economy is fundamentally wrong. Price is a good indicator of productivity in a free-market economy. When hydrocarbon and nuclear electricity plants are replaced by wind and solar plants that provide green energy at twice the price, the productivity of the energy industry *declines*, impacting all commerce that uses energy. Introduction of negative productivity into a wide sector of the energy industry will result in negative national economic growth.

Climatist proposals for a renewable revolution create no new economic value. Implementation of ACES will transform U.S. the economy from a low-cost, efficient hydrocarbon base to a high-cost, inefficient renewable energy base, using massive subsidies taken by taxes from the rest of the economy. Since greenhouse gas emissions are not causing climate change, there is no net economic gain from these policies, only increased regulation, higher costs, and economic dislocation impacting all U.S. industries.

As a hypothetical example, suppose tomorrow we introduce a new green manufacturing process into the U.S. auto industry. Suppose also that the new process eliminates greenhouse gas emissions from all automotive factories, but the number of automobiles produced with the new process per hour of labor is only two-thirds that of the previous process. Such a program may eliminate emissions and create some jobs, but auto industry productivity would decrease 33%, resulting in negative growth for the U.S. economy.

It's easy to directly create jobs. The government can simply pay people to walk on treadmills connected to electric generators to replace coal-fired electricity. But this would obviously reduce the productivity of the electrical power industry, retard economic growth, and cause job loss in other parts of the economy.

Toll roads are an example of our illogical modern-day society. The State of Illinois has constructed an expensive highway toll system that collects revenue of over $400 million per year. The system employs about 1,700 people to collect the tolls.[18] Yet, the same revenue could be collected by a less than 1% increase in sales tax or other levy, without traffic congestion and creation of the large bureaucracy. The system has created 1,700 jobs, but brings no net value to society.

Waxman-Markey will create thousands of government jobs in every state as well as federal agencies to enforce the massive regulations of the W-M legislation. The Act will create jobs in the same way that collection of fees from a toll road "creates jobs." In both cases, there is no increased output of goods or services in the economy. There is only the cost of over-regulation and additional taxation that reduces economic productivity. As a result, *there will be a net job loss and reduced economic growth from Waxman-Markey* or any other cap-and-trade legislation.

An August 2009 study commissioned by the American Council for Capital Formation and the National Association of Manufacturers projects significant losses from ACES in both Gross National Product and employment. The study finds that ACES, if passed, will result in an inflation-adjusted GDP reduction of between 1.8% ($419 billion) to 2.4% ($571 billion) by the year 2030. For comparison, the amount the U.S. Government spent on social security payments to retirees in 2008 was $612 billion. Cumulative GDP losses from 2012 to 2030 are projected to be between $2.2 trillion and $3.1 trillion. The study also estimates employment will drop by 1,790,000 to 2,440,000 by 2030.[19] Contrary to the claims of advocates, Waxman-Markey will stunt U.S. economic growth and reduce employment.

FALSE CLAIM #3: MINIMAL COST

Co-author of ACES, Representative Ed Markey, cites a Congressional Budget Office (CBO) analysis regarding the cost of the Act:

Americans know that building a clean energy economy has real value, and this CBO analysis proves it...for the cost of about a postage stamp a day, all American families will see a return on their investment as our nation breaks our dependence on foreign oil, cuts dangerous carbon pollution and creates millions of new clean energy jobs that can't be shipped over seas.[20]

Mr. Markey's statement includes several exaggerated claims, including claims about foreign oil and clean-energy jobs that are nonsense, as we have already discussed. He refers to a CBO report issued on June 19, 2009 that estimates the cost to households from the cap-and-trade provisions of the ACES bill.

The CBO estimates a cost to households of $22 billion in 2020, about $175 per year or 48 cents per day, hence the "postage stamp per day" quoted by Markey. The study estimates $110 billion in costs to businesses in 2020 for purchasing 30% of allowances planned to be auctioned. The study expects these costs to be passed to consumers in terms of higher prices. Of the $100 billion total, CBO estimates that $85 billion would be allocated back to U.S. households and after some tax adjustments, hence the $22 billion number.[21]

But the CBO study takes a very limited view, sticking to the costs due to allowances alone. The CBO study does not discuss:

- Electricity price increases due to the RPS-mandated substitution of wind and solar for hydrocarbon sources at twice the price, and the addition of carbon capture and storage to coal-fired plants, a 50% cost adder
- The costs of new transmission lines needed to connect to wind and solar facilities in remote areas
- Over $15 billion per year in renewable energy subsidies, paid for by U.S. taxpayers
- Job losses from businesses locating offshore to avoid carbon permit costs
- Purchasing power losses from ACES tariffs established to block imports from "polluting" nations

The $22 billion from the CBO analysis is just the tip of the ACES-cost iceberg.

A cost that few mention is the capital expense of new transmission lines to connect to wind and solar facilities in remote locations. The Department of Energy estimates that for the U.S. to obtain 20% of its electricity from wind, more than 19,000 miles of new high-voltage transmission lines will need to be constructed. This cost alone is estimated at $60 billion.[22]

A May 2009 study for the National Black Chamber of Commerce, conducted by the consulting group Charles River Associates, found significant economic costs from ACES. The study found for the year 2030 a 1.3% loss in GDP, job loss of 2.5 million, and purchasing power loss of $830 annually per U.S. household. A major part of these losses was a wealth transfer of $40 billion to $60 billion per year to developing nations for purchase of carbon offsets.[23]

The study we mentioned earlier by the American Council for Capital Formation and the National Association of Manufacturers estimates electricity prices to rise by 31% to 50% and household income losses of $730 to $1,248 by 2030.[24] Clearly, the costs of "postage" are very high. Myron Ebell of the Competitive Enterprise Institute summarizes:

> Waxman-Markey would put big government in charge of how much energy people can use. It would be the biggest government intervention in people's lives since the second world war, which was the last time people had to have rationing coupons in order to buy a gallon of gas.[25]

FALSE CLAIM #4: 20% RENEWABLES POSSIBLE BY 2020

The American Climate and Energy Security Act calls for an 83% reduction in greenhouse gas emissions from 2005 levels by 2050 and for 20% of U.S. energy to be provided from renewable sources by 2020. These are revolutionary changes. Has anyone in the U.S. Government seriously looked at the feasibility of these goals?

The United States emitted about 1.6 billion metric tons of carbon in 2005. An 83% reduction would limit annual emissions to about 280 million metric tons. The last time U.S. emissions were this low was 1905, according to the Carbon Dioxide Information Analysis Center at Oak Ridge National Laboratory.[26] This was before automobiles, airplanes, refrigerators, computers, washers and dryers, gas heating, gas water heaters, air conditioning, light bulbs for most people, and most other modern electronic or mechanical devices. The U.S. population in 1905 was also much smaller, about 80 million or less than one-fifth of today's 307 million.[27]

An average U.S. home uses about 11.5 kWh of electricity and about 67,000 cubic feet of natural gas during a year. In terms of energy use, residential heating is about 40%, water heating about 20%, and lighting and appliances about 25%.[28] A family won't be able to heat the house or power the lights after a cut of 83%.

So the key question is: Will renewable energy sources be able to replace hydrocarbon fuels and provide the power? Figure 125 shows that little progress has been made by renewables over the last 20 years, despite wind, solar, and biofuel subsidies over most of that period. The 2016 and 2020 goals are far above the current output from renewables.

We are talking about an economy that consumes huge amounts of energy. Just like the enormous quantity of annual oil imports, it's not clear U.S. political leaders understand the magnitude of the nation's energy usage. In 2008, the U.S. consumed almost 100 quads of energy. Remember that each quad is 1,000,000,000,000,000 Btu. Hydrocarbon fuels provided 84% of this energy usage. Nuclear power provided another 8.5%. Renewables delivered only 7.4% of the usage. Of the 7.4% delivered by renewables, hydroelectric power provided 2.5%, wood and waste provided 2.5%, and biofuels

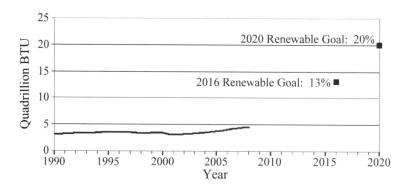

Figure 125. U.S. Energy from Renewable Sources and Goals. Energy from renewables is shown from 1990 to 2008, with ACES goals for 2016 and 2020. Existing hydropower is not counted since it is excluded from the ACES goals. Since annual U.S. energy usage is about 100 quads, the left axis also shows the percentage of U.S. energy from non-hydro renewables. (EIA, 2009)[29]

provided 1.4% (mostly ethanol vehicle fuel). Where were solar and wind? *Solar provided a microscopic 0.1%. Wind power, despite having about 20,000 wind turbines installed in the U.S., provided only 0.5% of the total energy usage.*[30]

Let's calculate the wind power needed to meet the ACES goal. The ACES calls for 20% renewables in 2020, while excluding current hydropower as a renewable. Energy from renewables was 4.8 quads (without hydropower) in 2008 and the 2020 goal requires 20 quads. If wind power accounts for 80% of the needed increase of 15.2 quads, this equals about 12 quads, or 407 gigawatts of average delivered power. If we generously assume the new wind turbines will be 35% efficient, an astounding 1,163 gigawatts of nameplate power will be required. At the end of 2008, the installed nameplate wind power in the U.S. was only 25.4 GW.[31] If each wind turbine is rated at 2 MW, 582,000 new wind turbines are required!

To meet the goal, 145 wind turbines must be installed and commissioned *each day over the next 11 years.* For comparison, 2008 was a huge building year for the heavily subsidized the wind industry, with 8,558 MW of nameplate power and 5,125 turbines installed.[32] But on average, only 14 turbines were installed per day. Get ready for a massive and costly build out as the U.S. Government strives to increase the use of wind power by a factor of 45!

Consider also that 600,000 300-foot-high steel and concrete towers will cause the annual death of about 300,000 birds. *This is more birds killed every year than by the 1989 Exxon Valdez oil spill.* Climatists point out that cats kill more birds. Probably true, but cats don't kill owls, hawks, and eagles.

On top of all this, the cost of U.S. wind energy is going up. These costs are likely to continue to rise for four reasons. First, the cost of wind towers is rising. Data from the U.S. Department of Energy shows that the installed cost of U.S. wind turbine towers climbed 65% from a low of $1,285 per kilowatt-hour in 2001 to $2,120 per kWh in 2009 (see Figure 126).[33] Many reasons are given for this, but state RPS mandates on electrical utilities must play a large part. Utilities are forced to buy wind power, boosting the price of wind turbines, and thereby raising the price of green energy for citizens and businesses. A 65% increase over eight years is quite a rise, since manufacturer prices should fall as the volume of turbines sold increases. This increase may be only a temporary trend, but the next

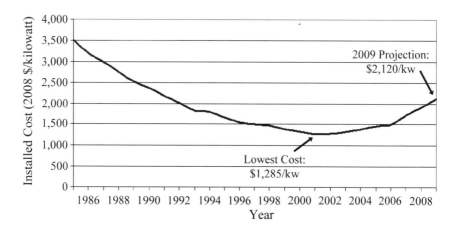

Figure 126. Installed U.S. Wind Project Costs 1985–2009. After falling for many years, U.S. wind turbine installed costs have increased since 2001. The average 2 MW wind tower now costs about $4.2 million. (DOE, 2009)[34]

three reasons for cost increases will only grow in importance.

The second reason for increasing wind turbine costs is that more systems will be offshore wind. Many southeastern states in the U.S. do not have good wind locations. Expensive offshore wind may be their only way to meet national mandates. In Europe, where citizens are increasingly opposing new onshore towers, most of the new systems will be offshore. The United Kingdom, Denmark, and Germany are emphasizing offshore wind. We have cited the U.K. House of Lords estimate showing offshore wind at 15% more expensive than onshore wind (Figure 102), but recent data shows the cost difference is much larger. The International Energy Agency surveyed 20 IEA Wind member nations and found for 2008 installations, offshore wind was *80% to 115% more expensive* than onshore wind.[35]

The third driver of rising wind energy costs is network balancing costs. These will rise as unpredictable wind power becomes a higher percentage of supply. Both maximum and minimum wind outputs must be handled by the electrical power grid, requiring higher levels of standby reserve power and other balancing measures. The consulting group Open Europe voices this concern:

The more overall generation capacity becomes dependant on the wind, the greater the risks posed by its intermittent nature—which leads to higher costs.[36]

Fourth, wind power efficiencies will eventually decline as more systems are installed. Average power output will be higher as the first wind systems are installed in optimal wind areas. Succeeding installations in sites with less desirable wind characteristics will suffer reduced average power output. Open Europe agrees:

> While renewable technologies are improving (like all other energy technologies), expanding renewables may still suffer declining marginal efficiency, as the best sites to install new plant are gradually used up.[37]

In addition to rising wind power costs, taxpayer-financed subsidies will explode. U.S. Government agencies spent $5 billion in renewable subsidies in 2008 to install 14 wind turbines per day. If the unlikely 140-per-day installation rate can be achieved, U.S. taxpayers will be pouring over $50 billion per year into a renewable energy black hole.

In summary, *the goals of United States renewable mandates are not realistic and will be prohibitively expensive to achieve.* The ACES calls for an explosion in installed wind turbine towers from 20,000 to about 600,000 in eleven years. If common sense enters the discussion, look for a reassessment of these goals in the near future.

FOOLISHNESS OF U.S. ENERGY POLICY

Over the last 40 years, United States energy policy has moved from the rational to the irrational. Environmental-become-Climatist groups have mounted a continuous and successful attack on the U.S nuclear and hydrocarbon energy industries. Licensing and construction of nuclear power plants has been halted nationwide. Oil drilling has been banned in many states and is discouraged by the national government. Oil refineries, liquefied natural gas facilities, and gas-fired electrical power plants are opposed. Coal-fired power plants are under full-scale assault by Climatism as the worst polluters. How does the U.S. find itself in this situation?

Nuclear Power
On December 2, 1942, Enrico Fermi and a team of scientists

gathered on the floor of a squash court beneath the stadium at the University of Chicago. Following Fermi's direction, cadmium control rods were withdrawn from the experimental reactor Chicago Pile-1, successfully triggering the world's first self-sustaining nuclear reaction. The nuclear age was born.[38]

World War II ended in 1945 with the detonation of atomic bombs at Hiroshima and Nagasaki, Japan, shocking the world with the devastating power of nuclear weapons. Four years later, the USSR exploded its first nuclear bomb.[39] For the next 40 years until the fall of the Berlin Wall in 1989, the U.S. and USSR engaged in confrontational global politics with the threat of thermonuclear war always at hand.

Against this background, the commercial nuclear industry began to grow. The first commercial nuclear power plant for generation of electricity was commissioned in Shippingport, Pennsylvania in 1957. By 1971, 22 nuclear power plants were in operation in the U.S.[40]

At the same time, the world environmental movement became a rising force. Along with Rachel Carson's assault on pesticides, anti-nuclear opposition provided a foundation for environmental groups. Rex Wyler, a founder of Greenpeace, credits the start of environmentalism with the founding of the U.K. organization Campaign for Nuclear Disarmament in 1957. Greenpeace itself was formed in Vancouver, Canada, in 1971, as a result of a protest of U.S. underground nuclear testing at Amchitka Island in Alaska.[41]

Then on March 28, 1979, a failure occurred at Three Mile Island reactor in Pennsylvania that would damage the U.S. nuclear industry for 30 years (and counting). A chain of events including both mechanical failure and human error caused a loss of coolant water, resulting in a meltdown of half of the reactor core before it was brought under control. The containment vessel held, but some radioactive gas was vented into the atmosphere.[42] Governor Dick Thornburgh, on recommendation of the Nuclear Regulatory Commission (NRC), advised evacuation of pregnant women and small children from the vicinity. Within days, over 100,000 people had left the area.[43]

In one of the most amazing coincidences in history, the Three Mile Island accident occurred only 12 days after the release of the movie *The China Syndrome*, starring Jane Fonda. The movie plot is about a fictional California nuclear plant that had a near-fatal accident from a cooling system failure. In the film, a prophetic nuclear safety expert warns that a reactor meltdown could make an area "the size of Pennsylvania" uninhabitable.[44]

No one died in the Three Mile Island incident. Studies by the NRC and independent groups estimate the average radiation exposure to local residents was less than from a chest X-ray.[45] Nevertheless, media reaction was one of alarm. The NRC introduced tighter reactor safety regulations and inspection procedures the same year. Licenses for nuclear plants were also put on hold for the short term. The Three Mile Island clean up cost totaled almost $1 billion.[46]

On April 26, 1986, a power surge at the nuclear plant at Chernobyl, Ukraine, in the former USSR, destroyed the reactor and released large amounts of radiative material into the environment. Of 600 workers at the plant, 134 received high radiation doses, with 30 dying within four months of the incident. More than 500,000 people were exposed to radiation contamination. It's estimated about 4,000 of these people may eventually die from cancer-related illnesses related to the event. The U.S. Nuclear Regulatory Commission analyzed the event and believes "the causes of the accident have been adequately dealt with in the design of U.S. commercial reactors."[47]

As a result of the Three Mile Island and Chernobyl accidents and growing public opposition, licensing for U.S. power plants ended in the early 1990s (Figure 127). However, many U.S. citizens may

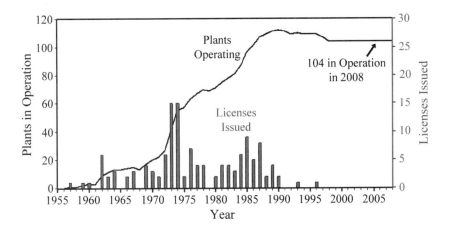

Figure 127. U.S. Nuclear Plants Operating and Licenses Issued 1955–2008.
One hundred and four U.S. nuclear plants remain in operation today, but no new plants have been licensed since 1996. (EIA, 2009)[48]

not know that 104 nuclear plants remain in operation today. Electricity generated from U.S. nuclear plants has continued to climb as the average plant capacity factor has increased from about 50% in 1973 to 92% in 2008.[49] But, not a single new U.S. nuclear plant license has been granted since 1996.

Nuclear power is the most concentrated form of energy currently available to mankind. According to IOR Energy, a U.K.-based energy consulting group, black coal has an energy density of about 29 gigajoules per ton, crude oil about 45 GJ per ton, and uranium oxide 470,000 GJ per ton.[50] For the same weight, there is more than 10,000 times as much energy available from uranium as the concentrated chemical energy in crude oil or coal. The power density of nuclear fuel provides the aircraft carrier U.S.S. Enterprise with the ability to travel 400,000 miles without refueling.[51]

But, along with high energy density come the problems of safe control of high power and temperature, shielding from radioactivity, waste disposal, and proliferation of nuclear weapons. Chernobyl caused a tragic loss of life and Three Mile Island was the most serious accident in the history of the U.S. Nuclear industry. There have been many other lesser incidents across the world over the last 50 years, but are these a reason to ban nuclear power?

Much of the world has decided nuclear power should be pursued. The 104 nuclear plants in the U.S. are about one-quarter of the 436 nuclear plants in operation in 31 nations around the world. Fifty-two plants are now under construction outside of the U.S. The 59 nuclear plants in France generated 76% of the country's electricity in 2008.[52] Japan, the nation which suffered the only war-time use of nuclear weapons, gets 30% of its electricity from 53 nuclear facilities. Japan is planning to increase its electricity from nuclear to 40% by 2017.[53] Yet, the U.S. continues to shun nuclear power, foolishly pursuing dilute, intermittent, and expensive wind and solar power.

Is nuclear power too dangerous for commercial use? There has never been a U.S. death from the use of nuclear power. Including Chernobyl, there have been fewer than 500 deaths worldwide over a fifty-year history that now totals over 13,000 reactor-years of operation.[54] In comparison, the Aircraft Crashes Record Office in Geneva, Switzerland reports over 121,000 aircraft fatalities since 1918.[55] Should we ban aircraft? Nuclear opponents warn that nuclear plants have become terrorist targets in today's world. Granted,

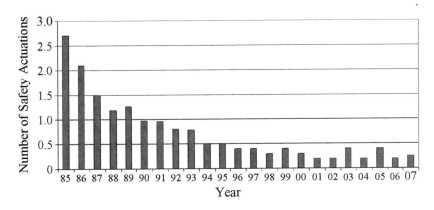

Figure 128. Safety System Actuations at U.S. Nuclear Plants 1985–2007. The graph shows safety actuations per plant per year. Safety systems are triggered either automatically or manually to deal with a problem in the reactor. Actuations have declined to one for every four plants per year. (NRC, 2009)[56]

but so are tall buildings. Should we now ban tall buildings?

Figure 128 shows the number of automatic and manual actuations of safety systems at U.S. nuclear power stations over the last 23 years. In 1985, safety systems were triggered at a rate of more than 2.5 times per plant per year. By 2008, the number of actuations had been reduced by a factor of ten. The lessons of Three Mile Island have been put to good use in the U.S. nuclear industry. It's time for the United States to again license nuclear power.

Petroleum

On January 29, 1969, a "blowout" occurred at a Union Oil platform five miles off the coast of Santa Barbara, California. Under extreme pressure from natural gas buildup, the drilling casing ruptured, spilling an estimated three million gallons of petroleum, creating an oil slick over an 800-square-mile area, and coating 35 miles of coastline with up to six inches of oil. The oil spill fouled local beaches and killed seals, dolphins, and an estimated 3,600 birds.[57]

The oil platform ruptured because of inadequate protective casing. Union Oil had been granted a waiver by the United States Geological Survey to use shorter casing on the pipe than was called for by federal standards. Casing is the reinforcing metal sheeting inside the well designed to prevent blowouts. Investigators later determined that more steel sheeting, as required by the standard, would have prevented the spill.[58]

The Santa Barbara oil spill ranks with *Silent Spring* and Three Mile Island as an early foundation of the environmental movement. The organization Get Oil Out (GOO) was founded in Santa Barbara only days after the incident. GOO quickly gathered 100,000 signatures on a petition to ban offshore drilling in California. The Santa Barbara spill also energized grass roots support for the first Earth Day, a nationwide demonstration on April 22, 1970.[59]

On March 24, 1989, the supertanker Exxon Valdez collided with Bligh Reef in Alaska's Prince William Sound, rupturing its hull and spilling an estimated 11 million gallons of crude oil into the sea.[60] The Exxon Valdez disaster remains the largest single oil spill in U.S. coastal waters. The oil slick eventually covered more than 1,000 miles of pristine Alaskan shoreline. Impact on marine life, including orcas, sea otters and shellfish was severe, and an estimated 100,000 to 300,000 birds were killed.[61] At the time of the accident, the tanker was piloted by the ship's third mate, who was not certified to be piloting the ship in such waters. Exxon paid over $2.1 billion for oil clean up and another $1 billion in civil and voluntary payments.[62]

The Exxon Valdez disaster produced a high level of media hysteria. The incident strengthened environmental demands for banning of oil and gas production in U.S. territories. A federal judge in the Exxon case called the incident "worse than Hiroshima."[63] To this day, environmental groups claim lasting damage from the incident. Margaret Williams of the World Wildlife Foundation refers to the scene of the disaster at Prince William Sound: "If it's lost, it's lost forever."[64] But is this really the case?

The National Oceanic and Atmospheric Administration found a surprisingly rapid recovery at Prince William Sound:

> Surface oil at our study sites had all but disappeared by 1992, 3 years after the spill...Recovery based on parallelism between oiled and unoiled intertidal populations occurred for most study species by 1992–1993.[65]

In other words, surface oil was gone and most species had recovered within three-to-four years after the disaster. NOAA still finds oil at the site, but its effect on wildlife is small:

However, today there is still residual oil to be found in the impacted areas we study...Interestingly, despite the fresh appearance of oil at these sites, chemical analysis and biological observations indicate that the oil is actually highly weathered and of such reduced toxicity that many intertidal species can tolerate its presence...[66]

Clearly, humanity can do a better job to protect the environment and minimize the chances of another such disaster. The Oil Pollution Act was made law in August 1990, largely in response to the Exxon Valdez incident. The act set aside funds for clean up of future spills, established requirements for contingency planning by governments and industry, and toughened penalties for future spills.[67] A 1992 amendment to the International Convention for the Prevention of Pollution from Ships established a double hull requirement for all oil tankers built in 1994 and later, as well as a phase-in period for existing ships.[68] While a double hull would not have prevented the Exxon Valdez oil spill, it could have reduced spillage by 60%.[69]

Tanker and oil platform spills claim the headlines, but most oil that enters the ocean is from natural seepage. A study by the National Academies in 2003 found that only 5% of the petroleum in U.S. waters was due to oil drilling or oil transportation (see Figure 129). *Sixty-three percent was due to natural seepage from the bottom of the sea,* like the previously discussed petroleum leak from the sea floor off Santa Barbara, California. U.S. communities should focus on land

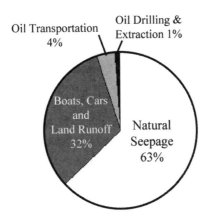

Figure 129. Petroleum in U.S. Waters. Only 5% of oil in the sea off U.S. shores is from oil drilling or transportation. (Adapted from NOIA and NAS, 2003)[70]

runoff into rivers and streams and discharges from cars and boats to reduce the 32% that is the largest man-made contribution.

In any case, in the wake of the Santa Barbara and Exxon Valdez oil spills, and 40 years of environmental and now Climatist activism, the United States energy policy has become strongly anti-oil. The U.S. Geological Survey estimates U.S. oil potential at 22 billion barrels of proven reserves, with an additional 163 billion barrels of unproven reserves, compared to current annual consumption of 7 billion barrels.[71] But about 85% of those 163 billion barrels are in areas where exploration and drilling are banned.[72] From 1982 to 2008, Congress did not provide funds for oil leasing on the Federal Outer Continental Shelf (OCS), effectively prohibiting leasing for offshore drilling and exploration. In 1990, President George H. W. Bush placed an additional executive ban on OCS drilling through 2012. George W. Bush removed the ban in September, 2008 during the height of the spike in oil and gasoline prices. Congress also allowed its ban to expire.[73] Secretary of the Interior Ken Salazar is currently considering a proposal to expand offshore drilling, but it appears that the anti-petroleum lobby will force the ban to be re-established.

The last U.S. oil refinery was built in 1976. Environmental restrictions, local community opposition, and the high cost of construction have precluded any new facilities. Instead, the industry has upgraded and expanded existing refineries, which are now running at close to 100% capacity.[74] Growing U.S. demand for refined petroleum products will need to be met by imports.

On the positive side for U.S. energy security, in August 2009, the U.S. State Department approved construction of a $3.3 billion pipeline to bring crude oil south from Canadian oil sands in the province of Alberta. Named the "Alberta Clipper," the project will bring up to 450,000 barrels a day by mid-2010.[75] While it won't reduce U.S. oil imports, the pipeline has the potential to shift 20% of the oil imported from OPEC to friendly neighbor Canada. The non-profit legal group Earthjustice has filed suit against the decision of the State Department on behalf of several Climatist groups. Attorney Sarah Burt of Earthjustice states:

> The Alberta Clipper will mean more air, water and global warming pollution, particularly in communities near refineries that process tar sands oil.[76]

Hyperion, a Dallas Texas-based oil company, plans to build a new $10 billion refinery in South Dakota to process the oil from the Alberta Clipper pipeline, the first new U.S. refinery in 43 years. The refinery project is opposed by the Sierra Club, Citizens Opposed to Oil Pollution and other groups.[77] The Alberta Clipper pipeline and Hyperion refinery show examples of strong opposition facing efforts to bring petroleum energy to U.S. citizens.

The Arctic National Wildlife Range was established on the north slope of Alaska in 1960 under President Dwight Eisenhower, one year after Alaska became a state. In 1980, President Jimmy Carter signed the Alaska National Interest Lands Conservation Act, a bill that expanded the Range to 18 million acres and renamed it the Arctic National Wildlife Refuge (ANWR). The Act also called for an assessment of the oil and gas potential in the 1.5 million acres of the refuge coastal plain. Total ANWR land area today is 19.3 million acres, the largest refuge in the U.S. system, and home to caribou, grizzly and polar bears, and many other species of arctic wildlife.[78]

Studies by the U.S. Geological Survey in 1999 and later provide a mean estimate of 10.4 billion barrels of recoverable oil in ANWR.[79] Since the 1980 Act, the debate has raged regarding whether ANWR should be opened for drilling. Opposing groups warn drilling would harm wildlife, showing picturesque landscapes and endearing wildlife scenes, while proponents argue the oil can be recovered with minimal impact on the ecology.

Figure 130 shows a summer photograph of the ANWR coastal plain where the oil is located. The plain is a 1.5 million-acre area of tundra and marsh, forming the upper 8% of the 19 million-acre refuge. Modern horizontal drilling technology should allow recovery of the oil using 2,000 acres, or less than 2% of the coastal plain, minimizing the impact on wildlife. Since most of the oil is located in the marshy coastal plane and only a small part of the plain is needed for access, it appears that ANWR should be opened for drilling.

The benefits to drilling in ANWR are large. Oil from the refuge would allow the U.S. to replace oil imports from Saudi Arabia for 18 years. The combination of ANWR and oil through the Alberta Clipper pipeline from Canada could replace all oil imports from the Persian Gulf for a period of 20 years. Maybe by that time, electric cars can become serious choices for consumers.

As an alternative to reducing oil imports from the Persian Gulf,

Ethanol from Indiana	Petroleum from Arctic National Wildlife Refuge
18.4 Million Acres (80% of Entire State of Indiana)	2,000-Acre Drilling Area[81] (1/20 of Washington, D.C.)
6.9 Billion Gallons Ethanol per Year (Energy Equivalent to 4.55 Billion Gallons Gasoline)	198 Billion Gallons of Gasoline[82] (from 10.4 Billion Barrels of Oil) **Equal to 43.5 Years of Indiana Ethanol**
Other Costs and Benefits	Other Costs and Benefits
- Over 200 Billion Gallons of Water per Year (assumes 4% irrigation)[80] - Pesticides and Fertilizers - Gasoline and Farm Energy **- No Food Exports to Developing Nations**	**- Some Impact on Arctic Wildlife** + 95 Billion Gallons Diesel Fuel + 40 Billion Gallons of Jet Fuel + 129 Billion Gallons of Other Products

Figure 130. Comparison of Ethanol from Indiana and ANWR Oil. The entire state of Indiana would need to grow 100% corn for Ethanol for 43 years, with high uses of water, pesticides, fertilizers, and gasoline to equal the oil energy from ANWR. But no food could be produced for developing nations. Refining costs are not included. The photo shows a typical view of the ANWR coastal plain.[83] (After Driessen, 2008)[84]

the U.S. could use ANWR oil to reduce ethanol production. Figure 130 compares the benefits of using oil from the refuge compared to production of ethanol in Indiana. To equal the gasoline energy available in ANWR, the entire state of Indiana would need to be devoted to grow corn for ethanol production for the next 43 years. But in addition, over 200 billion gallons of water and large volumes of pesticides, fertilizers, and farm energy would be needed *every year* to grow the corn. Yet, even with all this, there is no way for corn ethanol production to match the diesel fuel, jet fuel, and other petroleum products that are refined along with gasoline from each barrel of crude oil.

But the biggest loss in using corn ethanol from Indiana versus oil from ANWR is the opportunity cost of not producing food for the world. As we discussed in Chapter 13, the United States is well down the road of ethanol production for vehicles, rather than using oil reserves for vehicles and using agriculture to feed people around the world. Whether 18.4 million acres are used from Indiana or other states for ethanol, 2.3 billion bushels (60 million metric tons) of food corn would not be available for export. Sixty million metric tons is roughly equal to total 2008 U.S. corn exports.[85] By opening ANWR, the U.S. could reduce ethanol production and double annual corn exports. At least indirectly, Climatist opposition to domestic U.S. petroleum production has resulted in emphasis on ethanol, and less food for hungry people of the world.

Coal

Coal fuel provides 40% of the world's electricity.[86] Coal was the fuel for 49% of United States electrical power in 2008.[87] Coal continues to be the world's lowest-cost fuel for generating electricity.

The history of coal usage is one of large benefit, such as being fundamental to the Industrial Revolution (see Chapter 13), but also a history of struggle with air pollution from the burning of coal. As early at the 12th century, Londoners began to burn inexpensive "sea-coal" to heat their homes. By the 1800s, more than a million London citizens were burning coal. Winter fogs lingered for months at a time and became more than a nuisance. The term smog (smoke and fog) was believed to be first used in London in the early 20th century.[88]

Episodes of smog in Europe and the U.S. became major health problems in the first half of the 20th century, resulting in many air pollution-related deaths. A body of warm stagnant air sat over Donora Valley, Pennsylvania, in October 1948, trapping a thick cloud of yellow, acrid smog from local steel and metal fabrication plants. Twenty people died and many town residents became ill from the polluted air.[89]

The great London smog of 1952 is probably the worst pollution event in history. In December 1952, thick smog settled over the city and remained for several weeks. An estimated 13,500 deaths occurred due to breathing-related illnesses and influenza. The air pollution was driven by smoke containing sulfur dioxide emissions.[90] In response, Parliament enacted the Clean Air Act in 1956 as the foundation of air pollution reform in Britain.[91]

Smoke is composed primarily of fine particles of carbon (soot), nitrogen oxides, and sulfur oxides. Smoke is created by the incomplete combustion of carbon-based fuels. In contrast, CO_2 is an odorless, colorless gas created by the complete combustion of hydrocarbon fuel. **Carbon dioxide does not cause smoke.** Carbon particulates, carbon monoxide, sulfur dioxide, nitrogen dioxide, and other products of incomplete combustion are pollutants harmful to people, while CO_2 is neither a pollutant nor harmful.

Despite past problems, the ingenuity of mankind has made major progress in reducing air pollution from coal, especially in developed nations. The largest measure of progress came from the substitution of clean-burning natural gas in place of coal for residential and business heating and cooking. This transition in developed nations occurred in the last half of the 20th century and effectively eliminated smog in the U.S. and Europe. But as we saw in Chapter 14, much of the developed world is still plagued by air pollution from residential burning of wood, charcoal, coal, and dung.

The U.S. Congress passed the Air Pollution Control Act in 1955, followed by other measures, including the Clear Air Act of 1990.[92] Since 1980, *coal-fired generation of electricity in the United States has increased 72%, but air pollution has declined significantly.* According to the Environmental Protection Agency (EPA), U.S. air quality is much improved and continues to improve. The EPA reports that since 1980, the airborne concentration of carbon monoxide is down 79%, lead is down 92%, nitrogen dioxide is down 46%, ozone is down 25%, and sulfur dioxide is down 71%. Particulates have been tracked for fewer years, but PM10 particulates are down 31% since 1990 and PM2.5 particulates are down 19% since 2000.[93] We showed a graph of this declining air pollution in Figure 78, back in Chapter 8.

Yet, despite the value of coal in generating low-cost electrical power and significant U.S. progress in improving air quality, coal has become the number one U.S. energy target for Climatism (see Figure 131). Climatism has redefined air pollution to include carbon

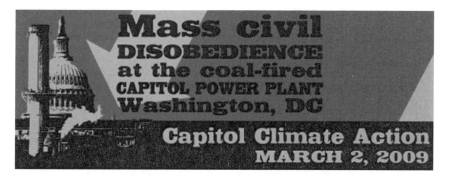

Figure 131. Climatist Attack on Coal-Fired Energy. Flyer distributed for the climate protest in Washington D.C. in March, 2009. The colors are red, brown and black in the best tradition of revolutionary flyers. (WUWT, 2009)[94]

dioxide. Since coal-fired power plants are the largest source of man-made CO_2 emissions, they must be protested, halted, and eliminated.

Natural Gas

Natural gas provided 21% of U.S. energy needs in 2008 and about the same percentage of world energy usage. It occurs naturally from decomposition of organic molecules in marshes, animal digestive systems and landfills. Most of the gas used by mankind is obtained as a byproduct from petroleum fields or stand-alone natural gas fields.

Refining of natural gas produces methane (commercial natural gas), propane, ethane, butane, and other gas products. Gas products burn cleanly, producing no ash and emitting very low levels of carbon particulates and oxides of nitrogen and sulfur. Even though combustion of natural gas produces about 30% less CO_2 than oil and 45% less CO_2 than coal per unit of energy,[95] Climatism opposes the use of natural gas. The IPCC estimates gas usage will grow to 31% of world primary energy demand by 2030, making gas a large CO_2 emitter.[96] Therefore, although coal and petroleum are larger "problems," natural gas usage must also be curtailed.

U.S. Climatist organizations have answered the call. An example of opposition was the February 2008 human blockade of construction of two new gas-fired electrical power plants in Palm Beach, Florida. The protest was organized by the groups Rising Tide North America and Earth First! Brian Sloan, an on-site protester and organizer with Rising Tide stated:

Gas-fired power is not a clean or sustainable energy. It is a dirty and dwindling fossil fuel.[97]

ROAD TO BLACKOUT PARADISE?

In March 2009, after a long struggle, Alliant Energy cancelled plans for a 649 MW coal-fired power plant in Marshall, Iowa. The company cited environmental, legislative, and regulatory uncertainty as reasons for cancelling the planned facility. Ken Kuyper, executive vice president of the Corn Belt Power Cooperative, stated:

> It's regrettable that this plant won't be built. This new source of generation was an important part of our plan for meeting the growing demand for electricity from the people of rural Iowa.[98]

Climatist groups consider the cancellation a major victory in their fight against greenhouse gas emissions. Neila Seaman, spokesperson for the Iowa Chapter of the Sierra Club, remarked:

> Environmental organizations were poised to encourage all Iowans to attend hearings and make informed comments on the damage a new coal plant could have on our state...We believe Alliant saw the writing on the wall that the time has come for new coal plants to finally account for their global warming pollution.[99]

The Sierra Club web site has a map of the U.S. titled "Stopping the Coal Rush" that identifies over 200 coal-fired power plant projects. The Club claims to have stopped over 100 of the projects.[100] Al Gore presents a slide showing the number of halted U.S. coal plants. In 2007, 59 out of 151 proposed U.S. coal plants were denied licenses by state governments or abandoned. Many of the others are being contested in court.[101] California, Florida, Hawaii, and Michigan have bans on new coal-fired electrical plants, and other states are considering bans. If Waxman-Markey passes, it's likely that construction of new coal plants will grind to a halt.

Is the United States on the road to blackout paradise? Coal delivers 49% of electricity production, but the nation is not going to build any more coal plants. Natural gas provides 21% of U.S. electricity, but gas plants are also frequent targets of protesting groups. Nuclear meets 20% of U.S. electricity needs, but nuclear is out. Hydropower is 6%, but few good undammed rivers remain and new dams are also opposed. *This means 96% of what works for generating electricity is now precluded.* Wait—here's the answer! Let's put all bets on zephyrs (currently 1% of electricity generation) and sunbeams (currently 0.02%).

The blackouts in California may be an indication of what is in store for the United States (see text box). California suffered a series of rolling blackouts in 2000 and 2001 due to a shortage of capacity, state-fixed retail prices, unexpectedly warm temperatures, and a reduction in available electricity imports from neighbor states. Nationwide, U.S. electrical power capacity margin has been declining slowly over the last 20 years. Capacity margin is the amount of electrical generating capacity available above peak demand, which for

---

2000–2001 Blackouts in California

California experienced severe shortages in electrical power in 2000 and 2001. State officials had been convinced by the "zero energy growth" theorists (Chapter 12) that electricity demand had peaked and did not invest in new power plants. From 1990 to 1999, generating capacity declined by 1.7%, but demand increased by 11.3% as the state's economy grew. California became increasingly dependent upon hydropower imports from neighbor states in the Pacific Northwest.

In the summer of 2000, lack of rain in the Pacific Northwest led to shortages of available hydroelectric power, and Northern California experienced "rolling blackouts." Rolling blackouts are controlled shutdowns of electrical power initiated by electrical grid operators. When supply is unable to meet demand, operators shut down different communities in succession to keep the entire system from collapsing. The blackouts caused millions of dollars in lost income to California businesses and industries.

The power shortage caused the price of wholesale electricity to skyrocket, well above the retail price, which was fixed by the state. Southern California Edison and Pacific Gas & Electric, the two largest electrical utilities, were forced to sell electricity at a retail price well below their cost and were unable to pay independent California power providers. Out-of-state power companies became hesitant to sell power to Edison and PG&E as the financial position of the companies deteriorated.

In March of 2001, a number of in-state power producers were forced to shut down because they had not been paid for several months. Due to unexpectedly warm temperatures and a shortage of capacity, rolling blackouts resumed throughout California. Pacific Gas & Electric declared bankruptcy in April. Hot temperatures led again to rolling blackouts in May, before state and national governments were able to get the situation under control.[102]

the U.S. occurs during the summer due to air conditioner use. The capacity margin in 2008 was 16.1%, down from 21.6% in 1990.[103] While 16% is still a good cushion, this margin will disappear if the nation stops building power plants.

CLIMATIST DEMANDS OF THE U.S.

Professor John Schellnhuber, the director of the Potsdam Institute for Climate Change, criticizes the unwillingness of the U.S. to sign up to ambitious greenhouse gas reduction targets. He states:

> In a sense the U.S. is climate illiterate. If you look at global polls about what the public knows about climate change even in Brazil, China you have more people who know about the problem and think deep cuts in emission are needed.[104]

Make no mistake. The United States, because of its affluence and energy usage, is the primary target of Climatism. The U.S. was the subject of many negative comments during the presidency of George Bush, and these continue today. See the text box that follows for just a few of these comments.

Much of the world believes that the U.S., the largest emitter of greenhouse gases for most of the 20th century, is responsible for destroying Earth's climate. Recall the United Nations crusade to end "unsustainable" consumption and production patterns from Chapter 9. The UN considers the U.S., as the largest consuming nation on Earth, to be the heart of the problem.

In addition to curbing its consumption patterns, the U.S. must make the largest transfer payments to developing nations to compensate for the effects of climate change. The international agency Oxfam recommends the U.S. be "responsible for meeting nearly 44% of developing country adaptation costs."[105] This means if the world agrees on a redistribution of 1% of Gross National Product, the U.S. annual bill will be over $100 billion.

Europe is desperate to have the U.S. implement cap-and-trade. U.S. gasoline and electricity prices are well below those of Europe and the gap is widening. Jürgen Thumann, president of BusinessEurope describes the European Trading System as the most "costly climate policy program in the world." Mr. Thumann wants the U.S. and

---

Climatist Criticism of the United States

"It's a pity the US is still very much unwilling to join the international community, to have a multi-lateral effort to deal with climate change."
  –Kenya's Emily Ojoo Massawa, chair of African group of nations, 2005[106]

"The US is responsible for 25 per cent of the world's greenhouse gas emissions. It should take responsibility for leading the way."
  –Tony Juniper, director of Friends of the Earth, 2005[107]

"Seeking Relief from Violations Resulting from Global Warming Caused by Acts and Omissions of the United States."
  –Suit filed by the Inuit people in the Arctic of Alaska and Canada, 2005[108]

"Europe cannot solve this alone. We need stronger U.S. activity on the climate change issue if we are going to move forward. We are trying to do what we can to convince them to be more active on this."
  –Anders Borg, Swedish Finance Minister, 2009[109]

---

other nations to adopt climate measures to allow European firms to continue to "compete internationally."[110]

If the U.S. Government passes Waxman-Markey, U.S. citizens and businesses will be faced with the same punitive measures in which the Europeans take such pride. After all, according to Secretary of State Clinton, the U.S. is to blame for causing the Earth to warm up:

> We acknowledge—now with President Obama—that we have made mistakes in the United States, and we along with other developed countries have contributed most significantly to the problem we face with climate change.[111]

THE SURRENDER TO ENERGY MADNESS

The U.S. has finally heard the call and begun to follow the Pied Piper of Climatism. The Waxman-Markey climate bill, if enacted, will punish the American people with much higher costs for using energy. Past access to low-cost energy has been a foundation for U.S. economic success, but such access is set to disappear.

U.S. energy policy has discarded economics and common sense in favor of the foolishness of environmental and Climatist ideology. Nuclear power is banned, petroleum, natural gas, and hydroelectric power are under attack, and coal-fired power is soon to be precluded. When the wind ceases, let the electricity blackouts begin. And the reduction in greenhouse gases from all this might be worth one thin Washington quarter.

CHAPTER 17

COMMON SENSE AND THE FUTURE

"Truth, crushed to earth, shall rise again."
William Cullen Bryant (1839)

Climatism is the belief that man-made greenhouse gases are destroying Earth's climate. Twenty-one years ago, the Intergovernmental Panel on Climate Change of the United Nations concluded that manmade greenhouse gas emissions were causing global warming. In so doing, the IPCC launched mankind on a renewable-energy crusade in a pointless effort to control Earth's climate.

Secretary-General of the UN, Ban Ki-moon, addressed the Global Environment Forum in Incheon, Korea in August, 2009:

> Climate change, as all previous speakers have already stated, is the fundamental threat to humankind...If we fail to act, climate change will intensify droughts, floods and other natural disasters. Water shortages will affect hundreds of millions of people. Malnutrition will engulf large parts of the developing world. Tensions will worsen. Social unrest - even violence - could follow. The damage to national economies will be enormous. The human suffering will be incalculable...We have just four months. Four months to secure the future of our planet.[1]

The Secretary-General appears to have completely lost touch with reality. Both Chicken Little and the Emperor without clothes would be proud of such comments.

WE CAN NO LONGER CONTROL THE SEAS

King Canute the Great ruled England from 1016 to 1035. According to legend, when his flattering courtiers claimed he could command the tides, he directed his throne to be carried to the seashore. But the waves did not obey his commands. Being a man who believed in God, he reportedly said:

> Let all men know how empty and worthless is the power of kings. For there is none worthy of the name but God, whom heaven and sea obey.[2]

Stefan Rahmstorf, of the Potsdam Institute in Germany, recently presented at the U.K.'s Oxford University and discussed sea level rise:

> The crux of the sea level issue is that it starts very slowly but once it gets going it is practically unstoppable...There is no way I can see to stop this rise, even if we have gone to zero emissions.[3]

With all respect to Dr. Rahmstorf, when has mankind ever been able to stop the rise of the sea? When have we ever been able to control the weather at a single location on Earth? Yet, Climatism now demands that we switch to renewable energy, forego economic growth and consumption, and accept thousands of regulations on energy use for our home, transportation, business, and recreation. If we do so, *then* we'll be able to command the tides and control the weather. Climatism is a belief system—not science.

In September 2009, Mojib Latif of the Leibniz Institute of Marine Sciences at Kiel University, Germany, addressed 1,500 scientists at the UN's World Climate Conference in Geneva, Switzerland. Dr. Latif is one of the world's leading climate modelers and a true believer in man-made global warming. However, Latif was surprised by the recent cooling in global temperatures:

> And then you see right away [that] it may well happen that you enter a decade or maybe even two, when the temperature cools relative to

the present level…I'm definitely not one of the skeptics…However, we have to ask the nasty questions ourselves, or some other people will do it.[4]

In fact, *not one* of the twenty multi-million-dollar supercomputers the IPCC trusts predicted the global cooling since 2002.

WHAT'S ALL THE CONCERN ABOUT?

From a common sense point of view, one must question what all the climate fuss is about. *Global temperatures have increased only 1.3 °F (0.7 °C) in the last century.* Figure 132 shows temperature records and average daily maximum and minimums for Chicago, compiled over the last 136 years. The annual temperature variation in Chicago is

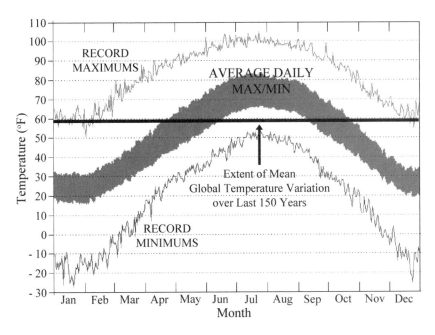

Figure 132. Chicago Temperatures and Global Temperature Change. Chicago record maximum and minimum temperatures by day of the year are shown in the top and bottom curves. The gray shaded curve in the center shows average daily maximum and minimum temperatures from 1872-2008. Total global average temperature change for the last 150 years is captured *within* the horizontal black line. (After Lindzen[5], adapted from Fisk,[6] 2009)

typically 100 degrees, changing from about -5°F to 95°F. In comparison, the extent of mean global temperature variation over the last 150 years is tiny, captured *within* the thickness of the horizontal black line in Figure 132. One degree is not much to a Chicagoan. For all the alarm, Climatism has only this historical 1.3°F temperature increase and computer model projections.

To put it another way, alarmists such as Mr. Ki-moon tell us global catastrophe will break loose with a 3°C (5.4°F) temperature rise by the year 2100. But these alarming forecasts are based solely on model predictions, not experimental evidence. As shown in Figure 133, the average annual temperature difference between Copenhagen and Paris is about 3°C (Figure 133). If the temperature rises by three degrees in Copenhagen over 100 years, won't the Danes be able to adapt? Are the people of Paris dying from heat, drought, and disease from the higher temperatures? The average temperature difference between Stockholm and Athens is 11.7°C. Does Sweden need to

Figure 133. Average Temperatures in Europe. Average annual temperatures for major cities in degrees Celsius. (Weatherbase)[7]

make payments to Greece to help them with their excessive temperatures?

GOOD NEWS FROM REALISTIC CLIMATE SCIENCE

There is no scientific evidence that carbon dioxide is in control of Earth's temperatures. As we discussed in Chapter 3, science increasingly shows that cloud formation, precipitation, water vapor (our largest greenhouse gas) and ocean temperature cycles are much greater climatic forces than carbon dioxide. Isotope analysis shows that only 4% of atmospheric CO_2 is due to man-made emissions (the rest is from natural causes), contrary to assertions by the IPCC. The Climatist assumption that increases in atmospheric carbon dioxide cause a positive feedback of increased atmospheric water vapor is looking more and more unlikely. The General Circulation Models unanimously predict an atmospheric warming signature above the equator, but *experimental data from satellites and weather balloons do not show this warming.*

In Chapter 4 we showed that Earth's climate is driven by long-term, medium-term, and short-term cycles. Proxy and historical evidence shows that Earth's climate was at least as warm during the Medieval Warm Period as it is today. The late 20th century warming is much better explained by a combination of long-term temperature recovery from the Little Ice Age and the temperature variation of the Pacific Decadal Oscillation, than by rising greenhouse gas emissions.

Chapter 5 showed evidence that the sun is the driver of Earth's climate. A combination of 1) long-term changes in solar radiation, and 2) short-term changes in low-level cloudiness driven by variation in sunspot activity, correlates well with changes in Earth's temperatures. The recent marked decline in solar activity matches the 8-year decline in global temperatures, while CO_2 levels continue to rise and *do not* match the temperature decrease.

In Chapter 6 we showed that, despite sensational claims from the alarmists, recent climate events are neither remarkable nor unprecedented in history. Seas continue to rise at a steady rate of seven inches per century. Even the IPCC projects a mean 100-year rise of only 38.5 cm (15 inches). Earth's icecaps are stable, with the Antarctic Icecap growing, the Greenland Icecap stable, and the Arctic

Icecap shrinking; all are within historical variation. Tropical storms, droughts, and floods are not stronger or more frequent and are within the range of historical patterns. Model projections of higher temperature-related deaths, increased vector-borne disease, and species extinction are not supported by experimental evidence. Glacier shrinkage has occurred for 200 years, beginning well before significant man-made greenhouse gas emissions. No evidence exists that the Gulf Stream is slowing down or connected to CO_2 emissions.

This is a book of good news. **The good news is that man-made greenhouse gases are an insignificant part of global climate change.** Global warming and global cooling are due to the natural cycles of Earth, which are ultimately driven by the sun. We do not yet understand Earth's climate well enough to control it. Until we do, mankind must continue to adapt to the climate of Earth as we have for thousands of years.

THE BAD NEWS IS CLIMATISM ITSELF

As we have discussed, the bad news is that, regardless of the science, the world is adopting "snake-oil remedies" from the prophets of Climatism. Cap-and-trade systems, renewable portfolio standards, automobile carbon emissions standards, and feed-in tariffs are all poor economics, but increasingly endorsed by Climatist-driven political leaders. The world is spending $170 billion per year, including over $50 billion in subsidies for wind power, solar power, and biofuels. **But there is no economic gain from all this effort.** These measures only switch from efficient low-cost hydrocarbon fuels to unreliable and expensive renewable energy in a fruitless quest to reduce greenhouse gas emissions. Almost every citizen on Earth is paying the price in higher energy costs, reduced economic growth, and loss of freedoms.

COMMON SENSE FOR THE 21st CENTURY

The road to improved prosperity is through sound science, advances in technology, and economic growth. Economic growth is best driven by free market competition and global trade, not by bureaucratic control. Policies restricting consumption, production, and use of energy in the name of sustainability and global responsibility will only promote economic stagnation in developed nations and prolong disease and suffering in the developing world. Here are a few common sense ideas for policy in the 21st century.

Air Pollution

Carbon dioxide should be removed from the list of pollutants. Despite what the media tells you, there is no scientific logic for classifying CO_2 as a pollutant based on the greenhouse effect. As we discussed in Chapter 8, if CO_2 is a pollutant because it's a greenhouse gas, then water vapor is also a pollutant.

Carbon monoxide, carbon particulates, lead, ozone, nitrogen dioxide, and sulfur dioxide are gases harmful to humans, and rightly classified as pollutants. Mankind should continue to reduce industrial emissions of these compounds.

Deforestation

As we discussed in Chapter 8, forests are expanding in all developed nations. The best way to stop deforestation in poor nations is to help them develop electricity and hydrocarbon fuel sources, such as propane and natural gas, so they stop burning their forests for fuel. Efforts to stop the use of hydrocarbons will only prolong the use of wood for fuel and resulting deforestation.

Subsidies for Energy

All subsidies for commercial production of energy should be eliminated. Government subsidies should only be used for research, to stimulate development of improved energy technology. Today, the world promotes wind, solar, and biofuel industries that exist solely because of government mandates and payments. These industries are now huge constituent groups that lobby for indefinite continuation of government subsidies and favored treatment. A prime example is the U.S. ethanol industry, which receives more than $13 billion per year in payments, and is one of the most powerful lobby forces in the U.S.

Coal for Electricity

Coal is the world's lowest-cost fuel for generation of electricity. The world has over 100 years of proven coal reserves. Humanity should continue to use coal where economically the best choice. Technology should be applied to further reduce sulfur dioxide and other traditional pollution from coal-fired plants. Carbon Capture and Storage provides *zero economic value* and will be prohibitively expensive. Research on CCS should be discontinued.

Nuclear Power
Nuclear power offers highly concentrated energy for electrical power generation. After decades of experience, the world has learned to use commercial nuclear power safely. Nuclear power should be used for peaceful purposes where economically viable.

Renewable Energy
The wind and solar power industries would be only a small fraction of their current size without government mandates and subsidies. They should be required to compete on an even footing with any other energy source. What will happen when the average person realizes that greenhouse gases are not the driver of global warming? To paraphrase Ronald Reagan, the day may come when the citizens of Britain say, "Mr. Prime Minister—tear down this windmill!"

Oil Drilling
As we have discussed, a continuous stream of petroleum is naturally leaking into U.S. coastal waters in the Gulf of Mexico and near California. Why do U.S. citizens choose to buy oil from Saudi Arabia or Libya and at the same time let it leak from the ocean floor into coastal waters? A common-sense energy policy would lift the ban on new oil production from the U.S. offshore continental shelf.

Drilling on only 2,000 acres of the 19.3-million-acre Arctic National Wildlife Refuge can yield oil that will replace U.S. oil imports from Saudi Arabia for 18 years. At $100 per barrel, such oil is worth over $1 trillion to the U.S. economy. It makes sense for the U.S. to access ANWR energy.

Biofuels
As we discussed in Chapter 13, the world is currently subsidizing biofuel production to the tune of $20 billion, and this bill is quickly rising. But because ethanol has less energy density than gasoline and because biodiesel yield per acre is small, these fuels have only a small

effect on replacing usage of petroleum. Biofuel usage will also not stop global warming. But the conversion of farmland from food crops to motor fuel does raise world food prices to millions in global poverty. Like other energy sources, subsidies for biofuel production should be eliminated. Suppose we use our agriculture to grow food and our oil industry to fuel vehicles?

Research and Development

Science and technology has been a foundation of our modern information and energy age. New sources of energy will be delivered by breakthroughs in technology. Government funding of energy research and development is appropriate public policy. Nuclear fusion, advanced nuclear power, energy storage, fuel cells for electric vehicles, and pollution reduction are all reasonable topics for research. Research into advanced solar cells and wind power is also appropriate, but such research won't help when the wind doesn't blow or the sun doesn't shine. However, government-directed research can also skew science, as in the case of today's climate science. The best course may be financial support for research with a minimum of government direction.

In our humble opinion, these are some of the policies that should be followed. But, given the current world view, climate madness is likely to continue for decades.

TRENDS FOR THE FUTURE

The physicist Neil Bohr reportedly said "Prediction is very difficult, especially if it is about the future." One way to predict the future is to project current trends. Here are some thoughts about the future, based on current trends.

Trend #1: The Earth is Cooling

Natural climatic forces appear to be cooling the Earth. Both the Pacific Decadal Oscillation and the Atlantic Multidecadal Oscillation have entered a cool phase. This means the Atlantic and Pacific Oceans are showing cooler temperatures. As we discussed in Chapter 5, the sun also continues to be in a quiet phase. As of late 2009, Solar

Cycle 24 has begun, but remains at a low level. This low level of solar activity is being compared to periods of solar minimums in the early 20th century and during the Little Ice Age.[8] Because of the combination of cooling PDO and AMO cycles and low solar activity, it appears we are in for 10 to 20 years of global cooling.

However, the forces of Climatism are not interested in data that shows Earth is cooling. Like the advice of the Wizard of Oz to "pay no attention to that man behind the curtain,"[9] Climatism marches onward with single-minded demands for energy revolution. According to Climatism, it really doesn't matter what temperatures are doing because the science is settled. We must forego hydrocarbon energy and adopt renewables, regardless of experimental data.

Trend #2: Continuing Growth of Carbon Constituencies

In May, 2009, Sir Muir Gray, Public Health Director of the Campaign for Greener Healthcare, wrote in the *UK Times*:

> Climate change is the cholera of our era–fear of the havoc that climate change will wreak should stimulate a new public health revolution.[10]

Paul Reiter, an expert in vector-borne diseases, labeled the article by Gray a "load of garbage." Dr. Reiter is a member of the Pasteur Institute of Paris, a member of the World Health Organization Expert Advisory Committee on Vector Biology and Control, and formerly of the Centers for Disease Control and Prevention. Reiter denounces the article by Sir Gray:

> They have cherry picked without remorse…these peddlers of garbage quote a 1998 model by two activists whose work is ridiculed by those of us who work in this field…I am flabbergasted that this can go on, and on, and on.[11]

But it is going to go on and on. The world now has millions of people who rely on Climatism and the myth of man-made global warming for their careers and their livelihood. From the science community, we have men like Dr. James Hansen, modeling scientists and their multi-million dollar supercomputers, and thousands of researchers who are studying global warming in every part of the globe. We have environmental editors at newspapers, deans of sustainability at colleges, and environmental vice presidents at thousands of companies. Local and national governments are building huge bureaucracies of climate regulators, enforcers, and consultants.

Climatism is big business. It's the basis for the growing wind power, solar, and biofuel industries. Even the nuclear industry views global warming mania as key to its revitalized growth. Global multinational companies strive to be "greener" than their competition to gain favor with environmental groups and the mislead public. Let's not forget the carbon traders and their artificial carbon markets, or the environmental attorneys planning to earn millions in suits against carbon "polluters."

Like the IPCC, we now have a great many people standing on many levels of the global warming house of cards. But climate science is inexact and the real cause of global warming will never be definitively proven. Expect the global warming crusade to go forward for many years to come.

Trend #3: Climatism-Driven Conflict between Nations

The level of conflict between nations is rapidly increasing. Conflict is escalating between Europe and the U.S., between the U.S. and China, and between the developed and developing nations. Driven by fear of climate disaster, the world is moving toward a breakdown in international relationships because of global warming delusion.

By taking the lead in reducing greenhouse gas emissions, Europe has painted itself into a corner. Aluminum, cement, chemicals, iron, and steel firms in Europe are looking to move their facilities to other world locations to avoid costly energy bills and ETS carbon trading costs. The European Union has agreed to provide free carbon permits to threatened industries to minimize "carbon leakage" of facilities out of Europe.[12] The cost of electricity and motor fuel in the U.S. is already much lower than in Europe. European governments and companies are desperate to force the U.S., China, and developing nations to accept the same onerous carbon costs and energy prices for their manufacturing industries. With the failure to achieve agreement at Copenhagen Summit, we may be on the verge of the biggest trade war since the 1930s.

Developed nations have accepted blame for carbon pollution, so developing nations now feel they are entitled to compensation for the coming "climate crisis." The Maldives Islands want compensation for rising sea levels. The Inuit people want compensation for melting arctic ice. In October, 2009 a mock "Asian Peoples' Climate Court was held in Bangkok. The Court found the G8 developed nations guilty of "planetary malpractice" in violation of the UN Framework Convention on Climate Change.[13] African nations are now demanding large payments to compensate for computer-model-projected climate disasters. Salifou Sawadogo, environmental minister for Burkina Faso states:

> We think 65 billion dollars are needed to deal with the effects of climate change on a continental scale. That is to say that our expectations are very high.[14]

Dr. Benny Peiser of Liverpool John Moores University summarizes:

> ...the climate hysteria created and perpetuated by Western government officials has opened Pandora's Box. What looked to be a valuable policy tool for green protectionism is now threatening to unleash political chaos and economic misery on its creators and their nations. Climate alarmism has turned into a Frankenstein monster that threatens to devour its own designers. I can't see why Africa and other developing nations should be ready to refrain from demanding hundreds of billions of dollars in reparations given that Nicholas Stern and other green campaigners and government officials claim that the West is liable for current and future climate disasters.[15]

Trend #4: Increasing Desperation of the Climatists

After hurricane Katrina in 2005, the widespread success of Al Gore's *Inconvenient Truth* in both book and movie versions, and the shrinkage of arctic ice in 2007, it appeared that the tsunami of global warming alarmism would sweep all in its path. But instead, the last two years indicate that Climatism momentum may have peaked in 2007. Hurricane activity in the Atlantic Ocean dropped to 30-year lows in 2009. The Arctic Icecap stopped shrinking and grew larger in both 2008 and 2009. Eight years of declining global temperatures are now recognized by the world. A growing number of scientists have announced skepticism of the theory of man-made global warming. Public opinion polls show that an increasing number of citizens in the U.K., the U.S. and other nations think that predicted global warming catastrophes are overstated.

In response to growing skepticism, Climatists are showing signs of desperation. Secretary-General Ki-moon, James Hansen, and Prince Charles all boosted the volume on apocalyptic warnings that we have "only a few years," pleading with listeners to change their carbon-emitting ways. If people won't act, the scenarios must get scarier.

In October of 2009, the British government released a prime-time £6 million advertising campaign rife with distortions that can only be regarded as *propaganda* by objective viewers. The video shows a father reading a story book to his daughter in bed. As the father reads about "awful heat waves" and "terrible storms and floods," the daughter sees pictures of a black CO_2 monster in the sky and people and animals drowning in flooded areas. The little girl asks in sadness "is there a happy ending?" The Ministers of Parliament approved the campaign because skepticism about climate change is growing among British citizens.[16] *The desperation of looming climate catastrophe demands the use of little children for propaganda.*

If punishing lifestyle-changing solutions are not enacted across the world, expect Climatist groups to engage in increasingly desperate and disruptive behavior. What can be more important than stopping global warming?

Trend #5: Growing Science Weight for Natural Causes, Not CO_2
The IPCC and climate modelers jumped to the conclusion that global warming was man-made at the end of the 1980s, despite the fact that much of Earth's climate was a mystery. Over the last 20 years, in part because of focus on global warming alarmism, many new aspects of Earth's climate have been researched. The weight of science continues to grow that climate change is caused by natural cycles, not man-made greenhouse gas emissions.

Scientists now have 30 years of satellite global temperature data, which refutes model-predicted atmospheric warming above the equator. The Pacific Decadal Oscillation was first named as recently as 1996. The PDO, the Atlantic Multidecadal Oscillation, and other ocean cycles have been recently characterized by scientists and shown to have been active for at least several thousand years.

Experts also better understand the role of the sun in driving Earth's climate. During the last 15 years, Svensmark developed his theory that sunspot activity and solar wind causes variations in

Earth's low-level cloudiness, with a significant impact on global temperatures. Willie Soon and others showed higher correlation of solar activity to global temperatures than CO_2 emissions. The IPCC will have a very difficult task if it again plans to discount the role of the sun in its next assessment report.

Much of the data on which the IPCC relies has also been shown to be flawed during the last 10 years. As we discussed in Chapter 2, the work of Anthony Watts showed significant Heat Island Bias in U.S. temperature station data. In 2009, the U.K. East Anglia Climate Research Unit admitted their original global temperature data had been erased.[17] This means the rise in global temperatures shown in Figure 6, Chapter 2 cannot be trusted.

In Chapter 8, we discussed the discrediting of the Mann "Hockey-Stick Curve." At the Senate hearings, those testifying in favor of Mann put forward the idea that, even if the Mann data were discredited, other studies showed unprecedented global warming in the 20th century. In 2009, after 10 years of trying, Steve McIntyre was able to obtain release of tree ring data from the Yamal Peninsula in Siberia. The data showed that several of these other studies were heavily based on only a small subset of overall tree-ring data. When the entire data set was considered, the "unprecedented 20th century warming" disappeared.[18] It's apparent that variations on Mann's hockey stick have also lost credibility.

In summary, the truth, crushed to earth, will rise again. It's only a matter of time before humanity realizes Climatism is based on false science.

SOME THINGS YOU CAN DO

It's a sad situation, but how can citizens trust their Congressperson, Minister of Parliament or other government official, the news media, or even many scientists to provide accurate information on climate change? How can we trust the leaders of our major scientific organizations, such as the Royal Society, the National Academy of Sciences, and the National Oceanic and Atmospheric Administration? National governments now provide the majority of funding for these organizations. To continue to receive this funding, the politically correct policy must be espoused.

If you have stayed with us throughout, you understand the real story. You can learn more from the Further Reading section that follows this chapter. The section provides a list of books that continue a realistic discussion of Earth's changing climate, energy economics, and the real state of the world.

We encourage you to challenge your government officials and Climatists in a public forum with some of the following questions:

- Why do you want to raise our energy prices?
- If this legislation is enacted, exactly how much will global temperatures be reduced?
- How do you know that greenhouse gases cause global warming, rather than natural cycles of Earth?
- Global temperatures have been cooling since 2002. Why didn't any climate models predict this?
- How much will our nation need to pay to developing nations in the name of climate change?

A FINAL NOTE: THE P-38 "GLACIER GIRL"

It was 1942 and the United States had entered the European war on the side of Britain against Germany and Italy. To provide needed aircraft to the U.K., the allies mounted Operation Bolero. Bolero ferried planes across the Atlantic Ocean from the U.S. to Britain using stops in Newfoundland, Greenland, Iceland, and Scotland.

On July 15, 1942, pilot Carl Rudder led a flight of six Lockheed P-38F Lightning fighters and two Boeing B-17 Flying Fortress bombers from Greenland toward Iceland on the third leg of one such ferry. Bad weather forced the flight to turn back. Low on fuel and disoriented by the storm, the eight planes crash-landed on the ice of southeast Greenland. Miraculously, all 25 men of the flight survived the landing and nine days on the icecap. They were finally rescued by a dogsled team from a U.S. Coast Guard cutter on July 24.[19]

Thirty-nine years later in 1981, Atlanta businessmen Pat Epps and Richard Taylor formed the Greenland Expedition Society and returned to Greenland to recover the planes. Taylor recalls:

Our thoughts were that the tails would be sticking out of the snow. We'd sweep off the wings and shovel them out a little bit, crank the planes up and fly them home. Of course, it didn't happen.[20]

The location of the 1942 crash site was well known, but the eight planes were nowhere to be found. Three more expeditions were mounted in the early 1980s, but were also unable to locate the planes.

In 1988, Epps and Taylor returned with a team equipped with high-technology sub-surface radar systems for a more thorough search. All eight aircraft were soon located about two miles from the original crash location. *The radar systems found the planes buried under 270 feet of solid ice.*[21] Forty-six years of snow and ice had accumulated on the lost aircraft.

It took Epps and Taylor more than $2 million, 11 years, and use of advanced technology to recover one of the P-38 fighters, which they christened the "Glacier Girl."[22] Figure 134 shows the Glacier Girl in an ice cave carved by the salvage team, 270 feet below the surface of the Greenland ice cap. Today, the Glacier Girl has been restored to working condition and periodically flies in air shows—rescued from under the "disappearing" Greenland icecap!

Figure 134. P-38 Glacier Girl in Greenland Ice Cave. After crash-landing in 1942, the Glacier Girl was rescued 50 years later under 270 feet of accumulated ice. (Photo by Lou Sapienza)[23]

FURTHER READING

Blue Planet in Green Shackles: What is Endangered: Climate or Freedom? by Václav Klaus (Competitive Enterprise Institute, 2007).

Climate Change Reconsidered: the Report of the Nongovernmental International Panel on Climate Change by Craig Idso and S. Fred Singer (Heartland Institute, 2009).

Climate Confusion: How Global Warming Hysteria Leads to Bad Science, Pandering Politicians and Misguided Policies that Hurt the Poor by Roy Spencer (Encounter, 2008).

Climate of Extremes: Global Warming Science They Don't Want You to Know by Patrick J. Michaels and Robert Balling, Jr. (Cato Institute, 2009).

CO_2, Global Warming and Coral Reefs: Prospects for the Future by Craig Idso (Science and Public Policy Institute, 2009).

Eco-Imperialism: Green Power, Black Death by Paul Driessen (Free Enterprise Press, 2003).

Green Hell: How Environmentalists Plan to Control Your Life and What You Can Do to Stop Them by Steve Milloy (Regnery, 2009).

Gusher of Lies: The Dangerous Delusions of Energy Independence by Robert Bryce (Public Affairs, 2008).

Heaven and Earth: Global Warming, the Missing Science by Ian Plimer (Connor Court, 2009).

Red Hot Lies: How Global Warming Alarmists Use Threats, Fraud, and Deception to Keep You Misinformed by Christopher Horner (Regnery,

2008).

Solar Fraud: Why Solar Energy Won't Run the World by Howard C. Hayden (Vales Lake, 2005).

The Bottomless Well: The Twilight of Fuel, the Virtue of Waste, and Why We Will Never Run Out of Energy by Peter Huber and Mark Mills (Basic Books, 2005).

The Chilling Stars: A New Theory of Climate Change by Henrik Svensmark and Nigel Calder (Totem, 2007).

The Deniers: The World Renowned Scientists Who Stood Up Against Global Warming Hysteria, Political Persecution, and Fraud–And those who are too fearful to do so by Lawrence Solomon (Richard Vigilante Books, 2008).

The Improving State of Our World by Indur Goklany (Cato Institute, 2007).

The Politically Correct Guide to Global Warming (and Environmentalism) by Christopher C. Horner (Regnery, 2007).

The Real Global Warming Disaster by Christopher Booker (Continuum, 2009).

The Really Inconvenient Truths: Seven Environmental Catastrophes Liberals Don't Want You to Know About–Because They Helped Cause Them by Iain Murray (Regnery, 2008).

The Skeptical Environmentalist: Measuring the Real State of the World by Bjorn Lomborg (Cambridge University, 2001).

The Ultimate Resource 2 by Julian L. Simon (Princeton University Press, 2008).

Unstoppable Global Warming: Every 1,500 Years by S. Fred Singer and Dennis Avery (Rowman & Littlefield, 2008).

NOTES

Introduction
1. Al Gore, Statement to the Senate Foreign Relations Committee, January 28, 2009.
2. "Coal-fired power stations are death factories," *The Observer*, February, 15, 2009, http://www.guardian.co.uk/commentisfree/2009/feb/15/james-hansen-power-plants-coal/print
3. "Tony Blair's foreword to climate change report," *Times Online*, January 30, 2006, http://www.timesonline.co.uk/tol/news/uk/article722952.ece?print=yes&randnum=123722

Chapter 1: The Tsunami for Global Warming Alarmism
1. Al Gore, invited address before the American Associate for Advancement of Science, February 14, 2009, web site: http://www.aaas.org/news/releases/2009/0215am_gore.shtml
2. Ibid.
3. Ibid.
4. The Oscars, web site: http://www.oscar.com/oscarnight/winners/?pn=detail&nominee=AnInconvenientTruthDocumentaryFeatureNominee
5. "Ingraham and Morano Expose Gore's Global Warming Profit Motive," *News Busters*, May 3, 2009, web site: http://newsbusters.org/blogs/noel- sheppard/2009/05/03/ingraham-morano-expose-gores-global-warming-profit-motive
6. "Gore Pulls Slide of Disaster Trends," *The New York Times*, February 23, 2009, http://dotearth.blogs.nytimes.com/2009/02/23/gore-pulls-slide-of-disaster-trends/?pagemode=print
7. Stuart Dimmock vs. Secretary of State for Education and Skills, Royal Courts of Justice, Strand, London, WC2A 2LL, October 10, 2007.
8. "Gore climate film's nine 'errors'", *BBC News*, October 11, 2007, http://news.bbc.co.uk/go/pr/fr/-/1/hi/education/7037671.stm
9. Al Gore website, http://www.algore.com
10. Ibid.
11. "Climate scientist sees cover-up," *The Washington Times*, March 20, 2007,

http://washingtontimes.com/news/2007/mar/20/20070320-120435-3136r/print/

12. "Global Warming Has Begun, Expert Tells Senate," *The New York Times*, June 24, 1988, http://www.nytimes.com/1988/06/24/us/global-warming-has-begun-expert-tells-senate.html

13. James Hansen, "Testimony by James Hansen: Political interference with Government Climate Change Science," U.S. House of Representatives Committee on Oversight and Government Reform, March 19, 2007, web site: http://www.spaceref.com/news/viewsr.html?pid=23642

14. "NASA Climate Scientist Honored by American Meteorological Society," Goddard Space Flight Center, Press Release No. 09-005, January 14, 2009, http://www.nasa.gov/centers/goddard/news/topstory/2009/hansen_ams.html

15. "Coal-fired power stations are death factories," *The Observer*, Sunday 15 February, 2009, http://www.guardian.co.uk/commentisfree/2009/feb/15/james-hansen-power-plants-coal/print

16. James Hansen, "Global Warming Twenty Years Later: Tipping Points Near," address to the National Press Club, June 23, 2008, www.columbia.edu/~jeh1/2008/TwentyYearsLater_20080623.pdf

17. *The Observer* (See no. 15)

18. Walter Cunningham, "In Science, Ignorance is not Bliss," *Launch Magazine*, web site: http://launchmagonline.com/walt-cunninghams-viewpoint/64-in-science-ignorance-is-not-bliss

19. *The Economics of Climate Change: The Stern Review* by Nicholas Herbert Stern, Great Britain, (Cambridge University Press, 2007).

20. "Cost of tackling global climate change has doubled, warns Stern," *The Guardian*, June 26, 2008, http://www.guardian.co.uk/environment/2008/jun/26/climatechange.scienceofclimatechange/print

21. "Mass Migrations and War: Dire Climate Scenario," *Associated Press*, February 21, 2009, http://abcnews.go.com/Technology/WireStory?id=6929425&page=2

22. *An Inconvenient Truth: The Planetary Emergency of Global Warming and What We Can Do About It* by Al Gore (Rodale, 2006), pp. 196-209.

23. Arthur Robinson, et al., "Environmental Effects of Increased Atmospheric Carbon Dioxide," *Journal of American Physicians and Surgeons* (2007) 12, pp. 79–90.

24. Intergovernmental Panel on Climate Change, 4th Assessment Report, Summary for Policy Makers, 2007, p. 13, www.ipcc.ch/ipccreports/ar4-syr.htm

25. *An Inconvenient Truth* (See no. 22), p. 97.

26. *The Hot Topic: What We Can Do about Global Warming* by Gabrielle Walker and David King, (Houghton Mifflin Harcourt, 2008).

27. *An Inconvenient Truth* (See no. 22), pp. 42-59.

28. Robinson (See no. 23).

29. *CO2, Global Warming and Coral Reefs* by Craig Idso (Science & Public Policy Institute, 2009).

30. *The Day After Tomorrow*, by Roland Emmerich, (Twentieth Century-Fox Film Corporation, 2004).

31. "The Younger Dryas," National Oceanic and Atmospheric Administration, web site: http://lwf.ncdc.noaa.gov/paleo/abrupt/data4.html

32. United Nations Framework Convention on Climate Change text, http://unfccc.int/resource/docs/convkp/conveng.pdf

33. Kyoto Protocol, United Nations Framework Convention on Climate Change web site: http://unfccc.int/kyoto_protocol/items/2830.php

34. Vaclav Klaus, Presentation at the International Conference on Climate Change, March 8, 2009.

Chapter 2: The Thin Science for Man-Made Global Warming

1. Chris Landsea, Atlantic Oceanographic and Meteorological Laboratory web site: http://www.aoml.noaa.gov/hrd/tcfaq/C5c.html

2. *A World of Weather: Fundamentals of Meteorology, 4th Edition* by Lee Grenci and John Nese, p. 134 (Kendall/Hunt, 2006).

3. National Lightning Safety Institute web site: http://www.lightningsafety.com/nlsi_lhm/lpts.html

4. Illinois Department of Natural Resources web site: http://dnr.state.il.us/Lands/Landmgt/PARKS/I&M/CORRIDOR/geo/geo.htm

5. *Global Warming: The Complete Briefing*, 3rd Edition by John Houghton, p.16, (Cambridge University Press, 2004).

6. Jean-Baptiste Joseph Fourier, "Remarques générales sur les températures du globe terrestre et des espaces planétaires," *Annales de Chimie et de Physique*, (Paris) 2nd ser. 27, 136–167, 1824.

7. John Tyndall, "On the absorption and radiation of Heat by Gases and Vapours, and the Physical Connection of Radiation, Absorption, and Conduction," *Philosophical Magazine and Journal of Science*, ser. 4, vol. 22, 169–194, 273–285, 1861.

8. Svante Arrhenius, "On the Influence of Carbonic Acid in the Air upon the Temperature of the Ground," *Philosophical Magazine and Journal of Science*, ser. 5, vol. 41, 237–276, 1896.

9. G. S. Callendar, "The Artificial Production of Carbon Dioxide and Its Influence on Temperature," *Quarterly Journal Royal Meteorological Society* vol. 64, pp. 223–240, 1938.

10. U.S. Department of Commerce, National Oceanic & Atmospheric Administration, Earth System Research Laboratory web site: http://www.esrl.noaa.gov/gmd/ccgg/trends/co2_data_mlo.html

11. Climatic Research Unit (UK) web site: http://www.cru.uea.ac.uk/cru/info/warming/

12. Joseph D'Aleo, "United States and Global Data Integrity Issues," web site: http://icecap.us/images/uploads/US_AND_GLOBAL_TEMP_ISSUES.pdf

13. Ibid.

14. J. D. Goodridge, "Comments on Regional Simulations of Greenhouse Warming, Including Natural Variability," *Bulletin of American Meteorology Society* vol. 77, pp. 1588–1599, 1996.

15. Ibid.

16. United States Historical Climatology Network, web site: http://cdiac.ornl.gov/epubs/ndp/ushcn/background.html

17. What's Up with That web site: http://wattsupwiththat.com/test/

18. Climate Reference Network (CRN) Site Information Handbook, National Oceanographic and Atmospheric Administration, December 10, 2002.

19. See web site: http://www.surfacestations.org/
20. What's Up with That (See no. 17).
21. D'Aleo (See no. 12).
22. Roger Revelle and Hans Seuss, "Carbon Dioxide Exchange between Atmosphere and Ocean and the Question of an Increase of Atmospheric CO2 during the Past Decades," *Tellus* IX, pp. 18–27, 1957.
23. John Lewis, "Clarifying the Dynamics of the General Circulation: Phillips's 1956 Experiment," *Bulletin of the American Meteorological Society*, vol. 79, No. 1, pp. 39-60, January, 1998.
24. Syukuro Manabe and Richard Wetherald, "Thermal Equilibrium of the Atmosphere with a Given Distribution of Relative Humidity," *Geophysical Fluid Dynamics Laboratory*, ESSA, Wash. D.C., pp. 241–259, Nov. 2, 1966.
25. John Sununu, Presentation at the International Conference on Climate Change, March 10, 2009.
26. James Hansen et al., "Climate Impact of Increasing Atmospheric Carbon Dioxide," *Science*, vol. 23, pp. 957-966, August 28, 2001.
27. Ibid.
28. J. M. Barnola et al., "Vostok Ice Core Provides 160,000-year Record of Atmospheric CO2," *Nature*, vol. 329, pp. 408-14, 1987.
29. J. R. Petit et al., "Climate and Atmospheric History of the Past 420,000 Years from the Vostok Ice Core, Antarctica," *Nature*, vol. 399, June 3, 1999, pp. 429–436.
30. S. S. Abysov et al., "Deciphering Mysteries of Past Climate from Antarctic Ice Cores," *Earth in Space*, vol. 8, no. 3, November, 1995, p. 9, http://www.agu.org/sci_soc/vostok.html
31. C. Lorius et al., "The Ice-Core Record: Climate Sensitivity and Future Greenhouse Warming," *Nature*, vol. 347, September 13, 1990, pp. 139–145.
32. Hubertus Fischer et al., "Ice Core Records of Atmospheric CO2 around the Last Three Glacial Terminations," *Science*, vol. 283, March 12, 1999, pp. 1712–1714.

Chapter 3: Carbon Dioxide: Not Guilty

1. Intergovernmental Panel on Climate Change, 4th Assessment Report, 2007, p. 515, www.ipcc.ch/ipccreports/ar4-syr.htm
2. Ibid, p. 514.
3. Ibid, p. 501.
4. "Henry's Law and the Solubility of Gases," Introductory University Chemistry I, web site: http://dwb4.unl.edu/Chem/CHEM869J/CHEM869JLinks /www.chem.ualberta.ca/courses/plambeck/p101/p01182.htm
5. Christopher Sabine, et al., "The Oceanic Sink for Anthropogenic CO2," *Science*, vol. 305, July 16, 2004, pp. 367-371.
6. Bert Bolin and Erik Eriksson, 1959, "Changes in the Carbon Dioxide Content of the Atmosphere and Sea Due to Fossil Fuel Combustion," *Rossby Memorial Volume*, (Rockefeller Institute Press), pp. 130-142.
7. Tom Segalstad, "Carbon Cycle Modeling and the Residence Time of Natural and Anthropogenic Atmospheric CO2: On the Construction of the 'Greenhouse Effect Global Warming' Dogma," web site: http://folk.uio.no/tomvs/esef/ESEF3VO2.htm

8. Ibid, pp. 13–14.
9. Ibid, pp. 15–18.
10. Conversation with Dr. Segalstad, April 1, 2009.
11. *Global Warming: The Complete Briefing*, 3rd Edition by John Houghton, p.37, (Cambridge University Press, 2004).
12. *Heaven and Earth: Global Warming, the Missing Science* by Ian Plimer, (Connor Court, 2009), p. 421.
13. Arthur Robinson et al., "Environmental Effects of Increased Atmospheric Carbon Dioxide," *Journal of American Physicians and Surgeons*, vol. 12, pp. 79–90, 2007.
14. Ibid.
15. Intergovernmental Panel on Climate Change, 4th Assessment Report, 2007, p. 133, www.ipcc.ch/ipccreports/ar4-syr.htm
16. Ibid, p. 38.
17. Richard Lindzen, "Understanding Common Climatic Claims," Proceedings of the 2005 ERICE Meeting of the World Federation of Scientists on Global Emergencies, 2005.
18. David Archibald, "Solar Cycle 24: Implications for the United States," presented at the International Conference on Climate Change, March, 2008
19. William Happer, statement before the U.S. Senate Environment and Public Works Committee, February 25, 2009.
20. William Gray, Presentation at the International Conference on Climate Change, March 10, 2009
21. Roy Spencer, "Statement to the Committee on Oversight and Government Reform of the United states House of Representatives," March 19, 2007.
22. Roy Spencer, "Global Warming and Nature's Thermostat," web site: http://www.mannkal.org/downloads/environment/sthermostat.pdf
23. Happer (See no. 19)
24. Intergovernmental Panel on Climate Change (See no. 15), pp. 590-662.
25. S. Fred Singer, "Nature, Not Human Activity, Rules the Climate," The Heartland Institute, 2008
26. Intergovernmental Panel on Climate Change (See no. 15) p. 40.
27. Ibid.
28. Houghton (See no. 11). p. 37.
29. Theodore Anderson et al., "Climate Forcing by Aerosols--a Hazy Picture," *Science*, vol. 300, May 16, 2003, pp. 1103–1104.
30. Craig Idso and S. Fred Singer, *Climate Change Reconsidered: 2009 Report of the Nongovernmental Panel on Climate Change (NIPCC)*, (The Heartland Institute, 2009).
31. David Randall et al., "Breaking the Cloud Parameterization Deadlock," *American Meteorological Society*, November 2003, pp. 1547–1563, web site: http://www.seas.harvard.edu/climate/seminars/pdfs/randall_etal_2003.pdf
32. S. Fred Singer (See no. 25).
33. Reconciling Observations of Global Temperature Change, Panel on Reconciling Temperature Observations, Climate Research Committee, Commission on Geosciences, Environment, and Resources, National Research Council (National Academy Press, 2000), http://www.nap.edu/catalog.php?record_id=9755

34. Thomas Karl et al., "Temperature Trends in the Lower Atmosphere: Steps for Understanding and Reconciling Differences," U.S. Climate Change Science Program, p. 116, April, 2006.
35. Intergovernmental Panel on Climate Change (See no. 14), p. 675.
36. USCCSP (See no. 33), p. III.
37. USCCSP (See no. 33), p. 112.

Chapter 4: Climate Change is Continuous

1. William McClenney, "The Sky is Falling or On Revising the Nine Times Rule, Part I of V," February 28, 2008, web site: http://icecap.us/index.php/go/joes-blog/the_sky_is_falling_or_revising_the_nine_times_rule/
2. US Global Change Research Program, web site: http://www.ncdc.noaa.gov/paleo/ctl/thc.html
3. *Canon of Insolation and the Ice-Age Problem*, by Milankovic Milutin, 1941, (Agency for Textbooks, 1998).
4. "The Seasons and the Earth's Orbit," U.S. Naval Oceanography web site: http://aa.usno.navy.mil/faq/docs/seasons_orbit.php
5. *Global Warming: The Complete Briefing*, 3rd Edition by John Houghton, p.71, (Cambridge University Press, 2004).
6. "The Seasons and the Earth's Orbit," U.S. Naval Oceanography web site: http://aa.usno.navy.mil/faq/docs/seasons_orbit.php
7. Richard Muller and Gordon MacDonald, "Spectrum of 100-kyr glacial cycle: Orbital inclination, not eccentricity," *Proceedings of the National Academy of Sciences of the United States of America*, Aug. 5, 1997, v. 94(16), pp. 8329-8334.
8. *Unstoppable Global Warming Every 1500 Years*, by S. Fred Singer and Dennis Avery, p. 239, (Rowman & Littlefield, 2007).
9. Al Gore, invited address before the American Associate for Advancement of Science, February 14, 2009, web site: http://www.aaas.org/news/releases/2009/0215am_gore.shtml
10. See the website: http://www.physics.sc.edu/~rjones/phys101/Fahrenheit'sThermometer.html
11. W. Dansgaard, "Summary of Activities at the Geophysical Isotope Laboratory, Copenhagen," at web site: http://www.cig.ensmp.fr/~iahs/redbooks/a118/iahs_118_0401.pdf
12. "Paleoclimatology: the Oxygen Balance," NASA Earth Observatory, web site: http://earthobservatory.nasa.gov/Features/Paleoclimatology_OxygenBalance/oxygen_balance.php
13. Intergovernmental Panel on Climate Change, 4th Assessment Report (2007), p. 9, web site: www.ipcc.ch/ipccreports/ar4-syr.htm
14. Official Greenland Travel Guide, http://www.greenland.com/content/english/tourist
15. *The Emperor's New Climate, Debunking the Myths of Global Warming* by Bruno Wiskel, p. 73, (Evergreen Environmental Company, 2006).
16. Official Greenland Travel Guide, http://www.greenland.com/content/english/tourist
17. D. L. Ashman, "Vikings in America," http://www.pitt.edu/~dash/vinland.html
18. Scott Mandia, "The Little Ice Age in Europe," web site: http://www2.sunysuffolk.edu/mandias/lia/little_ice_age.html

19. Lori Martinez, "Useful Tree Species for Tree-Ring Dating," web site: http://www.ltrr.arizona.edu/lorim/good.html

20. Q. Ge et al., "Winter half-year temperature reconstruction for the middle and lower reaches of the Yellow River and Yangtze River, China, during the past 2000 years," The Holocene v. 13, 6 (2003), pp. 933–940.

21. Q. Ge et al., "Winter half-year temperature reconstruction for the middle and lower reaches of the Yellow River and Yangtze River, China, during the past 2000 years," The Holocene v. 13, 6 (2003), pp. 933–940, adapted by CO_2 Science: http://www.co2science.org/data/mwp/studies/l1_easternchina.php

22. Lloyd Keigwin, "The Little Ice Age and Medieval Warm Period in the Sargasso Sea," Science, New Series, vol. 274, no. 5292, Nov. 29, 1996, pp. 1504–1508.

23. Ibid.

24. Arthur Robinson et al., "Environmental Effects of Increased Atmospheric Carbon Dioxide," Journal of American Physicians and Surgeons, vol. 12, pp. 79–90, (2007).

25. Haken Grudd, "Tornetrask tree-ring width and density AD 500–2004: a test of climatic sensitivity and a new 1500-year reconstruction of north Fennoscandian summers," Climate Dynamics, vol. 31, 30 January, 2008, pp. 842–857.

26. Ibid, adapted by CO_2 Science: http://www.co2science.org/data/mwp/studies/l1_tornetrask.php

27. Intergovernmental Panel on Climate Change, 4th Assessment Report, 2007, p. 9, www.ipcc.ch/ipccreports/ar4-syr.htm

28. CO_2 Science web site: http://www.co2science.org/data/mwp/mwpp.php

29. Unstoppable Global Warming Every 1,500 Years (See no. 8), p.1.

30. W. Dansgaard et al., "North Atlantic Climatic Oscillations Revealed by Deep Greenland Ice Cores," in Climate Processes and Climatic Sensitivity, ed. F. E. Hansen and T. Takahashi (Washington, D.C.: American Geophysical Union, 1984), Geophysical Monograph 29, 288-98.

31. Ibid, adapted by Dennis Avery, Presentation at the International Conference on Climate Change, Mar 9, 2009.

32. Climate: Past, Present and Future vol. 2 by H. H. Lamb, (Methuen, 1977).

33. The News Hour with Jim Lehrer, February 4, 1998, web site: http://www.pbs.org/newshour/bb/weather/jan-june98/nino_2-4.html

34. OceanWorld Web site: http://oceanworld.tamu.edu/print/students/elnino/elnino2.htm

35. Web site: http://www.aoml.noaa.gov/phod/amo_faq.php

36. "What is an El Niño?," U.S. Dept. of Commerce web site: http://www.pmel.noaa.gov/tao/elnino/el-nino-story.html

37. Ibid.

38. Fred Goldberg, Presentation at the International Conference on Climate Change, March 9, 2009.

39. "The Pacific Decadal Oscillation (PDO)," web site: http://jisao.washington.edu/pdo/

40. "What is the Pacific Decadal Oscillation?," National Weather Service Forecast Office, web site: http://www.wrh.noaa.gov/fgz/science/pdo.php?wfo=fgz

41. "Atlantic Multidecadal Oscillation Timeseries," 1856-2008, CDC NOAA, web site: http://www.cdc.noaa.gov/data/correlation/amon.us.long.data

42. Syun-Ichi Akasofu, Presentation at the International Conference on Climate Change, March 9, 2009.
43. Syun-Ichi Akasofu, "Is the Earth Still Recovering from the 'Little Ice Age?,' A Possible Cause of Global Warming," web site: http://www.iarc.uaf.edu /highlights/2007/akasofu_3_07/Earth_recovering_from_LIA.pdf
44. Ibid.
45. PDO (See no. 40).
46. Akasofu (See no. 43).
47. Roy Spencer, Presentation at the International Conference on Climate Change, March 10, 2009.

Chapter 5: The Sun is Our Climate Driver

1. "Climate Debate and FAQ," Danish National Space Center, web site: http://spacecenter.dk/research/sun-climate/climate-debate-and-faq
2. Solar Views, web site: http://www.solarviews.com/eng/sun.htm
3. J. Beer et al., "The Role of the Sun in Climate Forcing," *Quaternary Science Reviews,* vol. 19, pp. 403–415 (2000)
4. *Intergovernmental Panel on Climate Change, 4th Assessment Report* (2007), p. 691, www.ipcc.ch/ipccreports/ar4-syr.htm
5. Beer (See no. 3).
6. Solar and Heliospheric Observatory, NASA, web site: http://sohowww.nascom.nasa.gov/gallery/images/bigspotfd.html
7. What's Up with That web site: http://wattsupwiththat.com/test/
8. Sawyer Hogg, Out of Old Books (Schwabe's Determination of the Sun-spot Cycle), web site: http://articles.adsabs.harvard.edu/full/seri/JRASC/0042/0000042.000.html
9. US Air Force photo, web site: http://en.wikipedia.org/wiki/File:Polarlicht_2.jpg
10. John Eddy, "The Maunder Minimum," *Science,* vol. 192, no. 4245, June 18, 1976, pp. 1189–1202.
11. Ibid.
12. Douglas Hoyt and Kenneth Schatten, "Group Sunspot Numbers: A New Solar Activity Reconstruction," Solar Physics, vol. 181, iss. 2, pp. 491–512 (1998),
13. Alan Cheetham, "Global Warming – It's Not Anthropogenic CO2," web site: http://www.appinsys.com/globalwarming
14. E. Friis-Christensen, and K. Lassen, "Length of the Solar Cycle: an Indicator of Solar Activity Closely Associated with Climate," *Science,* vol. 192, pp. 1189–1202, (1991).
15. K. Lassen and E. Friis-Christensen, "Variability of the Solar Cycle Length during the Past Five Centuries and the Apparent Association with Terrestrial Climate," *Journal of Atmospheric Terrestrial Physics,* vol. 57, no. 8, pp. 835–845, (1995).
16. Joseph D'Aleo, Presentation at the International Conference on Climate Change, March 9, 2009.
17. Arthur Robinson et al., "Environmental Effects of Increased Atmospheric Carbon Dioxide," *Journal of American Physicians and Surgeons,* vol. 12, pp. 79–90, (2007).

18. Willie Soon, "Variable Solar Irradiance as a Plausible Agent for Multidecadal Variations in the Arctic-wide Surface Air Temperature Record of the Past 130 Years," *Geophysical Research Letters*, vol. 32, L16712, October 29, 2005.

19. Ibid.

20. Email discussion with Willie Soon, March 23, 2009.

21. Henrik Svensmark, "Influence of Cosmic Rays on the Earth's Climate," from web site: http://www.junkscience.com/Greenhouse/influence-of-cosmic-rays-on-the-earth.pdf

22. Eigil Friis-Christensen and Henrik Svensmark, "What do we really know about the sun-climate connection?," Advanced Space Research, vol. 20, no. 4/5, pp. 913–921, 1997.

23. What's Up with That web site: http://wattsupwiththat.com/2009/03/17/beryllium-10-and-climate/

24. Jürg Beer et al., "The Role of the Sun in Climate Forcing," *Quaternary Science Reviews*, vol. 19, (2000), pp. 403-415.

25. Ibid.

26. Nigel Marsh and Henrik Svensmark, "Cosmic Rays, Clouds, and Climate, *Space Science Reviews*, vol. 94, pp. 215–230, May 10, 2000.

27. Nir Shaviv, "Cosmic Rays and Climate," web site: http://www.sciencebits.com/CosmicRaysClimate

28. Henrik Svensmark and Eigil Friis-Christensen, "Variation of Cosmic Ray Flux and Global Cloud Coverage—a Missing Link in Solar-Climate Relationships," *Journal of Atmospheric and Solar-Terrestrial Physics*, v. 59, p. 1225-1232, (1997).

29. "The SKY Experiment," web site: http://spacecenter.dk/research/sun-climate/experiments/the-sky-experiment/sky.

30. "The CLOUD Experiment," web site: http://www.space.dtu.dk/English /Research/Research_divisions/Sun_Climate/Experiments_SC/CLOUD.aspx

31. *The Chilling Stars: A New Theory of Climate Change* by Henrik Svensmark and Nigel Calder, (Totem Books, 2007).

32. S. Fred Singer, "Nature, Not Human Activity, Rules the Climate," The Heartland Institute (2008)

33. Svensmark (See no. 21).

34. Dennis Avery, "Where's the Runaway Warming," *Environmental Views*, April 19, 2009.

35. Jack Langer, "A Bad Day for a Global Warming Protest," March 3, 2009, web site: www.humanevents.com/index.php

36. Joseph D'Aleo, "Global Temperature Trends," web site: http://icecap.us/index.php

37. Richard Harris, "The Mystery of Global Warming's Missing Heat," web site: http://www.npr.org/templates/story/story.php?storyId=88520025.

38. Josh Willis personal communication, 2009.

39. "Quiet Sun Baffling Astronomers," *BBC News*, April 21, 2009, web site: http://newsvote.bbc.co.uk/mpapps/pagetools/print/news.bbc.co.uk/2/hi/science /nature/8008473.stm?ad=1

40. David Archibald, "Solar Cycle 24: Implications for the United States," presented at the International Conference on Climate Change, March, 2008.

41. Ibid.

42. David Archibald, web site: http://anhonestclimatedebate.wordpress.com /2008/10/13/david-archibald%E2%80%99 s-elegant-illustration-of-how-late-and-weak-solar-cycle-24-is-proving/

43. Email conversation with David Archibald, December 5, 2009.

Chapter 6: Global Warming Disasters Debunked

1. "President 'has four years to save Earth,'" *The Observer*, January 18, 2009, web site: http://www.guardian.co.uk/environment/2009/jan/18/jim-hansen-obama/

2. "CBS: Global Warming Increasing; 'Vicious Cycle' That 'Must Be Broken,'" February 16, 2009, web site: http://newsbusters.org/node/28187/

3. "Prince Charles: World Must Act Now to Save Planet," *Associated Press*, March 12, 2009, web site: http://abcnews.go.com/International/wireStory?id=7069895

4. "Rachel Maddow thinks snow in Dubai caused by global warming," web site: http://www.ihatethemedia.com/Rachel-maddow-thinks-snow-in-dubai-caused-by-global-warming

5. Weiss and Overpeck, Environmental Studies Laboratory, University of Arizona, web site: http://www.geo.arizona.edu/dgesl/research/other/climate_change_and_sea_level/sea_level_rise/sea_level_rise_old.htm#images

6. *An Inconvenient Truth: The Planetary Emergency of Global Warming and What We Can Do About It* by Al Gore (Rodale, 2006), pp. 98-99.

7. NationMaster, web site: http://www.statemaster.com/encyclopedia/Polar-ice-cap

8. The Cryosphere Today, University of Illinois, web site: http://arctic.atmos.uiuc.edu/cryosphere/IMAGES/current.anom.jpg

9. "Arctic ice cap to melt faster than feared, scientists say," *The Seattle Times*, September 7, 2007, web site: http://seattletimes.nwsource.com/html/localnews/2003873003_arcticice07m.html

10. "Arctic summers ice-free 'by 2013,'" *BBC News*, December 12, 2007, web site: http://newsvote.bbc.co.uk/mpapps/pagetools/print/news.bbc.co.uk/2/hi/science/nature/7139797.stm

11. Torgny Vinje, "Anomalies and Trends of Sea-Ice Extent and Atmospheric Circulation in the Nordic Seas during the Period 1864-1998," *American Meteorological Society*, vol. 14, pp. 255–267, February 1, 2001.

12. Ibid.

13. NavSource Online: Submarine Photo Archive, web site: http://www.navsource.org/archives/08/08578.htm

14. British Antarctic Survey, web site: http://www.nerc-bas.ac.uk/icd/gjma/temps.html

15. Ibid.

16. British Antarctic Survey (See no. 14).

17. The Cryosphere Today, University of Illinois, web site: http://arctic.atmos.uiuc.edu/cryosphere/IMAGES/current.anom.south.jpg

18. Climate Change 2007: Synthesis Report, Intergovernmental Panel on Climate Change, p. 33, web site: www.ipcc.ch/ipccreports/ar4-syr.htm

19. "New Evidence From NSF-funded ANDRILL Demonstrates Climate Warming Affects Antarctic Ice Sheet Stability," National Science Foundation,

March 18, 2009, web site: http://www.nsf.gov/news/news_summ.jsp?cntn _id=114385&org=ANT&from=home

20. Ibid.

21. Ibid.

22. "First Evidence of Under-ice Volcanic Eruption in Antarctica," *Science News*, January 22, 2008, web site: http://www.sciencedaily.com/releases/2008/01/080120160720.htm

23. "Icebergs break away from Antarctic Iceshelf," web site: http://climatechange psychology.blogspot.com/2009/04/European-space-agency-wilkinson-ice.html

24. "Fire and Ice," *Investor's Business Daily*, April 8, 2009, web site: http://www.ibdeditorials.com/IBDArticles.aspx?id=324083725218923

25. Joseph D'Aleo, "Greenland Again," July 30, 2008, web site: http://icecap.us/index.php/go/joes-blog/Greenland_again/

26. Ibid.

27. Arthur Robinson et al., "Environmental Effects of Increased Atmospheric Carbon Dioxide," *Journal of American Physicians and Surgeons*, vol. 12, pp. 79–90, (2007).

28. Ibid.

29. Intergovernmental Panel on Climate Change, 4th Assessment Report (2007), p. 13, www.ipcc.ch/ipccreports/ar4-syr.htm

30. 2007 Synthesis Report, Intergovernmental Panel on Climate Change, 4th Assessment Report, p. 12, www.ipcc.ch/ipccreports/ar4-syr.htm

31. Ronald Parker et al., "Development Actions and the Rising Incidence of Disasters," Independent Evaluation Group, World Bank, June, 2007, web site: http://www.worldbank.org/ieg/

32. International Emergency Disasters Database (EMDAT), web site: http://www.emdat.be/

33. *An Inconvenient Truth* (See no. 6), pp. 102–106.

34. EMDAT (See no. 32).

35. Indur Goklany, email conversation April 11, 2009.

36. Parker (See no. 31).

37. Indur Goklany, "Death and Death Rates Due to Extreme Weather Events," Civil Society Coalition on Climate Change, November, 2007.

38. *Climate of Extremes: Global Warming Science They Don't Want You to Know*, by Patrick Michaels and Robert Balling, Jr., (Cato Institute, 2009).

39. Kerry Emanuel, "Increasing destructiveness of tropical cyclones over the past 30 years," *Nature*, vol. 436, August 4, 2005, pp. 686-688.

40. Christopher Landsea et al., "Can We Detect Trends in Extreme Tropical Cyclones?," *Science*, vol. 313, July 28, 2006, pp. 452–454.

41. Ibid.

42. Ibid.

43. William Gray, "Atlantic Seasonal Hurricane Frequency. Part I: El Niño and 30 mb Quasi-Biennial Oscillation Influences," *Monthly Weather Review*, vol. 112, September, 1984.

44. Philip Klotzbach, "Trends in global tropical cyclone activity over the past twenty years," *Geophysical Research Letters*, vol. 33, May 20, 2006, L10805.

45. Ibid.

46. Ryan Maue, "30-year lows: Global tropical cyclone activity at record low

levels," web site: http://www.coaps.fsu.edu/~maue/tropical/?p=8

47. Ibid.

48. "Senator Kerry Blames Tornadoes on Global Warming," *Business & Media Institute*, February 13, 2008, web site: http://www.businessandmedia.org/articles/2008/20080206170159.aspx

49. National Climatic Data Center, web site: http://www.ncdc.noaa.gov/oa/climate/severeweather/tornadoes.html

50. Stanley Goldberg, Presentation at the International Conference on Climate Change, March 9, 2009.

51. "Global warming prospects dire for Africa," *Independent Online*, April 10, 2007, web site: http://www.iol.co.za/index.php?set_id= 14&click_id=143&art_id=vn20070410091019256C800237

52. "Indian Drought, Floods Linked to Warming," UN Foundation, Aug. 6, 2002, web site: http://www.unwire.org/unwire/20020806/28141_story.asp

53. Al Gore, Testimony before the Energy and Environment Subcommittee of the U.S. House of Representatives, April 24, 2009.

54. EMDAT (See no. 32).

55. Henry Lamb, "Oxygen and carbon isotope composition of authigenic carbonate form an Ethiopian lake: a climate record of the last 2000 years," *The Holocene*, vol. 17, no. 4, pp. 517–526 (2007).

56. Ibid, adapted by CO_2Science: http://www.co2science.org/data/mwp/studies/l3_lakehayq.php

57. T. M. Shanahan et al., "Atlantic Forcing of Persistent Drought in West Africa," *Science*, vol. 324, no. 5925, pp. 377–380.

58. Ibid.

59. Masaki Sano et al., "Tree-ring based hydroclimate reconstruction over northern Vietnam from *Fokienia hodginsii*: eighteenth century mega-drought and tropical Pacific influence," *Climate Dynamics*, August 23, 2008.

60. "Global warming creates volatility. I feel it when I'm flying," Watts up with that?, August 8, 2009, web site: http://wattsupwiththat.com/2009/08/11 /global-warming-creates-volatility-i-feel-it-when-i%E2%80%99m-flying/

61. Warren Meyer, "No Trend in Drought or Floods," The Climate Skeptic, August 20, 2008, web site: http://www.climate-skeptic.com/2008/08/no-trend-in-dro.html, data from National Climatic Data Center, web site: http://www.ncdc.noaa.gov/oa/climate/research/2008/jul/uspctarea-wetdry-svr.txt

62. "Full text: Blair's climate change speech," *The Guardian*, September 15, 2004, web site: http://www.guardian.co.uk/politics/2004/sep/15/greenpolitics.uk

63. Intergovernmental Panel on Climate Change, 4th Assessment Report, Climate Change 2007: Impacts, Adaptation, and Vulnerability, Contribution of Working Group II, (2007), pp. 396–397, www.ipcc.ch/ipccreports/ar4-syr.htm

64. "Seasonal Influenza," Center for Disease Control, web site: http://www.cdc.gov/flu/about/qa/disease.htm

65. "Recommended composition of influenza virus vaccines for use in the 2006 influenza season," World Health Organization, web site: http://www.who.int/csr/disease/influenza/influenzarecommendations2006.pdf

66. W. R. Keatinge et al., "Heat related mortality in warm and cold regions of Europe: observational study," British Medical Journal, vol. 321, September 16, 2000, pp. 670–673.

67. Bjorn Lomborg, "Global warming will save millions of lives," Telegraph, March 13, 2009, web site: http://www.telegraph.co.uk/comment/personal-view/4981028/Global-warming-will-save-millions-of-lives.html

68. Oliver Deschenes and Enrico Moretti, "Extreme Weather Events, Mortality, and Migration," *National Bureau of Economic Research*, July, 2007.

69. Intergovernmental Panel on Climate Change, 4th Assessment Report, Climate Change 2007: Impacts, Adaptation, and Vulnerability, Contribution of Working Group II, (2007), pp. 396-397, www.ipcc.ch/ipccreports/ar4-syr.htm

70. Ibid, p. 404.

71. "Global warming causing disease to rise," *The Associated Press*, November 14, 2006, web site: http://www.msnbc.msn.com/id/15717706/

72. Ibid.

73. Paul Reiter, Presentation at the International Conference on Climate Change, March 9, 2009.

74. Paul Reiter, "From Shakespeare to Defoe: Malaria in England in the Little Ice Age," Center for Disease Control, web site: http://www.cdec.gov/ncidod/EID/vol6no1/reiter.htm

75. Ibid.

76. *The Hot Topic: What We Can Do About Global Warming* by Gabrielle Walker and Sir David King, (Houghton Mifflin Harcourt, 2008).

77. "Secretary Kempthorne Announces Decision to Protect Polar Bears under Endangered Species Act," U.S. Department of the Interior, May 14, 2008, web site: http://www.doi.gov/news/08_News_Releases/080514a.html

78. Intergovernmental Panel on Climate Change (See no. 69), p. 213.

79. *The Skeptical Environmentalist: Measuring the real state of the World* by Bjorn Lomborg, p. 250, (Cambridge University Press, 2001).

80. Ibid.

81. "The World at Six Billion," The United Nations Population Division, web site: http://www.un.org/esa/population/publications/sixbillion/sixbilpart1.pdf

82. Dennis Avery, "The Impact of Global Warming on the Chesapeake Bay," Testimony before the Senate Committee on Environment and Public Works, September 26, 2007.

83. Ian Murray, "Virtually Extinct," Competitive Enterprise Institute, January 9, 2004. web site: http://cei.org/gencon/019,03797.cfm

84. "Polar bear conference underway in Winnipeg," web site: http://greeningopseu.blogspot.com/2009/01/polar-bear-conference-underway-in.html

85. "Polar bear population is stable, Inuit say," *The Canadian Press*, January 17, 2009, web site: http://www.thespec.com/0117092359/utilities/todayPaper

86. Mitch Taylor, Presentation at the International Conference on Climate Change, March 10, 2009.

87. "NASA Examines Arctic Sea Ice Changes Leading to Record Low in 2007," NASA Press Release, October 1, 2007, web site: http://www.nasa.gov/vision/earth/lookingatearth/quikscat-20071001.html

88. "Ancient polar bear jawbone found," *BBC News*, October 12, 2007, web site: http://news.bbc.co.uk/2/hi/science/nature/7132220.stm

89. Ibid.

90. "Climate change lays waste to Spain's glaciers," *The Guardian*, February 23, 2009, web site: http://www.guardian.co.uk/environment/2009/feb/23/spain-

glaciers-climate-change

91. *An Inconvenient Truth* (See no. 6), pp. 42–59.

92. Georg Kaser et al., "Modern Glacier Retreat on Kilimanjaro as Evidence of Climate Change: Observations and Facts," International Journal of Climatology, Vol. 24, (2004), pp. 39–339.

93. John Christy, email correspondence, April 12, 2009.

94. *An Inconvenient Truth* (See no. 6), p. 58.

95. "Sizing up Earth's Glaciers," NASA Earth Observatory, web site: http://earthobservatory.nasa.gov/Features/GLIMS/

96. J. Oerlemans, "Extracting a Climate Signal from 169 Glacier Records," Science, vol. 308, April 29, 2005, pp. 657–677.

97. James Hansen, "Global Warming Twenty Years Later: Tipping Points Near," address to the National Press Club, June 23, 2008, www.columbia.edu/~jeh1/2008/TwentyYearsLater_20080623.pdf

98. Ove Hoegh-Guldberg, "Climate Change, coral bleaching and the future of the world's reefs," *Marine and Fresh Water Research*, vol. 50, pp. 839-866, (1999).

99. *CO_2, Global Warming and Coral Reefs: Prospects for the Future* by Dr. Craig D. Idso, (Center for the study of Carbon Dioxide and Global Change and Science and Public Policy Institute, 2009).

100. Glenn De'ath et al., "Declining Coral Calcification on the Great Barrier Reef," *Science*, vol. 2, January 2, 2009, pp. 116–119.

101. Idso (See no. 99)

102. Steve McIntyre, "'Unprecedented' in the past 153 Years," June 3, 2009, web site: http://www.climateaudit.org/?p=6189

103. *The Day After Tomorrow*, by Roland Emmerich, (Twentieth Century-Fox Film Corporation 2004).

104. *An Inconvenient Truth* (See no. 6), p. 149.

105. NASA photograph, web site: www.wonderquest.com/GulfStream.htm

106. Bryden et al., "Slowing of the Atlantic meridional overturning circulation at 25 degrees N.," *Nature*, December 1, 2005, vol. 438, pp. 655-657.

107. "Ocean Circulation Shut Down by Melting Glaciers After last Ice Age," Goddard Space Flight Center, November 19, 2001, web site: http://www.gsfc.nasa.gov/topstory/20011116meltwater.html

108. Friedrich Schott et al., "Variability of the Deep Western Boundary Current east of the Grand Banks," *Geophysical Research Letters*, vol. 33, Oct. 29, 2006.

109. Jennifer Marohasy, "Wrong 'Shot' at Atlantic Conveyor Belt Speed?," web site: http://www.jennifermarohasy.com/blog/archives/001755.html

Chapter 7: The IPCC and the Road to World Delusion

1. *Speaking of Earth: Environmental Speeches that Moved the World* by Alon Tal, (Rutgers University Press, 2006), p. 42.

2. 1972 Stockholm Conference on the Human Environment, United Nations Environment Program, web site: http://www.unep.org/Documents.Multilingual/Default.asp?DocumentID=97

3. Tal (See no. 1).

4. "Maurice Strong Dossier," National Center, web site: http://www.nationalcenter.org/DossierStrong.html

5. "The U.N.'s Man of Mystery," *The Wall Street Journal*, web site:

http://online.wsj.com/article/SB122368007369524679.html

6. Ibid.

7. Declaration of the World Climate Conference, web site: http://www.dgvn.de/fileadmin/user_upload/DOKUMENTE/WCC-3/Declaration_WCC1.pdf

8. Wendy Franz, "The Development of an International Agenda for Climate Change: Connecting Science to Policy," International Institute for Applied Systems Analysis, September, 1997, p. 8, web site: http://www.iiasa.ac.at/Admin/PUB/Documents/IR-97-034.pdf

9. "Report of the International Conference on the assessment of the role of carbon dioxide and of other greenhouse gases in climate variations and associated impacts," World Meteorological Organisation (WMO), Villach, Austria, 9-15 October, 1985, WMO no. 661.

10. Ibid.

11. "Digging up the roots of the IPCC," *Spiked*, web site: http://www.spiked-online.com/index.php?/site/article/3540/

12. "Hot Politics," Frontline, PBS, web site: http://www.pbs.org/wgbh/pages/frontline/hotpolitics/etc/script.html

13. Mario Molina and F. S. Rowland, "Stratospheric sink for chlorofluoromethanes: chlorine atom-catalysed destruction of ozone," *Nature*, vol. 249, no. 5460-, pp. 810-812, June 28, 1974.

14. "Reports to the Nation on Our Changing Planet: Our Ozone Shield," National Aeronautic and Atmospheric Association, web site: http://www.chymist.com/Our%20ozone%20shield.pdf

15. "Science Briefing--The ozone hole," British Antarctic Survey, web site: http://www.bas.ac.uk/bas_research/science_briefings/ozone_layer.php

16. Pope et al., "Ultraviolet Absorption Spectrum of Chlorine Peroxide, ClOOCl," *Journal of Physical Chemistry*, 111, May 3, 2007, pp. 4322–4332.

17. Q. B. Lu, "Correlation between Cosmic Rays and Ozone Depletion," *Physical Review Letters*, 102, 118501, March 20, 2009.

18. Frontline (See no. 12).

19. Franz (See no. 8).

20. *Global warming: are we entering the greenhouse century?* by Stephen H. Schneider, (Sierra Club, 1989), p. 194

21. Franz (See no. 8).

22. "Our Common Future: Report of the World Commission on Environment and Development," Center for a World in Balance, web site: http://worldinbalance.net/agreements/1987-brundtland.php

23. Henry Lamb, "Maurice Strong: The new guy in your future!," January, 1997, web site: http://sovereignty.net/p/sd/strong.html

24. Ibid.

25. Human Development Report 1996, United Nations Development Programme, web site: http://hdr.undp.org/en/reports/global/hdr1996/

26. *The Failures of American and European Climate Policy* by Loren Cass, (State University of New York, 2006), pp. 24, 70.

27. *A History of the Science and Politics of Climate Change* by Bert Bolin, (Cambridge University Press, 2007), p. 253.

28. Climate Change Damage and International Law by Roda Verheyen, (Martinus

Nijhoff, 2005), pp. 24–26.

29. Intergovernmental Panel on Climate Change, "Policymakers Summary of the Scientific Assessment of Climate Change: Report to IPCC from Working Group I," June 1990.

30. Timothy Ball, "Global Warming: The Cold, Hard Facts?," February 5, 2007, web site: http://www.canadafreepress.com/2007/global-warming020507.htm

31. John Zillman, "The IPCC: A View from the Inside," web site: http://www.apec.org.au/docs/zillman.pdf

32. Intergovernmental Panel on Climate Change, About IPCC, web site: http://www.ipcc.ch/about/index.htm

33. Zillman (See no. 31).

34. UN Conference on Environment and Development (1992), Earth Summit, web site: http://www.un.org/geninfo/bp/enviro.html

35. Agenda 21: The Report of the United Nations Conference on Environment and Development (Earth Summit - Rio de Janeiro, 3-14 June 1992), web site: http://www.un.org/esa/sustdev/documents/agenda21/english/agenda21toc.htm

36. Lamb (See no. 23).

37. United Nations Framework Convention on Climate Change, May 9, 1992, web site: http://unfccc.int/resource/docs/convkp/conveng.pdf

38. Ibid.

39. "Statement by Atmospheric Scientists on Greenhouse Warming," February 27, 1992, web site: http://www.sepp.org/policy%20declarations/statment.html

40. Intergovernmental Panel on Climate Change, Second Assessment Report, Summary for Policy Makers, 1995, web site: http://www.ipcc.ch/pdf/climate-changes-1995/spm-science-of-climate-changes.pdf

41. Intergovernmental Panel on Climate Change, Third Assessment Report, Stand-alone edition, 2001, p. 31, web site: http://www.ipcc.ch/pdf/climate-changes-2001/synthesis-spm/synthesis-spm-en.pdf

42. Intergovernmental Panel on Climate Change, Fourth Assessment Report, Working Group I, Summary for Policy Makers, 2007, p. 10, web site: http://www.ipcc.ch/pdf/assessment-report/ar4/wg1/ar4-wg1-spm.pdf

43. Statement of Dr. David Deming, U.S. Senate Committee on Environment & Public Works, December 6, 2006, web site: http://epw.senate.gov/hearing_statements.cfm?id=266543

44. Intergovernmental Panel on Climate Change, Second Assessment Report, 1995, Figure 22.

45. Intergovernmental Panel on Climate Change, Third Assessment Report, Working Group I, Technical Summary, Figure 5, p. 29, 2001, web site: http://www.ipcc.ch/ipccreports/tar/wg1/pdf/WG1_TAR-FRONT.PDF

46. Michael Mann et al., "Global-scale temperature patterns and climate forcing over the past six centuries," Nature, vol. 392, April 23, 1998.

47. Intergovernmental Panel on Climate Change, Third Assessment Report, Working Group I, Summary for Policymakers, p. 2, 2001, web site: http://www.ipcc.ch/ipccreports/tar/wg1/pdf/WG1_TAR-FRONT.PDF

48. Stephen McIntyre and Ross McKitrick, "Corrections to the Mann et. al. (1998) Proxy Data Base and Northern Hemispheric Average Temperature Series," Energy & Environment, vol. 14, no. 6, 2003.

49. Ibid.

50. Testimony of Edward Wegman before the Energy and Commerce Committee, United States House of Representatives, July 19, 2006 web site: http://archives.energycommerce.house.gov/reparchives/108/Hearings/0719200 6hearing1987/Wegman3108.htm

51. Christopher Monckton, "Climate chaos? Don't believe it," *Telegraph*, November 5, 2006, web site: http://www.telegraph.co.uk/news /uknews/1533290/Climate-chaos-Dont-believe-it.html

52. "Energy White Paper: Our energy future-creating a low carbon economy," DTI, web site: http://www.berr.gov.uk/files/file10719.pdf

53. Photograph copyright Kaihsu Tai, permission granted under GNU Fee Documentation License, Version 1.2, web site: http://en.wikipedia.org /wiki/File:JohnHoughtonHighWycombe20050226_CopyrightKaihsuTai.jpg

54. Intergovernmental Panel on Climate Change, Fourth Assessment Report, Working Group I, Technical Summary, 2007, p. 55, web site: http://www.ipcc.ch/pdf/assessment-report/ar4/wg1/ar4-wg1-ts.pdf

55. Timothy Ball, "How UN structures were designed to prove human CO_2 was causing global warming," *Canada Free Press*, April 30, 2008, web site: http://www.canadafreepress.com/index.php/article/2840

56. John McLean, "Prejudiced Authors, Prejudiced Findings: Did the UN bias its attribution of 'global warming' to humankind?," *Science & Public Policy Institute*, July, 2008, web site: http://scienceandpublicpolicy.org/originals /prejudiced_authors_prejudiced_findings.html

57. Ibid.

58. Zillman (See no. 31).

59. Paul Georgia, "IPCC report criticized by one of its lead authors," *The Heartland Institute*, June 1, 2001, web site: http://www.heartland.org/policybot /results/1069/IPCC_report_criticized_by_one_of_its_lead_authors.html

60. Peter Walsh, "President's Report 2008," The Lavoisier Group, 2008, web site: http://www.lavoisier.com.au/articles/climate-policy/science-and-policy/walsh2008-15.php

61. Zillman (See no. 31).

62. Frederick Seitz, "A Major Deception on Global Warming," *The Wall Street Journal Interactive Addition*, June 12, 1996, web site: http://www.eganinc.com/cslater/wsjgloba.htm

63. Intergovernmental Panel on Climate Change, Fourth Assessment Report, Working Group I, 2007, p. 203, web site: http://www.ipcc.ch/pdf/assessment-report/ar4/wg1/ar4-wg1-chapter2.pdf

64. Kyoto Protocol, United Nations Framework Convention on Climate Change, web site: http://unfccc.int/kyoto_protocol/items/2830.php

65. Kyoto Protocol to the United Nations Framework Convention on Climate Change, UN, 1992, web site: http://unfccc.int/resource/docs/convkp/kpeng.pdf

66. Ibid.

67. "Putin signals Russia will sign Kyoto Protocol for WTO membership," *Bellona*, May 23, 2004, web site: http://www.bellona.org/english_import_area/energy/34179

68. Marc Morano, Presentation at the International Conference on Climate Change, March 9, 2009.

69. Intergovernmental Panel on Climate Change, Fourth Assessment Report,

Synthesis Report, 2007, p. 46, web site: http://www.ipcc.ch/pdf/assessment-report/ar4/syr/ar4_syr.pdf

70. Intergovernmental Panel on Climate Change, Fourth Assessment Report, Working Group I, Technical Summary, 2007, p. 79, web site: http://www.ipcc.ch/pdf/assessment-report/ar4/wg1/ar4-wg1-ts.pdf

Chapter 8: Climatism: The Ideology Behind Global Warming Alarmism

1. Václav Klaus: "Climate Alarmism is about World Governance and Freedom," *Euromed*, Mar 17, 2008, web site: http://translate.google.com/translate?hl=en&sl=da&u=http://euro-med.dk/%3Fp%3D630&ei=UhssSsT2LIvoNNzK_NsJ&sa=X&oi=translate&resnum=2&ct=result&prev=/search%3Fq%3D%2522czech%2Bpresident%2BKlaus:%2B%2Bclimate%2522%26hl%3Den%26lr%3D

2. "WSI: Malthus Returns," web site: http://krusekronicle.typepad.com/kruse_kronicle/2008/08/wsi-malthus-returns.html

3. An Essay on the Principle of Population by Thomas Malthus, Library of Economics and Liberty, web site: http://www.econlib.org/library/Malthus/malPlong.html

4. "WSI: Technophysio Evolution," original data from J. Bradford Delong "Estimating World GDP, One Million B.C. - Present," web site: http://krusekronicle.typepad.com/kruse_kronicle/2008/07/wsi-technophysio-evolution.html

5. "World Grain Stocks Fall to 57 Days of Consumption," *Earth Policy Institute*, June, 2006, original data from the United States Department of Agriculture, web site: http://www.earth-policy.org/Indicators/Grain/2006.htm

6. *Inquires into Human Faculty and its Development* by Francis Galton, p. 219, (MacMillan, 1883).

7. "John D. Rockefeller, Jr. The Pursuit of Happiness," *Associated Content*, 2009, web site: http://www.associatedcontent.com/article/44035/john_d_rockefeller_jr.html?cat=9

8. "Gibson's Environmentalism," by William Kay, web site: http://www.ecofascism.com/review16.html

9. *Environmentalism: Ideology and Power* by Donald Gibson, (Nova Science, 2004), p. 37–38.

10. "Eugenics and Environmentalism: From quality control to quantity control," Old Thinker News, April 30, 2008, web site: http://www.oldthinkernews.com/Articles/oldthinker%20news/eugenics_and_environmentalism.htm

11. Population Council web site: http://www.popcouncil.org/

12. Kay (See no. 8).

13. Gibson (See no. 9).

14. *The Population Bomb* by Paul Ehrlich, (Ballantine Books, 1968). p. 5

15. Ibid p. 6.

16. *The End of Affluence* by Paul Ehrlich and Anne Ehrlich, (Ballantine Books, 1974), p. 21

17. John Tierney, "Betting on the Planet," *NYTimes Magazine*, December, 1990, web site: http://www.nytimes.com/1990/12/02/magazine/betting-on-the-planet.html?scp=5&sq=%22John+Tierney%22+Betting+the+Planet&st=nyt

18. Gretchen Daily and Paul Ehrlich, "Population, Sustainability, and Earth's Carrying Capacity: A framework for estimating population sizes and lifestyles that could be sustained without undermining future generations," from

BioScience, November, 1992, web site: http://www.dieoff.org/page112.htm

19. Holdren's Letterman Warning is No Laughing Matter, *Bloomberg*, February 10, 2009, web site: http://www.bloomberg.com/apps/news?pid=20601109 &sid=awOp50bE4Tko&refer=news

20. Daily and Ehrlich (See no. 18).

21. Donella Meadows et al., "The Limits to Growth: A Report to the Club of Rome," 1972, web site: http://www.clubofrome.org/docs/limits.rtf

22. Ibid.

23. The First Global Revolution, A Report by the Council of the Club of Rome, by Alexander King and Bertrand Schneider, (Pantheon Books, 191), p. 115

24. Klaus (See no. 1).

25. "'Club of Rome' member warns against council amalgamations," *ABC News*, June 5, 2007, web site: http://www.abc.net.au/news/newsitems/200706/s1942343.htm

26. The Canadian Association for the Club of Rome, web site: http://www3.sympatico.ca/drrennie/CACORhis.html

27. U.S. Assn. for the Club of Rome. Membership List., web site: http://www.namebase.org/sources/TL.html

28. Mikhail Gorbachev, web site: http://www.nndb.com/people/416/000023347/

29. "WSI: Worldwide per Capital Income," Kruse Kronicle, July 24, 2008, web site: http://krusekronicle.typepad.com/kruse_kronicle/

30. "WSI: Regional Variation in the Global Economic Expansion," Kruse Kronicle, July 25, 2008, web site: http://krusekronicle.typepad.com/kruse_kronicle/

31. *The Skeptical Environmentalist, Measuring the Real State of the World* by Bjorn Lomborg, (Cambridge University Press, 2001), p. 50.

32. "WSI: Life Expectancy," *Kruse Kronicle*, July 22, 2008, web site: http://kruse kronicle.typepad.com/kruse_kronicle/2008/07/wsi-life-expectancy.html

33. "Life Expectancy over Human History," *Masterliness, The World Knowledge Library*, web site: http://www.masterliness.com/a/Life.expectancy.htm

34. Earth Trends, World Resources Institute, web site: http://earthtrends.wri.org/

35. Ibid.

36. Martin Wolf, "Is Globalisation Causing World Poverty?," Presentation to Nottingham University, February 25, 2002.

37. *The Improving State of Our World* by Indur Goklany, (Cato Institute, 2007), p.47

38. The Millennium Development Goals Report 2009, United Nations, web site: http://www.un.org/millenniumgoals/pdf/MDG%20Report%202009%20ENG pdf

39. Ibid.

40. "Hunger and World Poverty," web site: http://www.poverty.com/index.html

41. Ibid.

42. International Energy Outlook 2009, Energy Information Administration, web site: http://www.eia.doe.gov/oiaf/ieo/

43. Paul Driessen, Presentation at the International Conference on Climate Change, March 2008.

44. United Nations (See no. 38).

45. United Nations (See no. 38).

46. *The Skeptical Environmentalist, Measuring the Real State of the World* by Bjorn Lomborg, (Cambridge University Press, 2001)

47. *The Improving State of Our World*, (See no. 37).

48. U.S. Environmental Protection Agency, web site: http://www.epa.gov/air/airtrends/2007/graphics/Figure_1.gif

49. Earth Trends Environmental Information, web site: http://earthtrends.wri.org/searchable_db/index.php?theme=2

50. Earth Trends Environmental Information, web site: http://earthtrends.wri.org/searchable_db/index.php?theme=5

51. Bruce Yandle et al., "Environmental Kuznets Curves: A Review of Findings, Methods, and Policy Implications," Property and Environment Research Center, April, 2004, web site: http://www.perc.org/articles/article207.php

52. Jesse Ausubel and Paul Waggoner, "The Jack Rabbit of Depression, or Do economic slumps benefit environment?," Apr. 18, 2009, web site: http://phe.rockefeller.edu/news/wp-content/uploads/2009/04/jackrabbitofdepressionfinal.pdf

53. Yandle (Same as no. 51).

54. Earth Trends Environmental Information, web site: http://earthtrends.wri.org/searchable_db/index.php?theme=5

55. "UN Report Says Deforestation Continuing at Alarming Rate," United Nations Press Release, May 12, 2005, web site: http://www.un.org/News/Press/docs/2005/envdev851.doc.htm

56. Intergovernmental Panel on Climate Change, Fourth Assessment Report, Working Group I, Summary for Policy Makers, 2007, p. 2, web site: http://www.ipcc.ch/publications_and_data/publications_ipcc_fourth_assessment_report_synthesis_report.htm

57. Earth Trends, World Resources Institute, web site: http://earthtrends.wri.org/

58. Pekka Kauppi et al., "Returning forests analyzed with the forest identity," PNAS, vol. 103, no. 46, November 14, 2006, web site: http://www.pnas.org/content/103/46/17574.short

59. "At a glance: Poverty," web site: http://www.ourplanet.com/imgversn/122/glance.html

Chapter 9: The Beliefs, Objectives, and Tactics of Climatism

1. Vinod Dar, "Climatology versus Climatism," *Right Side News*, July 24, 2008, web site: http://www.rightsidenews.com/200807231515/energy-and-environment/climatology-versus-climatism.html

2. "Senators Debate Global Warming Policy Despite Global Cooling Evidence," *CNSNews*, Wednesday, March 4, 2009, web site: http://www.cnsnews.com/public/content/article.aspx?RsrcID=44431

3. Noel Sheppard, "Former IPCC Member Slams UN Scientists' Lack of Geologic Knowledge," *NewsBusters*, July 9, 2007, web site: http://newsbusters.org/node/13971

4. Fox Business Network, June 30, 2008, web site: http://local-warming.blogspot.com/2008/07/reid-coal-makes-us-sick-oil-makes-us.html

5. Steve Conner, "US Climate Policy Bigger Threat to World than Terrorism," *Common Dreams News Center*, January 9, 2004, web site: http://www.commondreams.org/headlines04/0109-02.htm

6. *Climate Change and the Failure of Democracy* by David Shearman and Joseph Smith, (Greenwood Publishing Group, 2007)

7. C. J. Carnacchio, "Earth Day & Eco-totalitarians," *The Michigan Review*, April 22, 1998, web site: http://www.umich.edu/~mrev/archives/1998/4-22-98/pg6.htm

8. Charles Moore, "What you get when you mix Red and Green - a bad political climate," *Telegraph*, April 1, 2006, web site: http://www.telegraph.co.uk/comment/personal-view/3624047/What-you-get-when-you-mix-Red-and-Green—a-bad-political-climate.html

9. *Ecoscience: Population, Resources, Environment* by Paul Ehrlich and Anne Ehrlich, (W.H. Freeman, 1970), p. 323.

10. "Two Kids Should Be the Limit: Eco Expert," *Newser*, February 2, 2009, web site: http://www.newser.com/story/49694/two-kids-should-be-the-limit-eco-expert.html

11. Mike Morris, "Ted Turner: Global warming could lead to cannibalism," *Atlanta Journal Constitution*, April 3, 2008, web site: http://www.ajc.com/metro/content/news/stories/2008/04/03/turner_0404.html

12. Conor Clarke, "An Interview With Thomas Schelling," Atlantic Wire, July 13, 2009, web site: http://correspondents.theatlantic.com/conor_clarke/2009/07/an_interview_with_thomas_schelling_part_two.php

13. We Can Solve the Climate Crisis, web site: http://www.wecansolveit.org/content/about

14. Willie Soon, Presentation at the International Conference on Climate Change, June 2, 2009.

15. U.S. Environmental Protection Agency, "Endangerment and Cause or Contribute Findings for Greenhouse Gases under the Clean Air Act," Dec. 7, 2009, web site: http://www.epa.gov/climatechange/endangerment.html

16. John Broder, "Document is Critical of E.P.A. on Clean Air," *The New York Times*, May 13, 2009, web site: http://query.nytimes.com/gst/fullpage.html?res=9A01E2DA133FF930A25756C0A96F9C8B63

17. Henry Lamb, "Maurice Strong: The new guy in your future!," January, 1997, web site: http://sovereignty.net/p/sd/strong.html

18. United Nations International Conference on Population and Development, September, 1994, Cairo, Egypt, web site: http://www.iisd.ca/cairo.html

19. United Nations General Assembly, Programme for the Further Implementation of Agenda 21, September 19, 1997, web site: http://www.un.org/documents/ga/res/spec/aress19-2.htm

20. Ibid.

21. United Nations Environmental Programme, "Sustainable Lifestyles and Educations for Sustainable Consumption," web site: http://esa.un.org/marrakechprocess/pdf/Issues_Sus_Lifestyles.pdf

22. Tim Radford, "The whole world in our hands," The Guardian, September 16, 2000, web site: http://www.ecolo.org/lovelock/world-in-our-hands.html

23. Terrence Corcoran, "Global Warming: The Real Agenda," *Financial Post* (Canada), Dec. 26, 1998, web site: http://www.sepp.org/Archive/reality/realagenda.html

24. "UN REFORM - Restructuring for Global Governance," ECO-LOGIC, July/Aug 1997, web site: http://www.iahf.com/world/un-refm.html

25. Speech by Mr. Jacques Chirac, French President, to the VIth Conference for the Parties to the United Nations Framework Convention on Climate Change, The Hague, November 20, 2000.

26. "Section of Environment, Energy, and Resources, Sustainable Development, Ecosystems, and Climate Change Committee - Newsletter Archive," January, 2002, *American Bar Association*, web site: http://www.abanet.org/environ /committees/climatechange/newsletter/jan02/goldberg.html

27. *Ecodefense: A Field Guide to Monkey Wrenching* by David Foreman, (Abzug Publishing, 1987)

28. Pete Harrison, "Never waste a good crisis, Clinton says on climate," *Reuters*, March 7, 2009, web site: http://in.reuters.com/article/environmentNews/idINTRE5251VN20090306

29. "Paulson's rescue plan is called 'TARP,'" *Politico*, September 19, 2008, web site: http://www.politico.com/news/stories/0908/13609.html

30. "Bush Announces Bailout for GM, Chrysler," web site: http://www.gmbailout .com/2008/12/20/bush-announces-bailout-for-gm-chrysler/

31. "Obama signs stimulus into law," *MSNBC*, February 17, 2009, web site: http://www.msnbc.msn.com/id/29231790/

32. Václav Klaus, "President Klaus: Is Environmentalism a Bigger Threat to Humanity than Global Warming?," Mar. 7, 2009, web site: http://uddebatt .wordpress.com/2009/03/09/environmentalism-is-a-bigger-threat-to-humanity-than-global-warming-and-what-is-endangered-is-freedom-and-prosperity/

33. United States Climate Action Partnership, web site: http://www.us-cap.org/

34. Brian Merchant, "Report: Now 4 Climate Change Lobbyists for Every Member of Congress," *TreeHugger*, Feb. 26, 2009, web site: http://www.tree hugger.com/files/2009/02/4-climate-change-lobbyists-congress.php

35. Free Republic, web site: http://www.freerepublic.com/focus/news/845063/posts

36. *The Really Inconvenient Truths: Seven Environmental Catastrophes Liberals Don't Want You to Know About--Because They Helped Cause Them* by Iain Murray, (Regnery Publishing, 2008), pp. 50–51

37. Al Revere, "An interview with accidental movie star Al Gore," Grist, May 9, 2006, web site: http://www.grist.org/article/roberts2/

38. Piers Akerman, "Stern's report scare-mongering," *The Daily Telegraph*, November 5, 2006, web site: http://ff.org/centers/csspp/library/co2weekly/20061115/20061115_05.html

39. "Don't Bet All Environmental Changes Will Be Beneficial," *APS News Online*, Aug./Sep. 1996, web site: http://rpuchalsky.home.att.net/sci_env/sch_quote.html

40. "Greenpeace Just Kidding about Armageddon," The Washington Post, June 2, 2006, web site: http://www.washingtonpost.com/wp-dyn/content/article/2006/06/01/AR2006060101884.html

41. U.S. Senate Committee on Environment and Public Works, web site: http://epw.senate.gov/public/?CFID=1076364&CFTOKEN=60301686

42. Timothy Ball, "UN's IPCC preying on people's ignorance," *Canada Free Press*, May 27, 2008, web site: http://www.canadafreepress.com/index.php/article/3247

43. Greenpeace International, web site: http://www.greenpeace.org/international/

44. Jonathan Happ, web site: http://en.wikipedia.org/wiki/File:Gp-esso.jpg
45. Greenpeace (See no. 43).
46. "Breaking news: Kingsnorth Six found not guilty!," Greenpeace International, September 10, 2008, web site: http://www.greenpeace.org.uk/blog /climate/kingsnorth-trial-breaking-news-verdict-20080910
47. Greenpeace International Annual Report 2007, web site: http://www .greenpeace.org/usa/press-center/reports4/greenpeace-annual-report-2007
48. Nature Conservancy 2008 Annual Report, web site: http://www.nature.org/aboutus/annualreport/files/annualreport2008.pdf
49. "McKnight Foundation commits $100 million to combat climate change," *MinnPost*, March 19, 2009, web site: http://www.minnpost.com/ronway/2009 /03/19/7491/mcknight_foundation_commits_100_million_to_combat_climate _change
50. Suzanne Goldenberg, "Gore unveils $300m climate ads," *Guardian*, March 31, 2008, web site: http://www.guardian.co.uk/world/2008/mar/31/algore.uselections08.climate
51. Carbon Disclosure Project, web site: http://www.cdproject.net/aboutus.asp
52. "Conservative TM 14. The Church of the Intellectual," September 28, 2007, web site: http://conservativetm.blogspot.com/2007/09/14-church-of-intellectual.html
53. *The Down-to-Earth Guide to Global Warming* by Laurie David and Cambria Gordon, (Orchard Books, 2007), p. 17

Chapter 10: The Science is Not Settled

1. Rachel Waters, "Global Warming Movie Makes the Media Hot for Al Gore All Over Again," *Business and Media Institute*, web site: http://www .businessandmedia.org/specialreports/2006/SummerRerun/SummerRerun.asp
2. George Monbiot, "The media laps up fake controversy over climate change," *Guardian*, April 29, 2009, web site: http://www.guardian.co.uk/environment /georgemonbiot/2009/apr/29/george-monbiot-climate-change-scepticism
3. George Monbiot, "Monbiot's royal flush: Top 10 climate change deniers," *Guardian*, March 6, 2009, web site: http://www.guardian.co.uk /environment/georgemonbiot/2009/mar/06/climate-change-deniers-top-10
4. Sunny Hundal, "Stern: Climate change deniers are 'flat-earthers,'" *Guardian*, March 10, 2009, web site: http://www.guardian.co.uk/environment/2009 /mar/10/nicholas-stern-accuses-climate-change-deniers
5. *An Inconvenient Truth: The Planetary Emergency of Global Warming and What We Can Do About It* by Al Gore (Rodale, 2006), p. 264
6. Global Warming Skeptics: A Primer, *Environmental Defense*, web site: http://www.edf.org/documents/3943_paidskeptics.pdf
7. William Reilly, "Analysis: U.N. calls climate debate 'over,'" UPI, May 10, 2007, web site: http://www.upi.com/Security_Industry/2007/05/10/Analysis-UN-calls-climate-debate-over/UPI-64801178838911/
8. James Phogan, "Heretics Catalog," web site: http://www.jamesphogan.com/heretics/book.php?titleID=273
9. David Roberts, "An excerpt from a new book by George Monbiot," *Grist*, September 19, 2006, web site: http://www.upi.com/Security_Industry/2007 /05/10/Analysis-UN-calls-climate-debate-over/UPI-64801178838911/

10. Craig Offman, "Suzuki Exposed," *National Post*, Feb. 7, 2008, web site: http://torontochange.com/index2.php?option=com_content&do_pdf=1&id=113

11. George Monbiot, "I'm all for putting more vehicles on our roads. As long as they're coaches.," *Guardian*, December 5, 2006, web site: http://www.guardian.co.uk/commentisfree/2006/dec/05/comment.politics

12. "Climate denial ads to air on US national television," May 19, 2006, web site: http://www.marklynas.org/2006/5/19/climate-denial-ads-to-air-on-us-national-television

13. "Planet Gore," *National Review*, January 5, 2009, web site: http://planetgore.nationalreview.com/post/?q=ZjZkYTQ0MWFjOGExZDgxO WRhMzFmNjMwMDVhYzg4YTQ=

14. Alexander Cockburn, "I Am an Intellectual Blasphemer," Jan. 25, 2008, web site: http://www.spiked-online.com/index.php?/site/reviewofbooks_article/4357/

15. *Politicizing Science: The Alchemy of Policymaking*, edited by Michael Gough, pp. 283-297, (Hoover Institution Press, 2003)

16. Fred Singer et al., "What To Do about Greenhouse Warming: Look Before You Leap," *Cosmos: A Journal of Emerging Issues*, Vol. 5, No. 2, Summer 1992, web site: http://www.sepp.org/key%20issues/glwarm/cosmos.html

17. *Earth in Balance: Ecology and the Human Spirit* by Al Gore, pp. 307, 321, (Houghton Mifflin, 1992)

18. Gough (See no. 15).

19. Gough (See no. 15).

20. Gough (See no. 15).

21. "Public Policy," Marshall Institute, web site: http://www.marshall.org/category.php?id=7

22. *The Skeptical Environmentalist, Measuring the Real State of the World* by Bjorn Lomborg, (Cambridge University Press, 2001)

23. Jason Cowley, "The man who demanded a recount," *NewStatesman*, June 30, 2003, web site: http://www.newstatesman.com/200306300013

24. Lomborg (See no. 22), p. xix.

25. Lomborg (See no. 22), p. 352.

26. Lomborg (See no. 22), p. 323.

27. Cowley (See no. 23).

28. John Rennie, editor, "Misleading Math about the Earth," *Scientific American*, January 2, 2002, web site: http://www.scientificamerican.com/article.cfm?id=misleading-math-about-the

29. "Danish Committee Cites Lomborg For Scientific Dishonesty," *SpaceDaily*, January 7, 2003, web site: http://www.spacedaily.com/news/earth-03b.html

30. "Skeptical Environmentalist Vindicated," Kuro5hin, December 21, 2003, web site: http://www.kuro5hin.org/story/2003/12/20/193036/63

31. Holman Jenkins, Jr., "Al Gore Leads a Purge," *The Wall Street Journal [Eastern Edition]*, May 25, 1993, web site: http://www.off-road.com/green/gore.txt

32. Richard Lindzen, "Global Warming: The Origin and Nature of the Alleged Scientific Consensus," *Cato Institute*, Vol. 15, No. 2, Spring 1992, web site: http://www.cato.org/pubs/regulation/regv15n2/reg15n2g.html

33. Tom Wicker, "In the Nation; Time for Action," *The New York Times*, October 21, 1991, web site: http://www.nytimes.com/1991/10/24/opinion/in-the-

nation-time-for-action.html

34. *Earth in Balance: Ecology and the Human Spirit* by Al Gore, (Houghton Mifflin, 1992)

35. Ibid, p. 85.

36. Ronald Bailey, "Fired at DOE," *Reason*, December, 1993, web site: http://www.sepp.org/Archive/controv/controversies/happer.html

37. William Gray, "On the Hijacking of the American Meteorological Society and Recommendation that NO Large-Scale Programs to Reduce CO_2 be Undertaken," March 8, 2009

38. *The Deniers: The World-Renowned Scientists Who Stood Up Against Global Warming Hysteria, Political Persecution, and Fraud* by Lawrence Solomon, (Richard Vigilante Books, 2008), p. 178.

39. Peter Friedman, "Global warming debate should focus on science," *SouthCoastToday*, March 11, 2008, web site: http://www.southcoasttoday.com/apps/pbcs.dll/article?AID=/20080311/OPINION/803110318/-1/NEWS01

40. Lindzen (See no. 32).

41. *Climate of Extremes: Global Warming Science They Don't Want You to Know* by Patrick Michaels and Robert Balling, Jr., (Cato Institute, 2009), pp. x-xiii.

42. Letter from The Royal Society to ExxonMobil, Sep. 4, 2006, web site: http://image.guardian.co.uk/sys-files/Guardian/documents/2006/09/19/LettertoNick.pdf

43. Ibid.

44. Solomon (See no. 38).

45. Lawrence Solomon, "The ice-core man," *National Post*, May 4, 2007, web site: http://www.nationalpost.com/news/story.html?id=25526754-e53a-4899-84af-5d9089a5dcb6&p=1%20

46. Richard Lindzen, "Climate of Fear," *The Wall Street Journal*, April 12, 2006, web site: http://www.heartland.org/custom/semod_policybot/pdf/20143.pdf

47. "Another Inconvenient Truth?," Denmark.dk, web site: http://www.denmark.dk/en/servicemenu/news/focuson/anotherinconvenienttruth.htm

48. Funding for Global Change Research, web site: http://www.climatesciencewatch.org/file-uploads/USGCRP-CCSP_Budget_History_Table_2.pdf

49. Roger Pielke, Sr., "Short Circuiting the Scientific Process - A Serious Problem In The Climate Science Community," *Climate Science*, June 4, 2009, web site: http://climatesci.org/2009/06/04/short-circuiting-the-scientific-process-a-serious-problem-in-the-climate-science-community/

50. "Steps of the Scientific Method," *Science Buddies*, web site: http://www.sciencebuddies.org/science-fair-projects/project_scientific_method.shtml

51. Pielke (See no. 49).

52. "The Civil Heretic," The New York Times, April 12, 2009, web site: http://query.nytimes.com/gst/fullpage.html?res=9D00E7D91E3DF931A25757C0A96F9C8B63&scp=2&sq=freeman+dyson&st=nyt

53. Richard Lindzen, "Understanding Common Climatic Claims," Proceedings of the 2005 ERICE Meeting of the World Federation of Scientists on Global Emergencies (2005)

54. Roy Spencer, "Another NASA Defection to the Skeptics' Camp," January 29, 2009, web site: http://www.drroyspencer.com/2009/01/another-nasa-defection-to-the-skeptics-camp/

55. Ibid.

56. "NASA Chief Questions Urgency of Global Warming," *National Public Radio*, May 31, 2007, web site: http://www.npr.org/templates/story/story.php?storyId=10571499

57. "NASA Chief Not Sure Warming is a Problem," The Associated Press, May 31, 2007, web site: http://www.msnbc.msn.com/id/18964176/

58. Jennifer Marohasy, "Jennifer Marohasy Commentary on Ferene Miskolczi's Atmospheric Model," May 2, 2009, web site: http://www.sott.net/articles/show/183475-Jennifer-Marohasy-Commentary-on-Ferene-Miskolczi-s-Atmospheric-Model

59. "BBC Shunned Me for Denying Climate Change," *Daily Express*, web site: http://www.dailyexpress.co.uk/posts/view/69623

60. Solomon (See no. 38), pp. 205–206

61. *Daily Express* (See no. 59).

62. "John Theon," *Landshape*, Feb. 7, 2009, web site: http://landshape.org/enm/john-theon/

63. US Senate Committee on Environment and Public Works: Minority Page, January 27, 2009, web site: http://epw.senate.gov/public/index.cfm?FuseAction=Minority.Blogs&ContentRecord_id=1A5E6E32-802A-23AD-40ED-ECD53CD3D320

64. Katharine Sanderson, "Claude Allègre back in French government?," *Nature.com*, May 28, 2009, web site: http://blogs.nature.com/news/thegreatbeyond/2009/05/claude_allegre_back_in_frenchg.html

65. Marc Morano, "April Fool's Report? French Reversal on Climate Policy? Outspoken Skeptical Scientist May Be Tapped as Environmental Minister!," Climate Depot, April 16, 2009, web site: http://climatedepot.com/a/278/April-Fools-Report-French-Reversal-on-Climate-Policy-Outspoken-Skeptical-Scientist-May-Be-Tapped-as-Environmental-Minister

66. Claude Allègre, "The Snows of Mount Kilimanjaro," *L'Express*, September 21, 2006, web site: http://blog.nam.org/The%20Snows%20of%20Mount%20Kilimanjaro.pdf

67. *Business Week*, August 16, 2004, web site: http://www.businessweek.com/magazine/content/04_33/b3896002_mz001.htm

68. *Time*, April 3, 2006, web site: http://www.time.com/time/covers/0,16641,20060403,00.html

69. Julian Cribb, "Shock Factor - Public perceptions of risk in science," ATSE Focus, No. 134, November/December 2004, web site: http://www.atse.org.au/index.php?sectionid=668

70. "The Cooling World," *Newsweek*, April 28, 1975, web site: http://denisdutton.com/newsweek_coolingworld.pdf

71. "Ends of the Earth," Ann Curry video, *MSNBC*, October 29, 2007, web site: http://allday.msnbc.msn.com/archive/2007/10/29/436528.aspx70. Peter Gwynne, "The Cooling World," *Newsweek*, April 28, 1975, p. 64, web site: http://www.freerepublic.com/focus/f-news/993807/posts

72. "Global Warming Censored," Business & Media Institute, 2008, web site: http://www.businessandmedia.org/specialreports/2008/GlobalWarmingCensored/GlobalWarmingCensored_execsum.asp

73. Brian Montopoli, "Scott Pelley and Catherine Herrick on Global Warming

Coverage," CBS News, March 23, 2006, web site: http://www.cbsnews.com/blogs/2006/03/22/publiceye/entry1431768.shtl

74. Peter Sissons, "Peter Sissons: I drove out of Television Centre for the final time last month . . . and I don't have a pang of regret," *DailyMail*, July 15, 2009, web site: http://www.dailymail.co.uk/news/article-1199006/PETER-SISSONS -I-drove-Television-Centre-final-time-month--I-dont-pang-regret.html

75. "The Heidelberg Appeal," Science and Public Policy Institute, web site: http://www.sepp.org/policy%20declarations/heidelberg_appeal.html

76. Ibid.

77. "U.S. Senate Minority Report: More Than 700 (Previously 650) International Scientists Dissent Over Man-Made Global Warming Claims," March 16, 2009, web site: http://epw.senate.gov/public/index.cfm?FuseAction=Minority.Press Releases&ContentRecord_id=d6d95751-802a-23ad-4496- 7ec7e1641f2f&Region_id=&Issue_id=

78. Ibid.

79. "Global Warming Petition Project," *Oregon Institute of Science and Medicine*, web site: http://www.petitionproject.org/

80. "About NIPCC," *Nongovernmental International Panel on Climate Change*, web site: http://www.nipccreport.org/aboutNIPCC.html

81. *Climate Change Reconsidered: the Report of the Nongovernmental International Panel on Climate Change* by Craig Idso and Fred Singer, (The Heartland Institute, 2009), p. iii.

82. "Met Office Hadley Centre observations datasets," web site: http://hadobs.metoffice.com/crutem3/

83. Terry Hurlbut, "Hadley CRU hacked with release of hundreds of docs and emails," *Examiner.com*, Nov. 19, 2009, web site: http://www.examiner.com/x- 28973-Essex-County-Conservative-Examiner-y2009m11d19-Hadley-CRU- hacked-with-release-of-hundreds-of-docs-and-emails

84. "Climategate Document Database," web site: http://www.climate- gate.org/index.php

85. Intergovernmental Panel on Climate Change, Third Assessment Report, Summary for Policy Makers, 2001, p. 5, web site: http://www.grida.no/publications/other/ipcc_tar/

86. Climategate (See no. 84).

87. Climategate (See no. 84).

88. Climategate (See no. 84).

89. Climategate (See no. 84).

90. Climategate (See no. 84).

91. Stephen McIntyre, "A Collation of CRU Correspondence," May 30, 2008, web site: http://climateaudit.files.wordpress.com/2008/05/cru.correspondence.pdf

92. Climategate (See no. 84).

93. Intergovernmental Panel on Climate Change, Fourth Assessment Report, 2007, web site: http://www.ipcc.ch/publications_and_data/publications_and_data _reports.htm#1

94. Climategate (See no. 84).

95. David Douglass et al., "A comparison of tropical temperature trends with model predictions," *International Journal of Climatology*, October 11, 2007, web site: http://www.pas.rochester.edu/~douglass/papers/Published

%20JOC1651.pdf
96. Climategate (See no. 84).
97. Climategate (See no. 84).
98. Raphael Satter, "UK climate scientist to temporarily step down," Associated Press, December 1, 2009, web site: http://www.google.com/hostednews/ap /article/ALeqM5j_dt9Bjj5yVV7k1PAyDnVHKvKtgAD9CAOU800

Chapter 11: Snake Oil Remedies to "Save the Planet"
1. Audra Ang, "Pelosi appeals for China's help on climate change," *Associated Press*, May 27, 2009, web site: http://www.sfgate.com/cgi-bin/article.cgi ?f=/n/a/2009/05/24/international/i210141D03.DTL&type=printable
2. "Climate Change: Global Risks, Challenges & Decisions," University of Copenhagen, March 10-12, 2009, p. 16, web site: http://climatecongress.ku.dk/pdf/synthesisreport
3. IPCC Third Assessment Report, "Climate Change 2001," Working Group II, Figure SPM-2, 2001, web site: http://www.grida.no/publications/other /ipcctar/?src=/climate/ipcc_tar/wg2/figspm-2.htm
4. IPCC Fourth Assessment Report, "Climate Change 2007: Synthesis Report," Table 5.1, p. 67, 2007, web site: http://www.ipcc.ch/pdf/assessment-report/ar4/syr/ar4_syr.pdf
5. Stephen Pacala and Robert Socolow, "Stabilization Wedges: Solving the Climate Problem for the Next 50 Years with Current Technologies," *Science*, vol. 305, August 13, 2004, web site: http://carbonsequestration.us/Papers-presentations/htm/Pacala-Socolow-ScienceMag-Aug2004.pdf
6. Carbon Mitigation Initiative, Princeton University, web site: http://www .princeton.edu/wedges/presentation_resources/Stabilization_Triangle.jpg
7. Carbon Mitigation Initiative, Princeton University, web site: http://www.princeton.edu/~cmi/news/CMIinBrief.pdf
8. Judit Kawaguchi, "Professor Kunihiko Takeda," The Japan Times, July 22, 2008, web site: http://search.japantimes.co.jp/cgi-bin/fl20080722jk.html
9. Lindsay Meisel, "Dangerous Assumptions," *The Breakthrough*, April, 2008, web site: http://thebreakthrough.org/blog/2008/04/the_significance_of_dangerous.shtml
10. "Cost of tackling global climate change has doubled, warns Stern," *The Guardian*, June 26, 2008, http://www.guardian.co.uk/environment /2008/jun/26/climatechange.scienceofclimatechange/print
11. Craig Whitlock, "Environment Minister Criticizes U.S. Policy," *Washington Post*, September 2, 2005, web site: http://www.washingtonpost.com/wp-dyn/content/article/2005/09/01/AR2005090101533.html
12. International Energy Annual 2006, U.S. Energy Information Administration, U.S. Department of Energy, December 8, 2008, web site: http://www.eia.doe.gov/pub/international/iealf/tableh1co2.xls
13. "PM: Climate accord process must change," NZHerald.co.nz, December 21, 2009, web site: http://www.nzherald.co.nz /politics/news/article.cfm?c_id=280&objectid=10616714&pnum=1
14. Dominic Lawson, "Dominic Lawson: Roll up, roll up for the great Copenhagen emissions-fest," *The Independent*, December 8, 2009, web site: http://www. independent.co.uk/opinion/commentators/dominic-lawson/dominic-lawson-

roll-up-roll-up-for-the-great-copenhagen-emissionsfest-1836067.html

15. "Politicians must heed public on climate: Kofi Annan," *AFP*, June 23, 2009, web site: http://www.google.com/hostednews/afp/article /ALeqM5iJw5p0SAsKLU3ald79yKMMEkbrPA

16. George Russell, "Seal the Deal or Sell It? U.N.'s High-Pressure Green Game," *Fox News*, November 19, 2009, web site: http://www.foxnews.com/story/0,2933,575758,00.html

17. "UNESCO organizes first international conference on Broadcast Media and Climate Change," UNESCO.ORG, web site: http://portal.unesco.org/ci/en/ev.php-URL_ID=29062&URL_DO=DO_TOPIC&URL_SECTION=201.html

18. "Paris Declaration on Broadcast Media & Climate Change," web site: http://portal.unesco.org/ci/en/ev.php-URL_ID=29090&URL_DO=DO_TOPIC&URL_SECTION=201.html

19. "India offers hand to US on climate change," *AFP*, March 4, 2009, web site: http://www.google.com/hostednews/afp/article/ALeqM5j9qtIShoetXpa8eaJFaU 5LCp2UEA

20. "Copenhagen Accord," United Nations Framework Convention on Climate Change, web site: http://unfccc.int/files/meetings/cop_15/application/pdf/cop15_cph_auv.pdf

21. Nathanial Gronewold, "U.N. Chief Declares Climate Accord 'Significant Achievement,'" *The New York Times*, December 21, 2009, web site: http://www.nytimes.com/gwire/2009/12/21/21greenwire-un-chief-declares-climate-accord-significant-a-24845.html

22. Jim Sciutto and Sunlen Miller, "Obama Calls Global Warming Agreement an 'Unprecedented Breakthrough,'" *ABC News*, December 18, 2009, web site: http://blogs.abcnews.com/politicalpunch/2009/12/obama-calls-global-warming-agreement-an-unprecedented-breakthrough.html

23. Siobhán Dowling and Daryl Lindsey, "Copenhagen Was an All-Out Failure," December 21, 2009, *Spiegel OnLine*, web site: http://www.spiegel.de/international/world/0,1518,668352,00.html

24. "Mexico wants binding climate accord at 2010 summit," *The Washington Post*, December 24, 2009, web site: http://www.washingtonpost.com/wp-dyn/content/article/2009/12/24/AR2009122402671.html

25. "It's all about green energy," *GuardianWeekly*, March 30, 2009, web site: http://www.guardianweekly.co.uk/?page=editorial&id=1007&catID=17

26. "Dutch Government ditches eco ticket tax in effort to halt declining traffic," *Green Air News*, April 6, 2009, web site: http://www.greenaironline.com/news.php?viewStory=412

27. "United Kingdom: Climate issues may limit flights," *Belfast Telegraph*, Feb. 7, 2009, web site: http://climateark.org/shared/reader/welcome.aspx?linkid=1177 23&keybold=climate%20AND%20%20change%20AND%20%20economy

28. John Vidal, "Levy on international air travel could fund climate change fight," *The Guardian*, June 7, 2009, web site: http://www.guardian.co.uk /environment/2009/jun/07/international-flight-levy-un-climate-change

29. Jim Pickard et al., "Brown plans subsidy for electric cars," *Financial Times*, April 8, 2009, web site: http://www.ft.com/cms/s/0/eb0c45ae-247e-11de-9a01-00144feabdc0.html?ftcamp=rss

30. Christopher Booker, "Gordon Brown shows how green he really is," *Telegraph*, Apr. 11, 2009, web site: http://www.telegraph.co.uk/comment/columnists/christopherbooker/5141208/Gordon-Brown-shows-how-green-he-really-is.html

31. Lonnie Miller, "Short-Term Slump, Year-Over-Year Gains for Global Hybrid Market," *PolkView*, December, 2008, web site: http://www.polk.com/TL/PV_20081201_Issue004_Hybrids.pdf

32. "The big chill," *Economist*, January 15, 2009, web site: http://www.economist.com/displaystory.cfm?story_id=12926505

33. "Brown wants all cars electric by 2020," ETA, July 9, 2008, web site: http://www.eta.co.uk/node/10848

34. Charles Clover, "4x4s to be priced off the road," Telegraph, Jan. 12, 2007, web site: http://www.telegraph.co.uk/news/uknews/1539264/4x4s-to-be-priced-off-the-road.html

35. EU News Feed, Issue 42, Dec. 11, 2008, web site: http://www.iaaglobal.org/_backup/resources/RSRC/2/newsletter/EU_News-Issue_42-November-December_2008.pdf

36. "Summary of California Climate Change 'Scoping Plan' and Transportation Measures," The International Council on Clean Transportation, web site: http://www.theicct.org/documents/California_plan_summary_for_ICCT_website_4-24-09.pdf

37. Matt Nauman, "Bunch of hot air? California isn't banning black cars," *The Mercury News*, March 27, 2009.

38. Thomas Tanton, "California Targets Auto Emissions, Ethanol Gets No Break," *Pacific Research Institute*, July 1, 2009, web site: http://liberty.pacific research.org/press/california-targets-auto-emissions-ethanol-gets-no-break

39. "Appliance Efficiency Standards," Pew Center on Global Climate Change, June 3, 2009, web site: http://www.pewclimate.org/what_s_being_done/in_the_states/energy_eff_map.cfm

40. Steven Titch, "California Wants to Ban Your Big Screen TV," *Reason Foundation*, Apr. 9, 2009, web site: http://reason.org/news/show/1007289.html

41. Pew Center (See no. 39).

42. "Hamburgers are the Hummers of food in global warming: scientists," AFP, February 15, 2009, web site: http://www.google.com/hostednews/afp/article/ALeqM5iSFzzJTQlFBr3vjq1Nv2wixJXpmA

43. "Dinner Choices Can Curb Global Warming According to Veg Climate Alliance," March 9, 2009, web site: http://vegclimatealliance.org/veg-climate-alliances-media-premiere/

44. "Turn veggie to save the world, say Pachauri and McCartney," *ANI*, November 30, 2008, web site: http://www.thaindian.com/newsportal/india-news/turn-veggie-to-save-the-world-say-pachauri-and-mccartney_100125033.html

45. Martin Hickman, "McCartney urges 'meat-free days' to tackle climate change," *The Independent*, June 15, 2009, web site: http://www.independent.co.uk/life-style/food-and-drink/news/mccartney-urges-meatfree-days-to-tackle-climate-change-1705289.html

46. Elisabeth Rosenthal, "To Cut Global Warming, Swedes Study Their Plates," *The New York Times*, October 23, 2009, web site: http://www.nytimes.com/2009/10/23/world/europe/23degrees.html

47. Elizabeth Landau, "Thinner is better to curb global warming, study says,"

CNN, April 20, 2009, web site:
http://www.cnn.com/2009/HEALTH/04/20/thin.global.warming/index.html

48. John Guillebaud and Pip Hayes, "Population growth and climate change," *British Medical Journal*, July24, 2008, web site:
http://www.bmj.com/cgi/content/full/337/jul24_2/a576

49. Daniel Martin and Simon Caldwell, "Call for two-child limit on families from the Government's leading green adviser," *Daily Mail*, February 2, 2009, web site: http://www.dailymail.co.uk/news/article-1133316/Call-child-limit-families-Governments-leading-green-adviser.html

50. The Center for Environment and Population, web site: http://www.cepnet.org/

51. Victoria Markham, "U.S. Population, Energy and Climate Change," 2008, web site: http://www.cepnet.org/documents/USPopulationEnergyandClimate ChangeReportCEP.pdf

52. Memorandum submitted by The Society for the Protection of Unborn Children, *Publications.Parliament.UK*, July 9, 2009, web site: http://www.pu blications.parliament.uk/pa/cm200304/cmselect/cmfaff/389/389we14.htm

53. Alister Doyle, "China says one-child policy helps protect climate," *Reuters*, August 30, 2007, web site:
http://www.reuters.com/article/environmentNews/idUSKUA07724020070831

54. Paul Murtaugh and Michael Schlax, "Reproduction and the carbon legacies of individuals," *Global Environmental Change*, Vol. 19, Iss. 1, February 2009, pp. 14–20, web site: http://www.sciencedirect.com/science?_ob=Article URL&_udi=B6VFV-4V8FFCG-1&_user=10&_rdoc=1&_fmt=&_orig =search&_sort=d&_docanchor=&view=c&_acct=C000050221&_version=1&_ urlVersion=0&_userid=10&md5=411a4bfd17ad84a0d2fd4c0a9061c717

55. Edwin Mora, "NYT Environment Reporter Floats Idea: Give Carbon Credits to Couples That Limit Themselves to One Child," *CNSNews*, October 19, 2009, web site: http://cnsnews.com/news/article/55667

56. *Time to Eat the Dog?: The Real Guide to Sustainable Living* by Robert Vale and Brenda Vale, (Thames & Hudson, 2009).

57. George Monbiot, "Here's the Plan," October 31, 2006, web site:
http://www.monbiot.com/archives/2006/10/31/heres-the-plan/

58. Jenny Haworth, "Carbon ration cards demanded," *The Scotsman*, July 9, 2009, web site: http://news.scotsman.com/scotland/Carbon-ration-cards-demanded.4287280.jp

59. Jonathan Leake, "Environment Agency sets up green police," *Times On Line,* July 4, 2009, web site: http://business.timesonline.co.uk/tol /business/industry_sectors/public_sector/article6639289.ece

60. Carolyn Jones, "Berkeley delays vote on climate-change plan," *SFGate*, April 23, 2009, web site: http://www.sfgate.com/cgi-bin/article.cgi?f=/c/a/2009/04/22/BAHU1774OV.DTL&type=newsbayarea

Chapter 12: Energy: The Lifeblood of Economic Prosperity

1. *The Bottomless Well: The Twilight of Fuel, the Virtue of Waste, and Why we Will Never Run Out of Energy* by Peter Huber and Mark Mills, (Basic Books, 2005); Earth Trends Environmental Information, web site: http://earthtrends.wri.org

2. Anne Stott, "Europe 1700–1914, web site:
http://europetransformed.blogspot.com/2006/10/europe-in-1700.html

3. *The Industrial Revolution, 1760–1830* by T.S. Ashton (Oxford University Press, 1998)

4. Rhys Jenkins, "Savery, Newcomen and the early history of the steam engine," Transactions of the Newcomen Society, 3 (1924 for 1922–1923; 4 (1925 for 1923–1924), pp. 113-33.

5. Ibid.

6. "Electric Conversions," Energy Information Administration, U.S. Department of Energy, web site:
http://www.eia.doe.gov/cneaf/electricity/page/prim2/charts.html

7. "Energy," *Wikipedia*, web site: http://en.wikipedia.org/wiki/Energy (physics)

8. *The Early Development of the Steam Engine* by David K. Hulse (TEE Publishing, 1999)

9. Huber (See no. 1).

10. "Head of Steam", web site:
http://www.darlington.gov.uk/Culture/headofsteam/about/About.htm

11. "Physics for Future Presidents," web site: http://muller.lbl.gov/teaching/physics10/old%20physics%2010/chapters%20 (old)/1-Explosions.htm

12. "Internal Combustion Engine," *BookRags*, web site:
http://www.bookrags.com/research/internal-combustion-engine-woi/

13. Ibid.

14. "History of the Oil and Gas Industry," Business Reference Services, web site:
http://www.loc.gov/rr/business/BERA/issue5/history.html

15. "Karl Benz," About.com, web site:
http://inventors.about.com/library/inventors/blbenz.htm

16. Mary Bellis, "Gottlieb Daimler," About.com, web site:
http://inventors.about.com/od/dstartinventors/a/Gottlieb_Daimler.htm

17. "A brief history of con Edison," web site:
http://www.coned.com/history/electricity.asp

18. "History of Electricity," Intermediate Energy Infobook, web site:
http://www.need.org/needpdf/infobook_activities/IntInfo/Elec3I.pdf

19. "Statistical Abstract of the United States: 2009," U.S. Census Bureau, web site:
http://www.census.gov/compendia/statab/

20. U.S. Energy Information Administration, Department of Energy, web site:
http://www.eia.doe.gov/

21. Ibid.

22. "Coal-fired power stations are death factories," *The Observer*, February, 15, 2009, http://www.guardian.co.uk/commentisfree/2009/feb/15/james-hansen-power-plants-coal/print

23. "Key World Energy Statistics 2008," International Energy Agency, web site:
http://www.iea.org/textbase/nppdf/free/2008/key_stats_2008.pdf

24. Ibid.

25. "The Plowboy Interview with Amory Lovins," *Mother Earth News*, Nov.-Dec., 1977, web site: http://www.motherearthnews.com/Renewable-Energy/1977-11-01/Amory-Lovins.aspx

26. Paul Ciotti, "Fear of Fusion: What if it Works?," *Los Angeles Times*, April 19, 1989.

27. *Earth in Balance: Ecology and the Human Spirit* by Al Gore, p. 243, (Houghton Mifflin, 1992).

28. *Ecoscience: Population, Resources, Environment,* by Paul Ehrlich et al., (W.H. Freeman, 1970), p. 323.

29. "Energy Strategy," Union of Concerned Scientists, 1980.

30. Amory Lovins, "Energy Strategy, The Road Not Taken," Nov., 1977, web site: http://www.rmi.org/images/PDFs/Energy/E77-01_TheRoadNotTaken.pdf

31. Department of Energy (See no. 20).

32. U.S. Census Bureau (See no. 19).

33. International Energy Agency (See no. 23).

34. International Energy Agency (See no. 23); Department of Energy (See no. 20)

35. "International Energy Outlook 2009," U.S. Energy Information Administration, Department of Energy, web site: http://www.eia.doe.gov/

36. Iain Thomson, "IT energy use could scupper global warming plans," *BusinessGreen,* web site: http://www.businessgreen.com /vnunet/news/2242243/scupper-global-warming-plans%20

37. Werner Zittel and Jorg Schindler, "Crude Oil: The Supply Outlook," Energy Watch Group, October, 2007, web site: http://www.energywatchgroup .org/fileadmin/global/pdf/EWG_Oilreport_10-2007.pdf

38. Ibid.

39. Ibid.

40. The Association for the Study of Peak Oil and Gas, web site: http://www.peakoil.net/

41. Dave Cohen, "The Perfect Storm," The Association for the Study of Peak Oil and Gas, October 31, 2007, web site: http://www.aspo-usa.com/archives /index.php?option=com_content&task=view&id=243&Itemid=91

42. *The Bottomless Well* (See no. 1).

43. "Oil Drilling," Science Clarified, web site: http://www.scienceclarified.com/Mu-Oi/Oil-Drilling.html

44. "Today's Drilling Leaves a Small Footprint," Arctic National Wildlife Refuge, web site: http://www.anwr.org/Technology/Today-s-drilling-leaves-a-small-footprint.php

45. "Drilling Technology," Arctic National Wildlife Refuge, web site: http://www.anwr.org/techno/drilling.htm

46. Department of Energy (See no. 20).

47. NASA Goddard Space Flight Center, 2009 and "Oil in Ocean Shows up on NASA Images: Half of the Oil in the ocean Bubbles up Naturally from Seafloor," *ScienceDaily,* Feb. 20, 2009, web site: http://www.sciencedaily.com/releases/2009/02/090219101658.htm

48. Ibid.

49. Bruce Allen, "Testimony before the House Committee on Natural Resources 'Offshore Drilling: Environmental and Commercial Perspectives,'" Feb. 11, 2009, web site: http://resourcescommittee.house.gov/images /Documents/20090211/testimony_allen.pdf

50. Annual Energy Review 2008, Energy Information Administration, US Department of Energy, web site: http://www.eia.doe.gov/emeu/aer/pdf/aer.pdf

51. "Controlling the Grid," U.S. Department of Energy, web site: http://www.eere.energy.gov/de/grid_control.html

52. Ibid.

Chapter 13: Renewable Energy: Reality Far Short of Promises

1. "How Solar Energy Works," *Union of Concerned Scientists,* web site: http://www.ucsusa.org/clean_energy/technology_and_impacts/energy_technolo gies/how-solar-energy-works.html

2. "Solar Electric Power (PV)," U.S. Department of Energy, http://www.eere.energy.gov/de/solar_electric.html

3. F. H. Lancaster, "The Gold Content of Sea-water," web site: http://www.goldbulletin.org/assets/file/goldbulletin/downloads/Swann_4_6.pdf

4. NationMaster.com, web site: http://www.statemaster.com/encyclopedia/polar-ice-cap

5. The Global Water Crisis: Our Inevitable Fate?," web site: http://blogs.princeton.edu/chm333/f2006/water/2006/11/how_does_water_use_in_developing_countries_differ.html

6. Walter Mossberg, "Solar Power Seen Meeting 20% of Needs by 2000; Carter may Seek Outlay Boost," *The Wall Street Journal,* Aug. 22, 1978.

7. Energy Information Administration (EIA), U.S. Department of Energy, web site: http://www.eia.doe.gov/emeu/mer/pdf/pages/sec1_7.pdf

8. Energy Information Administration (EIA), U.S. Department of Energy, web site: http://www.eia.doe.gov/emeu/mer/pdf/pages/sec7_5.pdf

9. A. Hunter Fanney et al.," Measured Performance of a 35 Kilowatt Roof Top Photovoltaic System," National Institute of Standards and Technology, U.S. Dept. of Commerce, March, 2003, web site: http://www.bfrl.nist.gov/863/bipv/documents/35kw_HI.pdf

10. "Will renewables become cost-competitive anytime soon?," Institute for Energy Research, April 1, 2009, web site: http://www.instituteforenergyresearch .org/2009/04/01/will-renewables-become-cost-competitive-anytime-soon-the-siren-song-of-wind-and-solar-energy/

11. Robert Bradley, "Getting Real: The Oil Major Move Away from Political Energy," *MasterResource,* Apr. 9, 2009, web site: http://masterresource.org/?p=1755

12. Christopher Flavin and Nicholas Lenssen, "Beyond the Petroleum Age Designing a Solar Economy," Worldwatch Institute, December, 1990.

13. Peter Lang, "Solar Power Realities," August, 2008, web site: http://carbon-sense.com/wp-content/uploads/2009/07/solar-realities.pdf

14. Ibid.

15. Mar Pop, "The Value of Distributed Urban Residential Photovoltaic Electricity in the Australian National Electricity Market," Centre for Energy and Environmental Markets, June 25, 2005, web site: http://www.ceem.unsw .edu.au/content/documents/ReportMarc-ValueofPV-25Jun05.pdf

16. "Queanbeyan Solar Farm," Bonzle, web site: http://maps.bonzle.com/c/a?a=p&p=7&cmd=sp&d=pics

17. Desertek-UK, web site: http://www.trec-uk.org.uk/pictures.htm

18. "Solar Thermal Electric Technology in 2004," Electric Power Research Institute, March, 2005, web site: http://mydocs.epri.com/docs/public/000000000001011615.pdf

19. "Diablo Canyon Nuclear Power Plant, California," Energy Information Administration, U.S. Department of Energy, web site: http://www.eia.doe.gov/cneaf/nuclear/page/at_a_glance/reactors/diablo.html

20. Levelized Cost of New Electricity Generating Technologies, Institute for Energy Research, May 12, 2009, web site: http://www.instituteforenergy research.org/2009/05/12/levelized-cost-of-new-generating-technologies/

21. City of Madison, Wisconsin, web site: http://www.cityofmadison.com/

22. Kent Ninomiya, "About Milan," eHow, web site: http://www.ehow.com/about_4597425_milan.html

23. Solar plant photograph by Alan Radecki, licensed under GNU Free Documentation License, web site: http://en.wikipedia.org/wiki/File:Solarplant-050406-04.jpg

24. Diablo Canyon Nuclear Plant photograph by Marya, licensed under Creative Commons Attribution 2.0 License, web site: http://en.wikipedia.org/wiki/File:Diablo_canyon_nuclear_power_plant.jpg

25. Julie Gozan, "Solar Eclipsed," MultinationalMonitor, April, 1992, web site: http://multinationalmonitor.org/hyper/issues/1992/04/mm0492_07.html

26. "Report to Congress on Assessment of Potential Impact of Concentrating Solar Power for Electricity Generation," EERE, January, 2007, web site: http://www.seia.org/galleries/pdf/CSP_Plants_in_the_US_Final.pdf

27. "Briefing Report: Reconsidering Nuclear Power Plants," California State Senate Republican Caucus, July 18, 2007, web site: http://cssrc.us/publications.aspx?id=3638

28. Bernie Woodall, "Los Angeles Will End Use of Coal-Fired Power," CommonDreams.org, July 3, 2009, web site: http://www.commondreams.org/headline/2009/07/03-5

29. Matthew Heberger et al., "The Impacts of Sea-Level Rise on the California Coast," California Climate Change Center, May, 2009, web site: http://www.pacinst.org/reports/sea_level_rise/report.pdf

30. Pacific Institute Board of Directors, web site: http://www.pacinst.org/

31. "Our Changing Climate," California Climate Change Center, July, 2005, web site: http://meteora.ucsd.edu/cap/pdffiles/CA_climate_Scenarios.pdf

32. Institute for Energy Research (See no. 20).

33. Mick Dumke, "Green for Green," Chicago Reader, July 29, 2009, web site: http://www.chicagoreader.com/TheBlog/archives/2009/07/29/green-to-green

34. Internal Revenue Service, web site: http://www.irs.gov/pub/irs-pdf/f5695.pdf

35. Austin City Council, May 14, 2009, web site: http://www.ci.austin.tx.us/edims/document.cfm?id=128081

36. "Texas, Incentives/Policies for Renewables & Efficiency," DSIRE, March 17, 2009, web site: http://www.dsireusa.org/incentives/incentive.cfm ?Incentive_Code=TX11F&re=1&ee=1

37. Gerhard Stryi-Hipp, presentation to the Intersolar North American Conference, San Francisco, July 13, 2009, web site: http://www.semiconwest .org/cms/groups/public/documents/web_content/ctr_030839.pdf

38. "European Household Electricity Price Index for Europe," E-Control, June, 2009, web site: http://www.vaasaett.com/2009/06/household-energy-price-index-for-europe/

39. U.S. Solar Industries Year in Review 2008, SEIA, March 2009, web site: http://www.seia.org/galleries/pdf/2008_Year_in_Review-small.pdf

40. Ibid.

41. *The Solar Fraud: Why Solar Energy Won't Run the World* by Howard Hayden, (Vales Lake Publishing, 2005), p. VI and p. 12.

42. "Turning the Vision into Reality," Desertec Foundation, July 13, 2009, web site: http://www.desertec.org/en/press/press-releases/090713-01-assembly-desertec-industrial-initiative/

43. "World's Most Daring Solar Energy Project Coming to Fruition," Renewable Energy Jobs, web site: http://www.renewableenergyjobs.net/showthread.php?t=4954

44. Desertec Foundation, web site: http://www.desertec.org/

45. Desertec (See no. 42)

46. Ryan Wiser and Mark Bolinger, "2008 Wind Technologies Market Report," Berkeley National Laboratory, July, 2009, web site: http://www1.eere.energy.gov/windandhydro/pdfs/46026.pdf

47. London Array, web site: http://www.londonarray.com/

48. Photograph by London Array, web site: http://www.londonarray.com/about/offshore-components/turbines/

49. Photograph copyright David Bowen, licensed for reuse under Creative Commons License 2.0, web site: http://www.geograph.org.uk/photo/577086

50. Institute for Energy Research (See no. 20).

51. Institute for Energy Research (See no. 20).

52. "The Economics of Renewable Energy," House of Lords, November 25, 2008, web site: http://www.publications.parliament.uk/pa/ld200708/ldselect/ldeconaf/195/195i.pdf

53. Ibid.

54. Ibid.

55. William Hyde, "When the Wind Stops," Kentish Weald Action Group, web site: http://kwag.co.uk/whenthewindstops.htm

56. Ibid.

57. "The Dash for Wind, West Denmark's Experience and UK's Energy Aspirations," Incoteco (Denmark) ApS, web site: http://www.glebemountaingroup.org/documents/DanishLessons.pdf

58. House of Lords (See no. 52).

59. "The UK Low Carbon Transition Plan," Department of Energy and Climate Change, July, 2009, web site: http://www.decc.gov.uk/default.aspx

60. *Blue Planet in Green Shackles*, Václav Klaus, (Competitive Enterprise Institute, 2008), p. 86.

61. Sara Scatasta and Tim Mennel, "Comparing Feed-In-Tariffs and Renewable Obligation Certificates - the Case of Wind Farming, Center for European Economic Research (ZEW), February 20, 2009, web site: http://realoptions.org/openconf/data/papers/43.pdf

62. Gabriel Calzada, "Study of the effects on employment of public aid to renewable energy sources," Universidad Rey Juan Carlos, March 2009, web site: http://www.juandemariana.org/pdf/090327-employment-public-aid-renewable.pdf

63. "Green Energy Certificates," BIS Department for Business Innovation & Skills, web site: http://www.berr.gov.uk/energy/sources/sustainable/microgeneration/strategy/implementation/certificates/page39834.html

64. "Renewable energy: Options for scrutiny," National Audit Office, July 1,

2008, web site: http://www.nao.org.uk/publications
/0708/renewable_energy_options.aspx

65. Aaron Patrick, "Wind energy endures a gale of hostility," *Daily Telegraph*, March 26, 2005, web site: http://www.doggedpictures.co.uk/clash/pdfs/Telegraph050326.pdf

66. Ibid.

67. Database for State Incentives for Renewables and Efficiency, North Carolina State University, web site: http://www.dsireusa.org/

68. Ibid.

69. Wiser (See no. 46).

70. "Federal Financial Interventions and Subsidies in Energy Markets 2007," Energy Information Administration, U.S. Department of Energy, April, 2008, web site: http://www.eia.doe.gov/oiaf/servicerpt/subsidy2/index.html

71. Ibid.

72. "Energy Statistics 2007," Danish Energy Agency, October, 2008, web site: http://www.ens.dk/da-DK/Info/TalOgKort/Statistik_og_noegletal /Maanedsstatistik/Documents/energy%20statistics%202007%20uk.pdf

73. "Secretary Chu Announces $2.4 billion in Funding for Carbon Capture and Storage Projects," U.S. Department of Energy, May 15, 2009, web site: http://www.energy.gov/news2009/7405.htm

74. Juliette Jowit and Tim Webb, "Ed Miliband plans clean coal scheme worth millions," *The Guardian*, April 18, 2009, web site: http://www.guardian .co.uk/environment/2009/apr/18/coal-carbon-capture-storage

75. "Carbon Capture & Storage: Assessing the Economics," McKinsey & Company, September 22, 2008, web site: http://www.mckinsey.com /clientservice/ccsi/pdf/ccs_assessing_the_economics.pdf

76. Mohammed Al-Juaied and Adam Whitmore, "Realistic Costs of Carbon Capture," Harvard Kennedy School, July, 2009, web site: http://belfercenter.ksg.harvard.edu/files/2009_AlJuaied_Whitmore_Realistic_C osts_of_Carbon_Capture_web.pdf

77. Christine Ehlig-Economides and Michael Economides, "Sequestering carbon dioxide in a closed underground volume," *Journal of Petroleum Science and Engineering*, vol. 70, November 4, 2009, pp. 123-130.

78. Freeman Dyson, "Heretical thoughts about science and society," Edge, The Third Culture, August 8, 2007, web site: http://www.edge.org/3rd_culture/dysonf07/dysonf07_index.html

79. "President Bush's State of the Union Address," *The Washington Post*, January 31, 2006, web site: http://www.washingtonpost.com/wp-dyn/content/article/2006/01/31/AR2006013101468.html

80. *Gusher of Lies: The Dangerous Delusions of Energy Independence* by Robert Bryce, (Public Affairs, 2008), p. 1

81. "Energy Kid's Page Ethanol Timeline," Energy Information Administration, web site: http://www.eia.doe.gov/kids/history/timelines/ethanol.html

82. Ronald Steenblik, "Biofuels--At What Cost?," Global Subsidies Initiative of the International Institute for Sustainable Development, September, 2007, web site: http://www.globalsubsidies.org/en/research/biofuel-subsidies

83. European Biodiesel Board, web site: http://www.ebb-eu.org/stats.php

84. Directive 2003/30/EC of the European Parliament and of the Council of 8

May 2003 on the promotion of the use of biofuels or other renewable fuels for transport, web site: http://ec.europa.eu/energy/res/legislation/doc/biofuels/en_final.pdf

85. F.O. Licht's World Ethanol and Biodiesel Report, 2009, web site: http://www.foodnews.co.uk/portal/puboptions.jsp?Option=menu&pubId=ag07 2, and Sam Lines, "An Exploding Market? Utilizing Waste Glycerol from the Biodiesel Production Process," April 19, 2009, web site: http://snrecmitigation.wordpress.com/2009/04/19/an-exploding-market-utilizing-waste-glycerol-from-the-biodiesel-production-process/

86. "Suspend 10 percent biofuels target, says EEA's scientific advisory body," European Environment Agency, April 10, 2008, web site: http://www.eea.europa.eu/highlights/suspend-10-percent-biofuels-target-says-eeas-scientific-advisory-body

87. "Green biodiesel production starts," *BBC News*, April 4, 2005, web site: http://news.bbc.co.uk/2/hi/uk_news/scotland/4409369.stm

88. "Bush reportedly regrets global warming, energy policy decisions," EV World, March 30, 2006, web site: http://www.energybulletin.net/node/14511

89. IISD (See no. 82).

90. "Federal Financial Interventions and Subsidies in Energy Markets 2007," Energy Information Administration, April 2008, web site: http://www.eia.doe.gov/oiaf/servicerpt/subsidy2/index.html

91. "Renewable energy: Options for scrutiny," National Audit Office, July 2008, web site: http://www.nao.org.uk/publications/0708/renewable_energy_options.aspx

92. IISD (See no. 82).

93. IISD (See no. 82).

94. IISD (See no. 82).

95. Paul Wescott, "Full Throttle U.S. Ethanol Expansion Faces Challenges Down the Road," U.S. Department of Agriculture, September, 2008, web site: http://www.ers.usda.gov/

96. EIA (See no. 81).

97. "Liquid Fuels Supply and Disposition," Energy Information Administration, web site: http://www.eia.doe.gov/

98. Massachusetts Government Operational Services Division, web site: http://www.mass.gov/?pageID=afterminal&L=7&L0=Home&L1=Budget%2c+Taxes+%26+Procurement&L2=Procurement+Information+%26+Resources&L3=Procurement+Programs+and+Services&L4=Environmentally+Preferable+Products+(EPP)+Procurement+Program&L5=Find+Green+Products+and+Services+on+Statewide+Contracts&L6=View+Information+on+Specific+Products+and+Services&sid=Eoaf&b=terminalcontent&f=osd_epp_product_es_hybrids&csid=Eoaf

99. Shapouri et al., "The Energy Balance of Corn Ethanol: An Update," U.S. Department of Energy, July, 2002, web site: http://www.usda.gov/oce/reports/energy/aer-814.pdf

100. Tiffany Groode and John Heywood, "Biomass to Ethanol: Potential Production and Environmental Impacts," Massachusetts Institute of Technology, Feb., 2008, web site: http://dspace.mit.edu/handle/1721.1/43144

101. Shapouri (See no. 99).

102. Marcello Dias de Oliveira et al., "Ethanol as Fuel: Energy, Carbon Dioxide

Balances, and Ecological Footprint," BioScience, July 2005, Vol. 55, no. 7, web site: http://goliath.ecnext.com/coms2/gi_0199-4483890/Ethanol-as-fuel-energy-carbon.html

103. Groode (See no. 100).

104. Kirk Berge, "Ethanol and Biodiesel: the Good, the Bad and the Unlikely," December 5, 2008, web site: http://www.peakoil.net/files /Biofuels_%20the%20G_%20the%20B_the%20U.pdf

105. Dennis Avery, "Biofuels, Food, or Wildlife? The Massive Land Costs of U.S. Ethanol," *Competitive Enterprise Institute*, September 21, 2006, web site: http://cei.org/pdf/5532.pdf

106. Vivienne Walt, "The World's Growing Food-Price Crisis," *Time*, Feb. 27, 2008, web site: http://www.time.com/time/world/article/0,8599,1717572,00.html

107. U.S. Department of Agriculture, 2009, web site: http://www.usda.gov/wps/portal/usdahome

108. Avery (See no. 105).

109. USDA (See no. 107).

110. Lester Brown, "Starving People to Feed the Cars," The Washington Post, September 10, 2006, web site: http://www.washingtonpost.com/wp-dyn/content/article/2006/09/08/AR2006090801596.html

111. Dennis Avery, Presentation at the International Conference on Climate Change, March, 2008.

112. "President Obama Announces National Fuel Efficiency Policy," The White House, May 19, 2009, web site: http://www.whitehouse.gov/the_press_office /President-Obama-Announces-National-Fuel-Efficiency-Policy/

113. Wendell Goler, "President Obama Announces Gas Mileage Standards," *Fox News.com*, May 19, 2009, web site: http://whitehouse.blogs.foxnews.com /2009/05/19/president-obama-announces-gas-mileage-standards/

114. "National Transportation Statistics 2009," U.S. Department of Transportation, web site: http://www.bts.gov/publications/national_transportation_statistics/

115. Ibid.

116. *The Bottomless Well: The Twilight of Fuel, the Virtue of Waste, and Why we Will Never Run Out of Energy* by Peter Huber and Mark Mills, (Basic Books, 2005), pp. 111–112.

117. "Reducing CO2 emissions from light-duty vehicles," EUROPA, web site: http://ec.europa.eu/environment/air/transport/co2/co2_home.htm

118. "Report on costs and reduction potential of CO2-saving measures," European Climate Change Program, October 31, 2006, web site: http://www.acea.be/images/uploads/co2/co2_0004.pdf

119. Stanley Pignal, "Motor lobby attacks planned emission rules," *FT.com*, August 11, 2009, web site: http://www.ft.com/cms/s/0/82fd0bfe-8696-11de-9e8e-00144feabdc0,dwp_uuid=62398742-53ce-11db-8a2a-0000779e2340.html

120. Micheline Maynard et al., "Toyota Hybrid Makes a Statement, and That Sells," *The New York Times*, July 4, 2007, web site: http://query.nytimes.com /gst/fullpage.html?res=9C0DE0D8153EF937A35754C0A9619C8B63

121. "Hydrogen Fuel Technology: a Cleaner and More Secure Energy Future," The White House, web site: http://georgewbush-whitehouse.archives.gov

/infocus/technology/economic_policy200404/chap2.html

122. *The Bottomless Well* (See no. 116), p. 86.

123. Mathew Wald, "U.S. Drops Research into Fuel Cells for Cars," *The New York Times*, May 7, 2009, web site: http://www.nytimes.com/2009/05/08/science/earth/08energy.html

Chapter 14: Chains for the Developing Nations

1. Rajendra Pachauri, "This silent suffering," *The Guardian*, May 29, 2009, web site: http://www.guardian.co.uk/commentisfree/cif-green/2009/may/29/climate-change-poor

2. "International Energy Outlook 2009," Energy Information Administration, U.S. Department of Energy, May, 2009, web site: http://www.eia.doe.gov/oiaf/ieo/index.html

3. Ibid.

4. "Shanghai population tops 20m," China Daily, May 5, 2003, web site: http://www.chinadaily.com.cn/en/doc/2003-12/05/content_287714.htm

5. "International Travel—China Shanghai Transportation," Love to Eat and Travel, web site: http://www.lovetoeatandtravel.com /Site/Intl/China/Shanghai/Lodging/transportation.htm

6. China Statistical Yearbook, National Bureau of Statistics of China, web site: http://www.stats.gov.cn/english/

7. Alex Felsinger, "How You Can Buy Part of Greenpeace's Heathrow Plot," Planetsave, January 13, 2009, web site: http://planetsave.com/blog /2009/01/13/how-you-can-buy-part-of-greenpeaces-heathrow-plot/

8. "Climate protest on Heathrow plane," BBC News, February 25, 2008, web site: http://news.bbc.co.uk/2/hi/uk_news/7262614.stm

9. Alex Felsinger, "'Plane Stupid' Climate Activists Block Runway at UK Airport," December 8, 2008, web site: http://planetsave.com/blog/2008/12/08/plane-stupid-climate-activists-block-runway-at-uk-airport/

10. Angela Gittens, "Challenges and opportunities for airport infrastructure," Presentation to the Airports Council International, October 14, 2008, web site: http://www.icao.int/DevelopmentForum/Forum_08/Presentations/gittens.pdf

11. Annual Energy Review 2008, Energy Information Administration, U.S. Department of Energy, June, 2009, web site: http://www.eia.doe.gov/emeu/aer/pdf/aer.pdf

12. Ibid.

13. International Energy Outlook, (See no. 2).

14. "Key World Energy Statistics 2008," International Energy Agency, web site: http://www.iea.org/

15. "Urban Population, Development and the Environment 2007," United Nations Department of Economic and Social Affairs, Population Division, web site: www.unpopulation.org

16. International Energy Outlook, (See no. 2).

17. International Energy Agency, (See no. 14).

18. Urban Population (See no. 15).

19. "2007/2008 Human Development Report," UNDP, web site: http://hdrstats.undp.org/indicators/210.html

20. "Itaipu Dam," Woodward South America, web site:

http://www.southamerica.cl/general_information/Itaipu_Dam.htm

21. Urban Population (See no. 15).

22. "South America," MSN Encarta, web site:
http://encarta.msn.com/encyclopedia_761574914_3/South_America.html

23. Photo by Tom Oates, Licensed under the Creative Commons Attribution-ShareAlike 3.0 License, web site:
http://en.wikipedia.org/wiki/File:Itaipu_171.jpg

24. Human Development Report (See no. 19).

25. "Nuclear Power in Southeast Asia: Implications for Australia and Non-Proliferation," Lowy Institute for International Policy, April 2008, web site:
http://www.lowyinstitute.org/Publication.asp?pid=786

26. Urban Population (See no. 15).

27. Human Development Report (See no. 19).

28. Urban Population (See no. 15).

29. "Earth at Night," NightSkyNation, image from NASA, web site:
http://www.nightskynation.com/objects/earth-at-night/Africa

30. Itai Madamombe, "Energy key to Africa's prosperity," Africa Renewal, United Nations, web site:
http://www.un.org/ecosocdev/geninfo/afrec/vol18no4/184electric.htm

31. Vijaya Ramachandran, "Power and Roads for Africa," Center for Global Development, web site: http://www.cgdev.org/content/publications/detail/1421340/

32. Ibid.

33. Ibid.

34. Rodney Slater, "The National Highway System: A Commitment to America's Future," U.S. Department of Transportation Federal Highway Administration, web site: http://www.tfhrc.gov/pubrds/spring96/p96sp2.htm

35. Conversation with Janet Goreham, November 10, 2009.

36. "Malaria Facts," Centers for Disease Control and Prevention, web site:
http://www.cdc.gov/Malaria/facts.htm

37. *Silent Spring* by Rachel Carson, (Houghton Mifflin, 1962)

38. Ronald Bailey, "Silent Spring at 40," Reasononline, web site:
http://www.reason.com/news/show/34823.html

39. Environmental Defense Fund, web site: http://www.edf.org/home.cfm

40. Todd Seavey, "The DDT Ban Turns 30–Millions Dead of Malaria Because of Ban, More Deaths Likely," American Council on Science and Health, June 1, 2002, web site:
http://www.acsh.org/healthissues/newsID.442/healthissue_detail.asp

41. "DDT and its Derivatives," International Programme on Chemical Safety, IPCS Inchem, web site:
http://www.inchem.org/documents/ehc/ehc/ehc009.htm

42. "DDT Ban Takes Effect," US Environmental Protection Agency, web site:
http://www.epa.gov/history/topics/ddt/01.htm

43. "Stockholm Convention on Persistent Organic Pollutants," May 22, 2001, web site: http://chm.pops.int/

44. Pat Sidley, "Malaria epidemic expected in Mozambique," *British Medical Journal*, vol. 320(7236), March 11, 2000, web site:

http://www.bmj.com/cgi/content/extract/320/7236/66941.

45. Bailey (See no. 38).

46. Seavey (See no. 40).

47. *The Really Inconvenient Truths: Seven Environmental Catastrophes Liberals Don't Want You to Know About--Because They Helped Cause Them* by Iain Murray, (Regnery Publishing, 2008), pp. 40–41.

48. "WHO Malaria Head to Environmentalists: 'Help save African babies as you are helping to save the environment,' Press statement by R. Arata Kochi, World Health Organization, September 15th, 2006, web site: http://malaria.who.int/docs/KochiIRSSpeech15Sep06.pdf

49. "Sustainable Energy: A Framework for New and Renewable Energy in Southern Africa," United Nations Economic Commission for Africa, March 2006, web site: http://www.uneca.org/srdc/sa/publications /SRO_SA_SustEnergy.pdf

50. Ibid.

51. Mallika Nair, "Renewable Energy for Africa," Institute for Environmental Security, May 2009, web site: http://www.envirosecurity.org/espa/PDF/Renewable_Energy_for_Africa.pdf

52. Ibid.

53. Ibid.

54. Eric Youngren, "Off-grid solar electricity in Africa," Island Energy Systems Blog, March 24, 2009, web site: http://islandenergysystems.wordpress.com /2009/03/24/off-grid-solar-electricity-in-africa/

55. Sarah van Schagen, "Eco-celebrity tries reality TV show," *MSNBC*, January 4, 2007, web site: http://www.msnbc.msn.com/id/16472662/

56. Marc Morano, "Environmentalist Laments Introduction of Electricity," *CNSNEWS*, August 26, 2002, web site: http://www.freerepublic.com/focus/news/739362/posts

57. "Mozambique dam may ease energy woes in SA," *Business Report*, July 13, 2009, web site: http://www.busrep.co.za/index.php?fArticleId=5080733

58. International Rivers, web site: http://www.internationalrivers.org/

59. "Renewable Independent Power Producers (RIPP)s: Restructuring the Southeast Asian Electricity Sector using Sustainable Energy," Greenpeace, July 1999, web site: http://www.greenpeace.org/raw/content /seasia/th/press/reports/renenewable-ipps-in-sea.pdf

60. Tom Fawthrop, "Mekong river hydroelectric dam threatens livelihoods and endangered species in landlocked Laos," Guardian, March 13, 2009, web site: http://www.guardian.co.uk/environment/2009/mar/13/laos-hydroelectric-dam

61. Photograph owned by Zenman, licensed under the GNU Free Documentation License, version 1.2, March 16, 2008, web site: http://en.wikipedia.org/wiki/File:Cifemrurales.JPG

62. "Indoor air pollution and health," World Health Organization, June 2005, web site: http://www.who.int/mediacentre/factsheets/fs292/en/index.html

63. *Eco-Imperialism: Green Power, Black Death* by Paul Driessen, (Free Enterprise Press, 2003), p. 42.

64. "Special Report: Financial Sector Responsibility," Ethical Corporation, Nov. 2006, web site: http://www.ethicalcorp.com/content.asp?ContentID=4578&ContTypeI=

65. "The Equator Principles," July 2006, web site: http://www.equator-principles.com/principles.shtml
66. Ibid.
67. "International Finance Corporation's Standards on Social & Environmental Sustainability," International Finance Corporation, World Bank Group, Apr. 30, 2006, web site: http://www.ifc.org/ifcext/sustainability.nsf/AttachmentsBy Title/pol_PerformanceStandards2006_full/$FILE/IFC+Performance+Standards .pdf
68. Ethical Corporation (See no. 64).
69. Christopher Wright, "For Citigroup, Greening Starts with Listening," The Equator Principles, April 4, 2006, web site: http://www.equator-principles.com/City.shtml
70. Steven Milloy, "Bank Must Take Stand for Third World," FoxNews, March 28, 2005, web site: http://www.foxnews.com/story/0,2933,151446,00.html
71. Steven Milloy, "Turning Children Against Business," CSRWatch, January 26, 2005, web site: http://www.freerepublic.com/focus/f-news/1329751/posts
72. JPMorgan Chase & Co. Environmental Policy, web site: http://www.jpmorganchase.com/cm/cs?pagename=Chase/Href&urlname=jpmc /community/env/policy
73. Jim Carlton, "J.P. Morgan Adopts 'Green' Lending Policies," Wall Street Journal, April 25, 2005, web site: http://www.climateark.org/shared/reader/welcome.aspx?linkid=41225&keybold =rainforest%20AND%20%20oil%20AND%20%20drilling
74. "World Bank and China Exim bank team up on lending to Africa," Bank Information Center, June 4, 2007, web site: http://www.bicusa.org/en/article.3386.aspx
75. Steve Herz and Johan Frijns, "A Challenging Climate: What international banks should do to combat climate change," *BankTrack*, Dec. 12, 2007, web site: http://www.banktrack.org/show/news/banks_challenged _to_fight_climate_change
76. BankTrack, web site: http://www.banktrack.org/show/pages/home
77. Paul Driessen, "Live Earth - Dead Africans," Center for the Defense of Free Enterprise, July 18, 2007, web site: http://www.eskimo.com/~rarnold/live_earth.htm
78. "Climate Change responsible for 300,000 deaths a year," Global Humanitarian Forum, May 29, 2009, web site: http://ghfgeneva.org/Media/PressReleases/tabid/265/EntryId/40/Climate-Change-responsible-for-300-000-deaths-a-year.aspx
79. OECD Glossary of Statistical Terms, web site: http://stats.oecd.org/glossary/detail.asp?ID=2074
80. N. Satyanarayana Murthy et al., "CO_2 Emissions Reduction Strategies and Economic Development of India," Indira Gandhi Institute of Development Research, Mumbai, August 2006, web site: http://www.igidr.ac.in/pdf/publication/WP-2006-004.pdf
81. Tsefaye Tadesse, "Africa wants $67 bln a year in global warming funds," *Reuters*, web site: http://www.reuters.com/article/latestCrisis/idUSLO544093
82. Nitin Sethi, "190 giga tonnes: Planet's carbon budget till 2050," *The Times of India*, May 5, 2009, web site:

http://economictimes.indiatimes.com/Earth/Global-Warming/190-giga-tonnes-Planets-carbon-budget-till-2050/articleshow/4484302.cms

83. Al Gore testimony before the U.S. House of Representatives, Committee on Energy and Commerce, April 24, 2009, web site: http://energycommerce.house.gov/index.php?option=com_content&view=articl e&id=1593:energy-and-commerce-subcommittee-hearing-on-the-american-clean-energy-and-security-act-of-2009&catid=130:subcommittee-on-energy-and-the-environment&Itemid=71

84. "Global 2008 CO$_2$ emissions rose 2 pct- German institute," *Reuters*, August 10, 2009, web site: http://www.reuters.com/article/latestCrisis/idUSLA354495

85. Al Gore Testimony (See no. 83).

86. "Tariff Provision May Be Critical to Senate Climate Change Bill," CQ Politics, August 12, 2009, web site: http://blog.climateandenergy.org/2009/08/13/cq-politics-tariff-provision-may-be-critical-to-senate-climate-change-bill/

87. "EU environment ministers unite on climate change action," *Deutsche Welle*, July 25, 2009, web site: http://www.dw-world.de/dw/article/0,,4517921,00.html

88. "The EU Climate Action and Renewable Energy Package: Are we about to be locked into the wrong policy?," Open Europe, October, 2008, web site: http://www.openeurope.org.uk/research/carep.pdf

89. Ian Talley and Tom Barkley, "Energy Chief Says U.S. Is Open to Carbon Tariff," *The Wall Street Journal*, March 18, 2009, web site: http://online.wsj.com/article/SB123733297926563315.html

90. Timothy Gardner, "U.S. Big Steel pushes for carbon fees on China," Reuters, March 23, 2009, web site: http://www.reuters.com/article /environmentNews/idUSTRE52M70S20090323

91. *The World Depression, 1929-39* by Charles Kindleberger, (Berkeley: University of California Press, 1973, pp. 291–308.

Chapter 15: Climatism in Action: Debacles around the World

1. "Austin's Renewable Energy Program Reduces CO2 Emissions By 370,257 Tons a Year," Climate Summit, 2007, web site: http://www.nyc climatesummit.com/casestudies/energy/energy_renew_Austin.html

2. "Texas Incentives/Policies for Renewables & Efficiency," Database of State Incentives for Renewables & Efficiency, web site: http://www.dsireusa .org/incentives/index.cfm?re=1&ee=1&spv=0&st=0&srp=1&state=TX

3. Climate Summit (See no. 1).

4. "Austin Energy Wins DOE Wind Power Award," U.S. Department of Energy, October 25, 2005, web site: http://www.windpoweringamerica.gov/filter_detail.asp?itemid=1214

5. "Austin Energy (leading green power program in US) struggles with costs of more renewables," Environmental Research Web Blog, July 20, 2009, web site: http://environmentalresearchweb.org/blog/2009/07/austin-energy-leading-green-po.html

6. "Austin Renewable Energy Program Plagued by Low Demand," *Environmental Leader*, July 12, 2009, web site: http://www.environmentalleader.com /2009/07/12/austins-greenchoice-program-faces-uncertain-future/

7. Environmental Research (See no. 5).

8. Environmental Research (See no. 5).
9. Memo from Roger Wood, Freescale Semiconductor, to Austin Generation Resource Planning Task Force, August 12, 2009, web site: http://powersmack .org/wp/wp-content/uploads/2009/08/roger-wood-memo.pdf
10. "The Enforcer (1976)," IMDB, web site: http://www.imdb.com/title/tt0074483/
11. "Renewables Global Status Report, 2009," Renewable Energy Policy Network for the 21st Century, 2009, web site: http://www.ren21.net/pdf/RE_GSR_2009_Update.pdf
12. "Obama aims for a million green cars by 2015," *ABCNews*, March 20, 2009, web site: http://www.abc.net.au/news/stories/2009/03/20/2521920.htm
13. Gabriel Calzada Álvarez et al., "Study of the effects on employment of public aid to renewable energy sources," Universidad Rey Juan Carlos, March 2009, web site: http://www.juandemariana.org/pdf/090327-employment-public-aid-renewable.pdf
14. Ibid.
15. Calzada (See no. 13).
16. Calzada (See no. 13).
17. Calzada (See no. 13).
18. "Interview Transcript: Jose Luis Rodriguez Zapatero," *Financial Times*, June 4, 2008, web site: http://us.ft.com/ftgateway/superpage.ft?news _id=fto06042008181438273&page=2
19. "Boosting growth and jobs by meeting our climate change commitments," *Europa*, January 23, 2008, web site: http://europa.eu/rapid/pressReleasesAction.do?reference=IP/08/80
20. Calzada (See no. 13).
21. Calzada (See no. 13).
22. Calzada (See no. 13).
23. Calzada (See no. 13).
24. Calzada (See no. 13).
25. International Energy Annual 2006, U.S. Energy Information Administration, U.S. Department of Energy, December 8, 2008, web site: http://www.eia.doe.gov/pub/international/iealf/tableh1co2.xls
26. Asociación de la Industria Fotovoltaica, 2009, web site: http://www.asif.org/
27. Report for Selected Countries and Subjects, Philippines, International Monetary Fund, web site: http://www.imf.org/external/pubs/ft/weo/2006 /01/data/dbcoutm.cfm?SD=1980&ED=2007&R1=1&R2=1&CS=3&SS=2&O S=C&DD=0&OUT=1&C=566&S=NGDP_RPCH-NGDP-NGDPD-NGDPPC-NGDPDPC-PPPWGT-PPPSH-PPPEX&CMP=0&x=85&y=17
28. "Urban Population, Development and the Environment 2007," United Nations Department of Economic and Social Affairs, Population Division, web site: www.unpopulation.org
29. "2007/2008 Human Development Report," UNDP, web site: http://hdrstats.undp.org/indicators/210.html
30. "Ring of Fire, Plate Tectonics, Sea-Floor Spreading, Subduction Zones, Hot Spots," US Geological Survey, web site: http://vulcan.wr.usgs.gov /Glossary/PlateTectonics/description_plate_tectonics.html
31. "Welcome to our page with data for Philippines," International Geothermal

Association, web site: http://www.iga.1it.pl/286,welcome
_to_our_page_with_data_for_philippines_-_direct_uses.html

32. "IEA Energy Statistics—Electricity for Philippines," International Energy
    Agency, 2009, web site:
    http://www.iea.org/Textbase/stats/electricitydata.asp?COUNTRY_CODE=PH

33. "Challenges beset renewable energy projects," *Sun.star Network Online*, web
    site: http://www.sunstar.com.ph/network/challenges-beset-renewable-energy-
    projects

34. "German firm praises Philippines for embracing renewable energy," *Balita.Ph*,
    April 9, 2009, web site: http://balita.ph/2009/04/09/german-firm-praises-
    philippines-for-embracing-renewable-energy/

35. Lilybeth Ison, "Philippines' Renewable Energy Bill Passed," PR-inside.com,
    October 8, 2008, web site:
    http://www.topix.com/forum/energy/biomass/TJCM5D2831M3GUCTR

36. "Philippines Alliance for Mindanao Off-Grid Renewable Energy," USAID, web
    site:
    http://www.un.org/esa/sustdev/csd/casestudies/renewableEnergyy_USA.pdf

37. Joefel Ortega Banzon, "Infrastructure cost, volatility deter renewable energy
    projects," *Philstar.com*, May 29, 2009, web site: http://www.philstar.com
    /Article.aspx?articleId=472222&publicationSubCategoryId=108

38. Challenges (See no. 33).

39. Challenges (See no. 33).

40. Challenges (See no. 33).

41. CIA World Factbook, Central Intelligence Agency, web site:
    https://www.cia.gov/library/publications/the-world-factbook/

42. "Cebu braces for power deficit," *Manila Bulletin*, 2009, web site:
    http://www.mb.com.ph/articles/219714/cebu-braces-power-deficit

43. Elias Baquero, "Group asks gov't to mothball coal-fired power plants," *Sun.star
    Network Online*, web site: http://www.sunstar.com.ph/cebu/group-asks-
    gov%E2%80%99t-mothball-coal-fired-power-plants

44. Ibid.

45. "Wind power to combat climate change," Energinet.dk, web site:
    http://www.energinet.dk/NR/rdonlyres/3097FD4E-F82A-43D0-BBD9-
    8BF07C349474/0/Windpowermagazine.pdf

46. Ibid.

47. Energinet (See no. 45).

48. Energinet (See no. 45).

49. "The Economics of Renewable Energy," House of Lords, November 25, 2008,
    web site: http://www.publications.parliament.uk
    /pa/ld200708/ldselect/ldeconaf/195/195i.pdf

50. Private communication with Hugh Sharman regarding his upcoming
    publication: "Wind Energy—The Case of Denmark," CEPOS, Sep., 2009.

51. "Energy Statistics 2007," Danish Energy Agency, October, 2008, web site:
    http://www.ens.dk/da-DK/Info/TalOgKort/Statistik_og_noegletal
    /Maanedsstatistik/Documents/energy%20statistics%202007%20uk.pdf

52. John Goerten and Daniel Ganea, "Electricity prices for the second semester
    2008," Eurostat, 25/2009, web site: http://epp.eurostat.ec.europa.eu
    /cache/ITY_OFFPUB/KS-QA-09-025/EN/KS-QA-09-025-EN.PDF

53. Sharman (See no. 50).
54. Sharman (See no. 50).
55. Sharman (See no. 50).
56. Sharman (See no. 50).
57. Energinet.Dk., web site:
    http://translate.google.com/translate?hl=en&sl=da&u=http://www.energinet.dk
    /&ei=_ey2SvCtPIX2NengvdoO&sa=X&oi=translate&resnum=1&ct=result&pr
    ev=/search%3Fq%3D%2522energinet.dk%2522%26hl%3Den%26lr%3D
58. Danish Energy Agency, (See no. 51).
59. Energinet (See no. 45).
60. "Gov. Schwarzenegger Signs Landmark Legislation to Reduce Greenhouse Gas
    Emissions," Press Release from Office of the Governor of California, September
    27, 2006, web site: http://gov.ca.gov/press-release/4111/
61. "Executive Order S-14-08," Office of the Governor of California, November
    17, 2008, web site: http://www.gov.ca.gov/executive-order/11072/
62. "California Electricity Generation," Energy Almanac of the California Energy
    Commission, web site: http://www.energyalmanac.ca.gov/
63. Ibid.
64. "Geothermal Energy in California," The California Energy Commission, web
    site: http://www.energy.ca.gov/geothermal/index.html
65. "Impact of Past, Present and Future Wind Turbine Technologies on
    Transmission System Operation and Performance," PIER Project Report for
    the California Energy Commission, May, 2006, web site:
    http://www.energy.ca.gov/pier/project_reports/CEC-500-2006-050.html
66. Energy Almanac (See no. 62).
67. PIER Report (See no. 65).
68. "California Greenhouse Gas Inventory for 2000–2006," California Air
    Resources Board, web site: http://www.arb.ca.gov/cc/inventory/data/data.htm
69. "Renewable Resources Development Report," California Energy Commission,
    November, 2003, web site: http://www.energy.ca.gov/reports/2003-11-
    24_500-03-080F.PDF
70. Energy Almanac (See no. 62).
71. "Developing Methods to Reduce Bird Mortality in the Altamont Pass Wind
    Resource Area," PIER Report for the California Energy Commission, August,
    2004, web site: http://www.energy.ca.gov/reports/500-04-052/500-04-
    052_00_EXEC_SUM.PDF
72. "California Guidelines for Reducing Impacts to Birds and Bats from Wind
    Energy Development," California Energy Commission, October, 2007, web
    site: http://www.energy.ca.gov/windguidelines/index.html
73. Ryan Wiser and Mark Bolinger, "2008 Wind Technologies Market Report,"
    Berkeley National Laboratory, July, 2009, web site:
    http://www1.eere.energy.gov/windandhydro/pdfs/46026.pdf
74. "Iberdrola Renewables Opens 202 MW Texas Wind Farm," *Spain Business*,
    April 24, 2009, web site: http://www.us.spainbusiness.com/icex/cda
    /controller/pageInv/0,2958,35868_594951_1026487_4211860,00.html
75. PIER Report (See no. 71).
76. Wiser (See no. 73).
77. Elizabeth Shogren, "Wind Farm Called Threat to Condors," *Los Angeles Times*,

Sep. 14, 1999, web site: http://articles.latimes.com/1999/sep/14/news/mn-9946

78. "Integrated Energy Report," California Energy Commission, November, 2005, web site: http://www.energy.ca.gov/2005publications/CEC-100-2005-007/CEC-100-2005-007-CMF.PDF

79. Timm Herdt, "Solar subsidies could reach 70 percent," *Ventura County Star*, April 22, 2009, web site: http://www.venturacountystar.com/news/2009/apr/22/solar-subsidies-could-reach-70-percent/

80. "Large Solar Energy Projects," The California Energy Commission, web site: http://www.energy.ca.gov/siting/solar/index.html

81. "Climate Change Scoping Plan," California Air Resources Board, December 2008, web site: http://www.arb.ca.gov/cc/scopingplan/document/scopingplandocument.htm

82. "Summary of California Climate Change 'Scoping Plan' and Transportation Measures," International Council on Clean Transportation, 2009, web site: http://www.theicct.org/documents/California_plan_summary_for_ICCT_website_4-24-09.pdf

83. Ibid.

84. Scoping Plan (See no. 81)

85. Sanjay Varshney and Dennis Tootelian, "Cost of AB 32 on California Small Businesses—Summary Report of Findings," Varshney & Associates, June 2009, web site: http://ncwatch.typepad.com/files/ab-32-report-07-13-09.pdf

86. Amy Kaleita, "California Counts the Cost on Climate Change Legislation," *Pacific Research Institute*, July 22, 2009, web site: http://liberty.pacificresearch.org/press/california-counts-the-cost-on-climate-change-legislation-2

87. "California," Energy Information Administration, U.S. Department of Energy, web site: http://tonto.eia.doe.gov/state/state_energy_profiles.cfm?sid=CA

88. "Coal–Initial Scoping Note," PIU Energy Project, August 2001, web site: http://www.cabinetoffice.gov.uk/media/cabinetoffice/strategy/assets/coal.pdf

89. "The UK Low Carbon Transition Plan, Analytical Annex," Department of Energy and Climate Change, 2009, web site: http://www.decc.gov.uk/en/content/cms/publications/lc_trans_plan/lc_trans_plan.aspx

90. "The UK Low Carbon Transition Plan," Department of Energy and Climate Change, July 15, 2009, web site: http://www.decc.gov.uk/en/content/cms/publications/lc_trans_plan/lc_trans_plan.aspx

91. Ibid.

92. "UK Wind Energy Database," BWEA, web site: http://www.bwea.com/ukwed/index.asp

93. Low Carbon Transition Plan (See no. 90).

94. "Quarterly Wholesale/Retail Price Report," OFGEM, August, 2009, web site: http://www.ofgem.gov.uk/Pages/MoreInformation.aspx?docid=215&refer=Markets/RetMkts/ensuppro

95. Low Carbon Transition Plan (See no. 90).

96. Christopher Booker, "Why should we pay for the beliefs of others?," Telegraph, March 14, 2009, web site: http://www.telegraph.co.uk/comment/columnists http://www.independent.co.uk/life-style/food-and-drink/news/mccartney-urges-meatfree-days-to-tackle-climate-change-1705289.html /christopherbooker/4990827/Why-should-we-pay-for-the-beliefs-of-

others.html

97. Low Carbon Transition Plan (See no. 90).

98. Charles Clover, "4x4s to be priced off the road," Telegraph, January 12, 2007, web site: http://www.telegraph.co.uk/news/uknews/1539264/4x4s-to-be-priced-off-the-road.html

99. Kevin Schofield, "Flights face green tax in new bid to tackle climate change," *The Daily Record*, June 27, 2009, web site: http://www.dailyrecord.co.uk/news/uk-world-news/2009/06/27/flights-face-green-tax-in-new-bid-to-tackle-climate-change-86908-21475349/

100. Nick Clark, "Airlines cut flights in tax-hike protest," *The Independent*, July 22, 2009, web site: http://www.independent.co.uk/news/business/news/airlines-cut-flights-in-taxhike-protest-1755867.html

101. Ian Drury, "Ed Miliband's global warming law 'could cost £20,000 per family,'" *Mail Online*, May 5, 2009, web site: http://www.dailymail.co.uk/news/article-1177274/Ed-Milibands-global-warming-law-cost-20-000-family.html

102. Louise Gray, "Government pamphlet urges people to walk to work to stop climate change," *Telegraph*, Jun 26, 2009, web site: http://www.telegraph.co.uk/earth/environment/climatechange/5638670/Government-pamphlet-urges-people-to-walk-to-work-to-stop-climate-change.html

103. David Adam and Randeep Ramesh, "Tory frontbench signs up to 10:10 climate change campaign," *Guardian*, Sep. 2, 2009, web site: http://www.guardian.co.uk/environment/2009/sep/02/10-10-campaign-tory-frontbench

104. Ian Sample, "Leading scientist calls on religious leaders to tackle climate change," *Guardian*, Sep. 7, 2009, web site: http://www.reasonproject.org/newsfeed/item/Leading_scientist_calls_on_religious_leaders_to_tackle_climate_change/

105. Michael Shellenberger, "When Diplomats Boo: How Global Climate Talks Reached a New Nadir," The Breakthrough Institute, Dec. 17, 2007, web site: http://www.thebreakthrough.org/blog/2007/12/when_diplomats_boo_how_global.shtml

106. "Delivering on Climate Action–The Challenge Facing Europe," Institute for European Environmental Policy, June 29, 2009, web site: http://www.climnet.org/resources/Press%20Release%20Implementation%2029%20June%202009.pdf

107. Low Carbon Transition Plan (See no. 90).

108. "Less Greenhouse Gases," European Commission, web site: http://ec.europa.eu/climateaction/eu_action/less_greenhouse_gases/index_en.htm

109. "Annual Report and Resource Accounts 2008-09," Department of Energy and Climate Change, July 20, 2009, web site: http://www.decc.gov.uk/en/content/cms/publications/annual_reports/2009/2009.aspx

110. "Europe's Dirty Secret: Why the EU emissions Trading Scheme isn't working," Open Europe, August, 2007, web site: http://www.openeurope.org.uk/research/etsp2.pdf

111. Ibid.

112. Open Europe (See no. 110).

113. Danny Fortson and Georgia Warren, "Carbon-trading market hit as UN suspends clean-energy auditor," *Times OnLine*, September 13, 2009, web site:

http://business.timesonline.co.uk/tol/business/industry_sectors/natural_resourc es/article6832259.ece

114. "The EU Climate Action and Renewable Energy Package: Are we about to be locked into the wrong policy?," Open Europe, October, 2008, web site: http://www.openeurope.org.uk/research/carep.pdf

115. Open Europe (See no. 110)

Chapter 16: Energy Nonsense for the Good Old USA

1. "Secretary of State Hillary Rodham Clinton Announces Appointment of Special Envoy for Climate Change Todd Stern," U.S. Department of State, January 26, 2009, web site: http://www.america.gov/st/texttrans-english/2009/January/20090126163647eaifas0.8665125.html

2. "Remarks of President Obama–Address to Joint Session of Congress," White House Press Office, Feb. 24, 2009, web site: http://www.whitehouse.gov/the_press_office/remarks-of-president-barack-obama-address-to-joint-session-of-congress/

3. "Polluters pay in Obama's 'green' budget," *AFP*, February 25, 2009, web site: http://www.google.com/hostednews/afp/article/ALeqM5jHxfjIJ_y2QS7sK-iZGBiOnxLOpQ

4. "The American Clean Energy and Security Act (Waxman-Markey Bill)," Pew Center on Global Climate Change, web site: http://www.pewclimate.org/acesa

5. "US House committee approves climate change bill," *Oil & Gas Journal*, web site: http://www.waterworld.com/index/display/article-display/363616/s-articles/s-oil-gas-journal/s-volume-107/s-issue-21/s-general-interest/s-us-house-committee-approves-climate-change-bill.html

6. "Pew Center Summary of H.R. 2454: American Clean Energy and Security Act of 2009 (Waxman-Markey)," Pew Center on Global Climate Change, web site: http://www.pewclimate.org/acesa

7. "U.S. Energy Sec. Pledges to Fight Global Warming," *Reuters*, Mar 5, 2009, web site: http://www.theepochtimes.com/n2/content/view/13161/

8. "Petroleum," Energy Information Administration, U.S. Department of Energy, web site: http://www.eia.doe.gov/oil_gas/petroleum/info_glance/petroleum.html

9. "Annual Energy Review 2008," Energy Information Administration, U.S. Department of Energy, June, 2009, web site: http://www.eia.doe.gov/emeu/aer/pdf/aer.pdf

10. Petroleum (See no. 8)

11. Joe Carroll and Edward Klump, "Big Oil's Answer to Carbon Law May Be Fuel Imports (Update 2)," Bloomberg.com, June 26, 2009, web site: http://www.bloomberg.com/apps/news?pid=20601087&sid=a1ZiIqv3E4QE

12. Foon Rhee, "Kerry: Clean energy revolution coming," *Boston.com*, March 5, 2009, web site: Kerry: Clean energy revolution coming

13. "The big chill," *Economist*, January 15, 2009, web site: http://www.economist.com/displaystory.cfm?story_id=12926505

14. Annual Energy Review (See no. 9).

15. Annual Energy Review (See no. 9).

16. Brendan Bradley, "House Democrats Introduce Bill on Climate and Energy," Independent Petroleum Association of America, March 31, 2009, web site:

http://www.ipaa.org/news/wr/2009/WR-2009-03-31.pdf

17. Kerry (See no. 12).

18. "Illinois Tollway System Valuation," Credit Suisse, August 2006, web site: http://www.jschoenberg.org/tollway/Illinois_Report_Final.pdf

19. "Analysis of the Waxman-Markey Bill (H.R. 2454) Using The National Energy Modeling System (NEMS/ACCF-NAM 2)," A Report by the American Council for Capital Formation and the National Association of Manufacturers, Aug. 12, 2009, web site: http://www.accf.org/publications/126/accf-nam-study

20. Al Gore, "A Postage Stamp a Day," Al Gore Blog, June 23, 2009, web site: http://blog.algore.com/2009/06/a_postage_stamp_a_day.html

21. "The Estimated Costs to Households From the Cap-and-Trade Provisions of H.R. 2454," Congressional Budget Office, June 19, 2009, web site: https://www.cbo.gov/ftpdocs/103xx/doc10327/06-19-CapAndTradeCosts.pdf

22. "IEA Wind Energy Annual Report 2008," International Energy Agency, July 2009, web site: http://www.ieawind.org/AnnualReports_PDF/2008/2008%20AR_small.pdf

23. "Impact on the Economy of the American Clean Energy and Security Act of 2009 (H.R.2454), CRA International, May, 2009, web site: http://www.crai.com/uploadedFiles/Publications/impact-on-the-economy-of-the-american-clean-energy-and-security-act-of-2009.pdf

24. Analysis of the Waxman-Markey Bill (See no. 19).

25. Pete Du Pont, "Sapping America's Energy," *The Wall Street Journal*, April 16, 2009, web site: http://online.wsj.com/article/SB123980462156321035.html

26. "United States of America Fossil-Fuel Emissions," Carbon Dioxide Information Analysis Center, Oak Ridge National Laboratory, web site: http://cdiac.ornl.gov/

27. Matt Rosenberg, "US Population Through History," *About.com*, web site: http://geography.about.com/od/obtainpopulationdata/a/uspop.htm

28. "2005 Residential Energy Consumption Survey," Energy Information Administration, U.S. Department of Energy, web site: http://www.eia.doe.gov/emeu/recs/recs2005/hc2005_tables/detailed_tables2005.html

29. Annual Energy Review (See no. 9).

30. Annual Energy Review (See no. 9).

31. IEA Wind Energy (See no. 22).

32. IEA Wind Energy (See no. 22).

33. Ryan Wiser and Mark Bolinger, "2008 Wind Technologies Market Report," Berkeley National Laboratory, July, 2009, web site: http://www1.eere.energy.gov/windandhydro/pdfs/46026.pdf

34. Ibid.

35. IEA Wind Energy (See no. 22).

36. "The EU Climate Action and Renewable Energy Package: Are we about to be locked into the wrong policy?," Open Europe, October, 2008, web site: http://www.openeurope.org.uk/research/carep.pdf

37. Ibid.

38. "The History of Nuclear Energy," U.S. Department of Energy, web site: http://www.ne.doe.gov/pdfFiles/History.pdf

39. "The Soviet Nuclear Weapons Program," nuclearweaponarchive.org, web site: http://nuclearweaponarchive.org/Russia/Sovwpnprog.html

40. History of Nuclear Energy (See no. 38)

41. *Greenpeace: How a Group of Ecologists, Journalists, and Visionaries Changed the World* by Rex Weyler, (Rodale Books, 2004)

42. "Three Mile Island Accident," U.S. Nuclear Regulatory Commission, web site: http://www.nrc.gov/reading-rm/doc-collections/fact-sheets/3mile-isle.html

43. "Dick Thornburgh," *PBS Online*, web site: http://www.pbs.org/wgbh/amex/three/peopleevents/pandeAMEX97.html

44. "The China Syndrome (1979)," IMDB, web site: http://www.imdb.com/title/tt0078966/

45. Three Mile Island (See no. 42)

46. "14-Year Cleanup at Three Mile Island Concludes," *The New York Times*, August 15, 1993, web site: http://www.nytimes.com/1993/08/15/us/14-year-cleanup-at-three-mile-island-concludes.html

47. "Chernobyl Nuclear Power Plant Accident," U.S. Nuclear Regulatory Commission, web site: http://www.nrc.gov/reading-rm/doc-collections/fact-sheets/chernobyl-bg.html

48. Annual Energy Review (See no. 9).

49. Annual Energy Review (See no. 9).

50. "Useful Conversion Factors," IOR Energy, web site: http://www.ior.com.au/ecflist.html

51. History of Nuclear Energy (See no. 38).

52. "Nuclear power plants, world-wide," European Nuclear Society, web site: http://www.euronuclear.org/info/encyclopedia/n/nuclear-power-plant-world-wide.htm

53. "Nuclear Power in Japan," World Nuclear Association, web site: http://www.world-nuclear.org/info/inf79.html

54. World Nuclear Association, web site: http://www.world-nuclear.org/

55. Aircraft Crashes Record Office, web site: http://www.baaa-acro.com/

56. "Plant Safety Performance After the TMI-2 Accident," U.S. Nuclear Regulatory Commission, web site: http://www.nrc.gov/reading-rm/doc-collections/fact-sheets/fs-plant-sfty-after-tmi2.html

57. Keith Clarke and Jeffrey Hemphill, "Santa Barbara Oil Spill: A Retrospective," University of California, Santa Barbara, web site: http://www.geog.ucsb.edu/~kclarke/Papers/SBOilSpill1969.pdf

58. "1969 Oil Spill," University of California, Santa Barbara, web site: http://www.geog.ucsb.edu/~jeff/sb_69oilspill/69oilspill_articles2.html

59. Clarke and Hemphill (See no. 57).

60. "Exxon Valdez Oil Spill," National Oceanic and Atmospheric Administration, web site: http://response.restoration.noaa.gov/topic_subtopic_entry.php?RECORD_KEY%28entry_subtopic_topic%29=entry_id,subtopic_id,topic_id&entry_id%28entry_subtopic_topic%29=700&subtopic_id%28entry_subtopic_topic%29=2&topic_id%28entry_subtopic_topic%29=1

61. "America's Largest Spill," Boise State University, web site: http://www.boisestate.edu/history/ncasner/hy210/valdez.htm

62. "Exxon Shipping Co. et al. v. Baker et al.," Supreme Court of the United States, Oct., 2007, web site: http://www.supremecourtus.gov/opinions/07pdf/07-219.pdf

63. J.R. Dunn, "How the Greens Captured Energy Policy," American Thinker,

July 10, 2008, web site: http://www.americanthinker.com/2008
/07/how_the_greens_captured_energy_1.html

64. Bryan Walsh, "Remembering the Lessons of the Exxon Valdez," March 24, 2009, *Time*, web site:
http://www.time.com/time/health/article/0,8599,1887165,00.html

65. "Summary Points: 10 Years of Intertidal Monitoring After the Exxon Valdez Spill," National Oceanic and Atmospheric Administration, web site:
http://response.restoration.noaa.gov/topic_subtopic_entry.php?RECORD_KE Y(entry_subtopic_topic)=entry_id,subtopic_id,topic_id&entry_id(entry_subto pic_topic)=254&subtopic_id(entry_subtopic_topic)=13&topic_id(entry_subto pic)=1

66. "Is the Oil Gone?," National Oceanic and Atmospheric Administration, web site: http://response.restoration.noaa.gov/topic_subtopic_entry .php?RECORD_KEY(entry_subtopic_topic)=entry_id,subtopic_id,topic_id&e ntry_id(entry_subtopic_topic)=257&subtopic_id(entry_subtopic_topic)=13&t opic_id(entry_subtopic_topic)=1

67. "Oil Pollution Act Overview," U.S. Environmental Protection Agency, web site: http://www.epa.gov/OEM/content/lawsregs/opaover.htm

68. International Convention for the Prevention of Pollution from Ships, 1973, as modified by the Protocol of 1978 relating thereto (MARPOL), International Maritime Organization, web site:
http://www.imo.org/Conventions/contents.asp?doc_id=678&topic_id=258

69. Tom Kizzia, "Double-hull tankers face slow going," *Anchorage Daily News*, May 13, 1999, web site: http://www.adn.com/evos/stories/T99032456.html

70. "Oil in the Sea III: Inputs, Fates and Effects," National Ocean Industries Association, web site: http://www.noia.org/website/article.asp?id=129

71. "Assumptions to the Annual Energy Outlook 2009," Energy Information Administration, U.S. Department of Energy, 2009, web site:
http://www.eia.doe.gov/oiaf/aeo/assumption/oil_gas.html

72. Alan Caruba, "The New Dark Ages of Britain & The U.S.," *Climate Realists*, August 16, 2009, web site: http://climaterealists.com/index.php?id=3873

73. Assumptions (See no. 71).

74. John Schoen, "U.S. refiners stretch to meet demand," *MSNBC*, November 22, 2004, web site: http://www.msnbc.msn.com/id/6019739/ns/business-oil_and_energy/

75. "U.S. approves Alberta Clipper pipeline project," Reuters, Aug. 20, 2009, web site: http://ca.reuters.com/article/businessNews/idCATRE57J65X20090820

76. "Native & Green Groups Challenge State Dept. Permit for Dirty Oil Pipeline," *EarthJustice*, Sep. 3, 2009, web site: http://www.earthjustice.org/news/press/20 09/native-green-groups-challenge-state-dept-permit-for-dirty-oil-pipeline.html

77. Chet Brokaw, "SD board OKs air quality permit for oil refinery," ABCNews, Aug. 20, 2009, web site:
http://abcnews.go.com/Business/wireStory?id=8377878

78. "Arctic National Wildlife Refuge," U.S. Fish and Wildlife Service, web site:
http://arctic.fws.gov/

79. "Arctic National Wildlife Refuge, 1002 Area, Petroleum Assessment, 1998, Including Economic Analysis," U.S. Geological Survey, April, 2001, web site:
http://pubs.usgs.gov/fs/fs-0028-01/fs-0028-01.htm

80. Andy Aden, "Water Usage for Current and Future Ethanol Production," *Southwest Hydrology*, September/October 2007, web site: http://www.swhydro.arizona.edu/archive/V6_N5/feature4.pdf

81. Charli Coon, "Tapping Oil Reserves In a Small Part of ANWR: Environmentally Sound, Energy Wise," *The Heritage Foundation*, Aug. 1, 2001, web site: http://www.heritage.org/research/energyandenvironment/wm27.cfm

82. "Where Does My Gasoline Come From?," Energy Information Administration, U.S. Department of Energy, April, 2008, web site: http://www.eia.doe.gov/bookshelf/brochures/gasoline/index.html

83. Photo of the ANWR 1002 area coastal plain by U.S. Geological Survey, August 24, 2005, web site: http://pubs.usgs.gov/fs/fs-0028-01/fs-0028-01.htm

84. Paul Driessen, Presentation at the International Conference on Climate Change, March 4, 2008

85. U.S. Department of Agriculture, 2009, web site: http://www.usda.gov/wps/portal/usdahome

86. "Key World Energy Statistics 2008," International Energy Agency, web site: http://www.iea.org/textbase/nppdf/free/2008/key_stats_2008.pdf

87. Annual Review (See no. 9)

88. David Urbinato, "London's Historic 'Pea-Soupers,'" Environmental Protection Agency, 1994, web site: http://www.epa.gov/history/topics/perspect/london.htm

89. Sean Hamill, "Unveiling a Museum, a Pennsylvania Town Remembers the Smog That Killed 20," *The New York Times*, November 2, 2008, web site: http://www.nytimes.com/2008/11/02/us/02smog.html

90. Michelle Bell et al., "A Retrospective Assessment of Mortality from the London Smog Episode of 1952: The Role of Influenza and Pollution," *Environmental Health Perspectives*, vol. 112, no. 1, January, 2004

91. Urbinato (See no. 88).

92. "Nonpoint Source Pollution," National Oceanic and Atmospheric Administration, web site: http://oceanservice.noaa.gov/education/kits/pollution/02history.html

93. "Air Trends," Environmental Protection Agency, web site: http://www.epa.gov/airtrends/

94. "Hansen's Coal and Global Warming protest may get snowed out," WattsUpWithThat, March 3, 2009, web site: http://wattsupwiththat.com /2009/03/01/hansens-coal-and-global-warming-protest-may-get-snowed-out/

95. "Natural Gas and the Environment," NaturalGas.org, web site: http://www.naturalgas.org/environment/naturalgas.asp

96. Intergovernmental Panel on Climate Change, Fourth Assessment Report, Working Group III, 2007, p. 266, web site: http://www.ipcc.ch/pdf/assessment-report/ar4/wg3/ar4-wg3-chapter4.pdf

97. "Earth First! Blockades Florida Power Plant Construction, 27 Arrested," Environment News Service, February 19, 2008, web site: http://www.ens-newswire.com/ens/feb2008/2008-02-19-091.asp

98. Lynda Waddington, "Alliant nixes plan for Marshalltown coal plant," *The Iowa Independent*, March 5, 2009, web site: http://iowaindependent.com /12343/alliant-nixes-plan-for-marshalltown-coal-plant

99. Nelia Seaman, "Iowans Can Breathe Easier: Alliant Energy Abandons Marshalltown Coal Plant Proposal," Sierra Club, March 5, 2009, web site: http://iowa.sierraclub.org/Energy/Alliant%20abandons%20coal%20plant%20n ews%20release030509.pdf

100. "Beyond Coal," Sierra Club, web site: http://www.sierraclub.org/coal/local.aspx

101. Jackie Grom, "Coal Plant Ban May Not Curb Climate Change," *ScienceNOW*, May 1, 2009, web site: http://www.ecoearth.info/shared/reader/welcome.aspx?linkid=126360&keybold =coal%20AND%20%20plant%20AND%20%20ban

102. "The California Energy Crisis," U.S. Department of Energy, web site: http://www.eere.energy.gov/de/ca_elec_crisis.html

103. Annual Energy Review (See no. 9).

104. "Americans are 'illiterate' about climate change, claims expert," Telegraph, Sep. 28, 2009, web site: http://www.telegraph.co.uk/earth/earthnews /6240611/Americans-are-illiterate-about-climate-change-claims-expert.html

105. "Oxfam Warns G8 on Cost of Inaction on Climate Change," OxfamAmerica, June 8, 2007, web site: http://www.oxfamamerica.org/press/pressreleas es/oxfam-warns-g8-on-cost-of-inaction-on-climate-change/?searchterm=None

106. Andrew Buncombe, "US walks out of climate change talks as 150 nations move forward to adopt Kyoto," *The Independent*, December 10, 2005, web site: http://www.independent.co.uk/news/world/americas/us-walks-out-of-climate-change-talks-as-150-nations-move-forward-to-adopt-kyoto-518888.html

107. Ibid.

108. Kathrine McGrow, "Climate refugee nation Tuvalu ponders legal options against polluters," *Pacific.scoop.co.nz*, September 9, 2009, web site: http://pacific.scoop.co.nz/2009/09/climate-refugee-nation-tuvalu-ponders-legal-options-against-polluters/

109. Aoife White, "EU says US must join climate change funding," *Boston.com*, October 2, 2009, web site: http://www.boston.com/business/articles/2009 /10/02/eu_to_decide_on_climate_funding_for_poor_countries?mode=PF

110. James Kanter, "E.U. Alone and Lonely on Carbon," *The New York Times*, September 28, 2009, web site: http://www.nytimes.com/2009 /09/28/business/energy-environment/28green.html

111. Patrick Goodenough, "Clinton Accepts Blame for 'Global Warming' Role, Ponders Link Between Climate Change and Family Planning," *CNSNews*, July 20, 2009, web site: http://www.cnsnews.com/PUBLIC/Content/article.aspx?RsrcID=51260

Chapter 17: Common Sense and the Future

1. Secretary-General Ban Ki-moon, "Remarks to the Global Environment Forum," UN News Centre, August 11, 2009, web site: http://www .un.org/apps/news/infocus/sgspeeches/statments_full.asp?statID=557

2. Barrie Rhodes, "Canute (Knud) The Great," The Viking Network, web site: http://www.viking.no/e/people/e-knud.htm

3. Gerard Wynn, "Two meter sea level rise unstoppable: experts," Reuters, September 30, 2009, web site: http://www.reuters.com/article/science News/idUSTRE58S4L420090930?feedType=RSS&feedName=scienceNews

4. "Key excerpts from Mojib Latif's WCC presentation," Deep Climate, October 2, 2009, web site: http://deepclimate.org/2009/10/02/key-excerpts-from-mojib- latifs-wcc-presentation/

5. Richard Lindzen, "Deconstructing Global Warming," October 26, 2009, web site: http://www.globalwarming.org/2009/11/03/video-of-dr-richard-lindzens-deconstructing-global-warming/

6. Charles, Fisk, "Graphical Climatology of Chicago Temperatures, Precipitation, and Snowfall (1871-present), web site: http://home.att.net/~chicago_climo/

7. Weatherbase, web site: http://www.weatherbase.com/

8. James Dacey, "The Sun could be heading into period of extended calm," *Physics World*, Sep. 23, 2009, web site: http://physicsworld.com/cws/article/news/40456

9. "The Wizard of Oz (1939)," IMDB, web site: http://www.imdb.com/title/tt0032138/

10. Muir Gray, "Climate change is the cholera of our era," *TimesOnLine*, May 25, 2009, web site: http://www.timesonline.co.uk /tol/comment/columnists/guest_contributors/article6355257.ece

11. Marc Morano, "Claim that 'climate change is the cholera of our era' ridiculed as 'load of garbage' by renowned disease expert," May 26, 2009, web site: http://www.climatedepot.com/

12. "'Carbon leakage': A challenge for EU industry, *EurActiv*," October 12, 2009, web site: http://www.euractiv.com/en/climate-change/carbon-leakage-challenge-eu-industry/article-176591

13. "G8 states could face class actions on climate change," *Irish Times*, October 8, 2009, web site: http://www.irishtimes.com/newspaper/world/2009 /1008/1224256166892.html

14. "Africa needs $65 bln to meet climate change: minister," *AFP*, web site: http://www.google.com/hostednews/afp/article/ALeqM5jUzMxeWM_060ikgk AZEKOR91mtMw

15. Benny Peiser, editor's note, *CCNET-News*, September 4, 2009.

16. Ben Webster, " Ministers target climate change doubters in prime-time TV advert," TimesOnLine, October 9, 2009, web site: http://www.timesonline.co.uk/tol/news/environment/article6867046.ece

17. Patrick Michaels, "The Dog Ate Global Warming," *National Review*, September 23, 2009, web site: http://article.nationalreview.com/print /?q=ZTBiMTRlMDQxNzEyMmRhZjU3ZmYzODI5MGY4ZWI5OWM=

18. Andrews Orlowski, "Treemometers: A new scientific scandal," The Register, September 29, 2009, web site: http://www.theregister.co.uk/2009/09/29/yamal_scandal/

19. Karen Jensen, "'Glacier Girl' The Back Story," *Air & Space Magazine*, July 1, 2007, web site: http://www.airspacemag.com/history-of-flight/16046577.html

20. Ibid.

21. Glacier Girl (See no. 19).

22. Glacier Girl (See no. 19).

23. Photograph by Lou Sapienza, 1992, all rights reserved.

INDEX